CHAPMAN & HALL/CRC APPLIED MATHEMATICS
AND NONLINEAR SCIENCE SERIES

Advanced Differential Quadrature Methods

CHAPMAN & HALL/CRC APPLIED MATHEMATICS AND NONLINEAR SCIENCE SERIES

Series Editors *Goong Chen and Thomas J. Bridges*

Published Titles

Advanced Differential Quadrature Methods, Zhi Zong and Yingyan Zhang
Computing with hp-ADAPTIVE FINITE ELEMENTS, Volume 1, One and Two Dimensional Elliptic and Maxwell Problems, Leszek Demkowicz
Computing with hp-ADAPTIVE FINITE ELEMENTS, Volume 2, Frontiers: Three Dimensional Elliptic and Maxwell Problems with Applications, Leszek Demkowicz, Jason Kurtz, David Pardo, Maciej Paszyński, Waldemar Rachowicz, and Adam Zdunek
CRC Standard Curves and Surfaces with Mathematica®: *Second Edition*, David H. von Seggern
Exact Solutions and Invariant Subspaces of Nonlinear Partial Differential Equations in Mechanics and Physics, Victor A. Galaktionov and Sergey R. Svirshchevskii
Geometric Sturmian Theory of Nonlinear Parabolic Equations and Applications, Victor A. Galaktionov
Introduction to Fuzzy Systems, Guanrong Chen and Trung Tat Pham
Introduction to non-Kerr Law Optical Solitons, Anjan Biswas and Swapan Konar
Introduction to Partial Differential Equations with MATLAB®, Matthew P. Coleman
Introduction to Quantum Control and Dynamics, Domenico D'Alessandro
Mathematical Methods in Physics and Engineering with Mathematica, Ferdinand F. Cap
Mathematical Theory of Quantum Computation, Goong Chen and Zijian Diao
Mathematics of Quantum Computation and Quantum Technology, Goong Chen, Louis Kauffman, and Samuel J. Lomonaco
Mixed Boundary Value Problems, Dean G. Duffy
Multi-Resolution Methods for Modeling and Control of Dynamical Systems, Puneet Singla and John L. Junkins
Optimal Estimation of Dynamic Systems, John L. Crassidis and John L. Junkins
Quantum Computing Devices: Principles, Designs, and Analysis, Goong Chen, David A. Church, Berthold-Georg Englert, Carsten Henkel, Bernd Rohwedder, Marlan O. Scully, and M. Suhail Zubairy
Stochastic Partial Differential Equations, Pao-Liu Chow

CHAPMAN & HALL/CRC APPLIED MATHEMATICS
AND NONLINEAR SCIENCE SERIES

Advanced Differential Quadrature Methods

Zhi Zong

Dalian University of Tecnology
Dalian, China

Yingyan Zhang

National University of Singapore
Singapore

CRC Press
Taylor & Francis Group
Boca Raton   London   New York

CRC Press is an imprint of the
Taylor & Francis Group, an **informa** business

A CHAPMAN & HALL BOOK

Chapman & Hall/CRC
Taylor & Francis Group
6000 Broken Sound Parkway NW, Suite 300
Boca Raton, FL 33487-2742

© 2009 by Taylor & Francis Group, LLC
Chapman & Hall/CRC is an imprint of Taylor & Francis Group, an Informa business

No claim to original U.S. Government works
Printed in the United States of America on acid-free paper
10 9 8 7 6 5 4 3 2 1

International Standard Book Number-13: 978-1-4200-8248-7 (Hardcover)

Library of Congress Cataloging-in-Publication Data

Zong, Zhi.
    Advanced differential quadrature methods / Zhi Zong and Yingyan Zhang.
        p. cm. -- (Chapman & Hall/CRC applied mathematics and nonlinear
        science series)
    Includes bibliographical references and index.
    ISBN 978-1-4200-8248-7 (alk. paper)
    1. Differential equations--Numerical solutions. 2. Numerical integration. I.
Zhang, Yingyan. II. Title. III. Series.

QA372.Z645 2009
518'.6--dc22   2008050784

Visit the Taylor & Francis Web site at
http://www.taylorandfrancis.com

and the CRC Press Web site at
http://www.crcpress.com

In memory of Z. Y. Zong for his love

To S.Q.Wei, L. Zhou, Xuezhou and Xueting

Preface

Despite the rapid development over recent years, problems involving non-linearity, discontinuity, multiple scale, singularity and irregularity continue to pose challenges in the field of computational science and engineering. Very often, closed form theoretical solutions are unavailable for such complex problems and approximate numerical solutions remain the only recourse. Of the various numerical solutions, differential quadrature (DQ) methods have distinguished themselves because of their high accuracy, straightforward implementation and generality in a variety of problems. Not surprisingly, DQ methods have seen phenomenal increase in research interest and experienced significant development in recent years.

DQ is essentially a generalization of the popular Gaussian Quadrature (GQ) used for numerical integration of functions. GQ approximates a finite integral as a weighted sum of integrand values at selected points in a problem domain whereas DQ approximates the derivatives of a smooth function at a point as a weighted sum of function values at selected nodes. The application of this elegant approach to solve ordinary and partial differential equations gives rise to the so-called direct DQ methods.

The applicability of the direct DQ method in its original form is limited. It has been known to fail for problems with strong nonlinearity and material discontinuity as well as for problems involving singularity, irregularity and multiple scales. Researchers working in applied mathematics, computational mechanics and engineering have developed a variety of DQ-based methods to overcome these shortcomings. Although these methods have different formulations and may even look completely different from one another, all DQ methods share a common objective—they are tools for performing numerical differentiation.

The purpose of this book is to introduce readers to the limitations of the direct DQ methods—their origins and common strategies to remove them. The formulations of several new DQ methods are presented and applied to solve problems which are beyond the capabilities of the direct DQ method. The results in this book represent the latest important developments of DQ methods in recent years. In addition to gaining an insight into the dynamic changes in this field, the reader will quickly master the use of DQ methods to solve complex problems.

There is no necessity for the reader to be familiar with the physical problems used in the examples in this book. The only prerequisite is an understanding of the fundamentals of calculus, ordinary and partial differential equations and numerical methods. As such, this book may serve as a textbook for postgraduates and a reference for researchers.

This book is organized as follows. The first chapter is devoted to the basic introduction of the direct DQ method. In Chapters 2 to 7, a variety of DQ methods are presented. They are arranged independently, in that each chapter focuses on a particular topic and is complete in itself. Hence, the reader may proceed directly to any selected chapter once he/she is familiar with the background knowledge discussed in Chapter 1. For the convenience of the reader, a mathematical compendium is summarized in Chapter 8 although the contents of the chapter can be found elsewhere. Chapter 9 contains three codes written in FORTRAN Language for readers who are keen to acquire hands-on experience of DQ methods quickly.

The first author gratefully acknowledges the financial support from the Natural Science Foundation of China (grant 50579004) and the State Key Laboratory of Structrual Analysis for Industrial Equipment (grant S08108). Part of this work had been done during the author's stay in Singapore through collaborating with Professor K. Y. Lam, Nanyang University of Technology. Credits also go to our colleagues, friends and students for their endurance, help and support. Among them are Guo Hai, Dao Lin, M. Hong, S. L. Ni, Hong Wu, Xiao Fang, J. Wang and Huaqiang. Also Y. Y. Zhang would like to express her sincere gratitude to her Ph.D. supervisors, Associate Professor Tan Beng Chye, Vincent and Prof. Wang Chien Ming from National University of Singapore, for their strong supports in the writing of this book.

Z. Zong

DUT

Y. Y. Zhang

NUS

May 2008

Credits

1. Copyright with permission from Elsevier for partial or full use of the following materials:

Zong, Z. (2006). *Information-theoretic methods for estimating complicated probability distributions.*

Shu, C., Yao, Q., and Yeo, K.S. (2002). "Block-marching in time with DQ discretization: an efficient method for time-dependent problems," *Computer Methods in Applied Mechanics and Engineering* 191, 4587–4597.

Zhong, H. (2001). "Triangular differential quadrature and its application to elastostatic analysis of Reissner plates," *International Journal of Solids and Structures* 38(16), 2821-2832.

Wu, C.P., and Tsai,Y.H. (2004). "Asymptotic DQ solutions of functionally graded annular spherical shells," *European Journal of Mechanics A-Solids* 23(2)283-299.

Zong, Z. (2003b). "A variable order approach to improve differential quadrature accuracy in dynamic analysis," *Journal of Sound and Vibration* 266(2), 307-323.

Zhong, H., and Guo, Q. (2004). "Vibration analysis of rectangular plates with free corners using spline-based differential quadrature," *Shock and Vibration* 11, 119–128.

Zong, Z. (2005). "Comments on 'A variable order approach to improve differential quadrature accuracy in dynamic analysis'–Author's reply," *Journal of Sound and Vibration* 280 (3-5), 1151-1153.

Ma, H., and Qin, Q.H. (2005). "A second-order scheme for integration of one-dimensional dynamic analysis," *Computers and Mathematics with Applications* 49, 239-252.

Zong, Z., Lam, K.Y., and Zhang, Y.Y. (2005). "A multi-domain differential quadrature approach to plane elastic problems with material discontinuity," *Mathematical and Computer Modelling 41*, 539-553.

Zhong, H.Z., and Lan M.Y. (2006). "Solution of nonlinear initial-value problems by the spline-based differential quadrature method," *Journal of Sound and Vibration* 296(4-5), 908-918.

Zhong, H. (2004). "Spline-based differential quadrature for fourth order differential equations and its application to Kirchhoff plates," *Applied Mathematical Modelling* 28, 353–366.

2. Copyright with permission from Wiley for partial or full use of the following materials:

Wang, X.W. Liu, F., Wang, X.F., and Gan, L.F. (2005). "New approaches in application of differential quadrature method to fourth-order differential equations," *Communications in Numerical Methods in Engineering* 21, 61–71.

Zhong, H. (2000). "Triangular differential quadrature," *Communications in Numerical Methods in Engineering* 16(6), 401-408.

Zhong, H. (2002). "Application of triangular differential quadrature to problems with curved boundaries," *Communications in Numerical Methods in Engineering* 18, 633-643.

3. Copyright with permission from Springer for partial or full use of the following materials:

Zong, Z. (2003a). "A complex variable boundary collocation method for plane elastic problems," *Computational Mechanics* 31(3-4), 284-292.

Zhong, H., Li, X., and He, Y. (2005) "Static flexural analysis of elliptic Reissner–Mindlin plates on a Pasternak foundation by the triangular differential quadrature method," *Arch Appl Mech* 74, 679–691.

Zong, Z., and Lam, K.Y. (2002). "A localized differential quadrature method and it application to the 2D wave equations," *Computational Mechanics* 29, 382-391.

4. Copyright with permission from Taylor & Francis for partial or full use of the following materials:

Zhang, Y. Y., Zong, Z., and Liu, L. (2007). "Complex differential quadrature method for two-dimensional potential and plane elastic problems," *Ships and Offshore Structures* 2(1), 1-10.

Contents

List of Tables

List of Figures

Chapter 1

Approximation and Differential Quadrature

Differential Quadrature (DQ) is a numerical method for evaluating derivatives of a sufficiently smooth function, proposed by Bellman and Casti in 1971. The basic idea of DQ comes from Gauss Quadrature, a useful numerical integration method. Gauss quadrature is characterized by approximating a definite integral with a weighted sum of integrand values at a group of so-called Gauss points. Extending it to finding the derivatives of various orders of a sufficiently smooth function gives rise to DQ (Bellman and Casti, 1971; Bellman et al., 1972). In other words, the derivatives of a smooth function are approximated with weighted sums of the function values at a group of so-called nodes. It should be noted that node is also called grid point or mesh point by various authors. Throughout the book all these names are used without distinction.

Differential Quadrature can be formulated either through approximation theory or solving a system of linear equations. In their original paper, Bellman and Casti (1971) used the latter to derive DQ. Throughout the book, however, we will employ the former to formulate DQ and DQ methods for the sake of simplicity. Thus in this chapter, an introduction to function approximation theory is first briefed in sections 1.1 and 1.2, followed by the fundamentals of the direct differential quadrature method in the subsequent of the sections.

1.1   Approximation and best approximation

We will closely follow the presentation of Zong (2006) about function approximation. This is a brief introduction to the theory of approximation. The interested reader is referred to monographs on this topic.

Suppose $C[a, b]$ is a set of all continuous functions defined on the interval $[a, b]$. Define a real-valued function $f(x) \in C[a, b]$ whose form is either complicated or hard to explicitly write. A general approximation is to use another simpler function $A(f)$ to replace $f(x)$, given that $A(f)$ is very close to $f(x)$. The following example is used to illustrate how the approximation works.

Example 1.1 *Approximation of* $\sin(x)$

Let $f(x) = \sin(x)$, $x \in [0, 0.1]$. Based on Taylor expansion, we may approximate the function through the formula

$$f(x) \approx x, \quad x \in [0, 0.1] \tag{1.1}$$

We then obtain

$$A(f) = 1 - x \tag{1.2}$$

The errors within the definition domain on the selected points are

x	f	$A(f)$	$|f - A(f)|$
0.02	0.019999	0.02	0.000001
0.04	0.039989	0.04	0.000011
0.06	0.059964	0.06	0.000036
0.08	0.079915	0.08	0.000085
0.1	0.099834	0.1	0.000166

It is observed that the errors are proportional to x, i.e., the errors increase as x increases. But within the interval, the approximation has high accuracy.

Approximation theory is about determining $A(f)$ and how well it works as a replacement of $f(x)$. To answer these questions, we begin with vector space.

A vector space is a set of vectors and scalars, in which there exist two types of operations: addition denoted by "+" only for vectors, and multiplication denoted by "×" for a scalar and a vector. Note that we have zero vector **0** and zero scalar 0 in the space.

A normed linear space denoted by **V** is a vector space with the additional structure of a norm. The norm is a function $\| \ \|$ from **V** to **R**, where **R** is the set of all real numbers, with the following properties: for each $x, y \in \mathbf{R}$ and each scalar;

(1) $\|\alpha x\| \geq 0$ $\tag{1.3a}$

(2) $\|x\| = 0$ if and only if x = 0, $\tag{1.3b}$

(3) $\|\alpha x\| \leq |\alpha| \, \|x\|$, $\tag{1.3c}$

(4) $\|x + y\| \leq \|x\| + \|y\|$ $\tag{1.3d}$

Let $\mathbf{V}$ be a vector space. Let $\phi_1, \phi_2, \cdots, \phi_n$ be vectors in $\mathbf{V}$, and $a_1, a_2, \cdots, a_n$ be scalars in $\mathbf{V}$. The vector $a_1\phi_1 + a_2\phi_2 + \cdots + a_n\phi_n$ is called a linear combination of $\phi_1, \phi_2, \cdots, \phi_n$ with combination coefficients $a_1, a_2, \cdots, a_n$. The vector $a_1\phi_1 + a_2\phi_2 + \cdots + a_n\phi_n$ is the zero vector $\mathbf{0}$ if and only if $a_1, a_2, \cdots, a_n$ are all zeros.

If all vectors in $\mathbf{V}$ can be written as linear combinations of $\phi_1, \phi_2, \cdots, \phi_n$, then these vectors form a set called the span of $\phi_1, \phi_2, \cdots, \phi_n$. Or in other words, we say that $\phi_1, \phi_2, \cdots, \phi_n$ span the vector space.

Linear independence and span are two key elements of the so-called basis. A set of vectors $\phi_1, \phi_2, \cdots, \phi_n$ is a basis for $\mathbf{V}$ if

(1) They are linearly independent,
(2) They span $\mathbf{V}$.

It can be shown that two bases for $\mathbf{V}$ have the same number of vectors. The number of vectors is termed as the dimension of $\mathbf{V}$.

Example 1.2 *Polynomial vector space*

Let Π_n be the set of polynomials of degree $\leq n$ in x with real coefficients. It is a vector space. The dimension is $n+1$. If $p \in \Pi_n$ then $p = a_0 + a_1 x + \cdots + a_n x^n$. The $n+1$ constants $a_0, a_1, \cdots, a_n$ provide $n+1$ degrees of freedom. Or, more precisely, we claim that $1, x_1, \cdots, x_n$ are a basis for Π_n. To prove this we need to show that

(1) $1, x_1, \cdots, x_n$ are linearly independent,

(2) They span Π_n.

Property (2) is clear from the definition of Π_n. Property (1) is a direct conclusion from the assumption that $a_0, a_1, \cdots, a_n$ are scalars satisfying the condition $a_0 + a_1 x + \cdots + a_n x^n = 0$.

The following concepts at least point out the possibility of how to construct an approximant based on linear combination.

Theorem 1.1 *(Best approximation) Let $\mathbf{V}$ be a normed linear space with norm $\| \ \|$. Let $\varphi_1, \varphi_2, \cdots, \varphi_n$ be any linearly independent members of $\mathbf{V}$. Let y be any member of $\mathbf{V}$. Then there are coefficients $a_1, a_2, \cdots, a_n$ which solve the problem minimizing*

$$\left\| y - \sum_{i=1}^{n} a_i \varphi_i \right\| \tag{1.4}$$

A solution to the minimizing problem is usually called a "best approximant" to y from $\mathbf{V}_n$. How many solutions are there? The answer is connected with convexity conditions.

Let $\mathbf{V}$ be a vector space. A subset $\mathbf{S}$ of $\mathbf{V}$ is convex if for any two members φ_1, φ_2 of $\mathbf{S}$, the set of all members of form $\varphi = t\varphi_1 + (1-t)\varphi_2$, $0 \le t \le 1$ called the line segment from φ_1 to ϕ_2, also belongs to $\mathbf{S}$.

Theorem 1.2 *(Uniqueness) Let $\mathbf{V}$ be a normed linear space with strictly convex norm. Then best approximations from finite dimensional subspaces are unique.*

Approximation is transformed into the problem searching for the bases which define a normed linear space with strictly convex norm. Depending on the sense in which the approximation is realized, or depending on the norm definition in equation (1.4), there are three types of approximation approaches:

(1) Interpolatory approximation: The parameters a_i are chosen so that on a fixed prescribed set of points x_k $(k = 1,\ldots,n)$ we have

$$\sum_{i=1}^{n} a_i\varphi_i(x_k) = y_k \tag{1.5}$$

Sometimes, we even further require that, for each i, the first r_i derivatives of the approximant agree with those of y at x_k.

(2) Least-square approximation: a_i are chosen so as to minimize

$$\left\| y - \sum_{i=1}^{n} a_i\varphi_i \right\|_2 = \sum_{k=1}^{n} \left(y_k - \sum_{i=1}^{n} a_i\varphi_i(x_k) \right)^2 \to \min \tag{1.6}$$

(3) Min-Max approximation: the parameters a_i are chosen so as to minimize

$$\left\| y - \sum_{i=1}^{n} a_i\varphi_i \right\|_\infty = \max \left| y - \sum_{i=1}^{n} a_i\varphi_i \right| \to \min \tag{1.7}$$

Interpolatory approximation will play a crucial role throughout the book, for DQ methods are based on interpolation.

1.2   Interpolating bases

Two frequently employed bases are polynomial basis and Fourier expansion basis which are to be detailed in the following. They are the approximants often used in DQ methods as well.

1.2.1   Polynomial bases

Choosing $\varphi_i = x^i$, we have polynomials as approximants. Weierstrass theorem guarantees that this is at least theoretically feasible.

Theorem 1.3 *(Weierstrass approximation theorem) Let $f(x)$ be a continuous function on the interval [a,b]. Then for any $\varepsilon > 0$, there exists an integer n and a polynomial p_n such that*

$$\max_{x \in [a,b]} |f(x) - p_n(x)| < \varepsilon \qquad (1.8)$$

In fact, if [a, b] = [0, 1], the Bernstein polynomial

$$p_n(x) = \sum_{k=0}^{n} C_n^k x^k (1-x)^{n-k} f\left(\frac{k}{n}\right) \qquad (1.9)$$

converges to $f(x)$ as $n \to \infty$.

Weierstrass theorems (and in fact their original proofs) postulate existence of some sequence of polynomials converging to a prescribed continuous function uniformly on bounded closed intervals.

However, polynomial approximants are not efficient in some sense. Take Lagrange interpolation for instance. If $x_1, x_2, \cdots, x_n$ are n distinct numbers at which the values of the function f are given, then the interpolating polynomial p is found from the Lagrange interpolation formula

$$p(x) = \sum_{k=1}^{n} f(x_k) \lambda_k(x) \qquad (1.10)$$

where $\lambda_k(x)$ is

$$\lambda_k(x) = \prod_{i=1, i \neq k}^{n} \frac{x - x_i}{x_k - x_i} \qquad (1.11)$$

The error in the approximation is given by

$$p(x) - f(x) = \frac{f^{(n)}[\zeta(x)]}{n!} \prod_{i=1}^{n} (x - x_i) \tag{1.12}$$

where $\zeta(x)$ is in the smallest interval containing x, x_1, x_n. Introducing the Lebesque function in the form of

$$\tau_n(x) = \sum_{k=1}^{n} |\lambda_k(x)| \tag{1.13}$$

and a norm

$$\|f\| = \max_{a \le x \le b} |f(x)| \tag{1.14}$$

then we have

$$\|p\| \le \|\tau_n\| \, \|f\| \tag{1.15}$$

This estimate is known to be sharp, that is, there exists a function for which the equality holds. First of all, it should be pointed out that equally spaced points may lead to bad consequences because it can be shown that

$$\|\tau_n\| \ge Ce^{n/2} \tag{1.16}$$

As n increases, function value becomes larger and larger, and entirely fails to approximate the function f. In other words, polynomial interpolation can be so bad that it does not yield the correct approximation at all.

This situation can be avoided if we have freedom to choose the interpolation points for the interval [a, b]. Chebyshev nodes in the following are known to be a good choice.

$$x_k = \frac{1}{2} \left[a + b + (a - b) \cos \frac{(k-1)\pi}{n-1} \right] \tag{1.17}$$

The maximum value for the associated Lebesque function in this case is

$$\|\tau_n^c\| < \frac{2}{\pi} \log n + 4 \tag{1.18}$$

Using Chebyshev nodes, we therefore obtain the following error bounds for polynomial interpolation

$$\|f - p\| \leq 2 \left(\frac{b-a}{4}\right)^n \max_{a \leq x \leq b} \frac{\left|f^{(n)}(\xi_n)\right|}{n!} \tag{1.19}$$

where Π_n denotes the linear space of all polynomials of degree n on [a, b]. We may further show that

$$\frac{\|f - p\|}{1 + \|\tau_n^c\|} \leq \|f - p\| \tag{1.20}$$

Thus by using the best interpolation scheme, we can still only hope to reduce the error from interpolation using Chebyshev nodes by less than a factor

$$\frac{2}{\pi} \log n + 5 \tag{1.21}$$

In summary, nearly the best approximation we are able to obtain cannot be better than that specified by Eq. (1.21). The worst case, however, can be as bad as divergent as specified by Eq. (1.16). Therefore, the above analysis outlines the accuracy we are able to obtain if polynomial interpolation is employed.

Example 1.3 *Runge example*

This is a well-known numerical example studied by Runge when he interpolated data based on a simple function of

$$y = \frac{1}{1 + 25x^2} \tag{1.22}$$

on an interval of [-1, 1].

For example, take six equidistantly spaced points in [-1, 1] and find y at these points as given in Table 1.1. Now through these six points, we can pass a fifth order polynomial

$$f_5(x) = 0.56731 - 1.7308x^2 + 1.2019x^4, \quad -1 \leq x \leq 1 \tag{1.23}$$

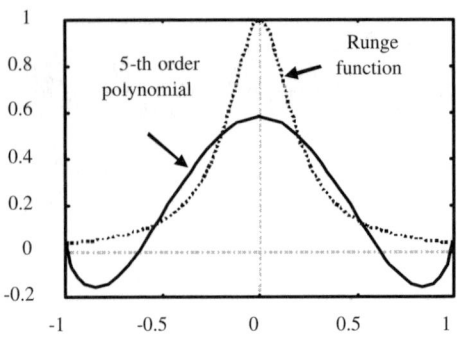

FIGURE 1.1:   The 5th-order interpolating polynomial vs exact function.

TABLE 1.1:   Six unequidistantly spaced points on $[-1, 1]$

x	$y = \frac{1}{1+25x^2}$
-1.0	0.03846
-0.6	0.1
-0.2	0.5
0.2	0.5
0.6	0.1
1.0	0.03846

TABLE 1.2:   Six unequidistantly spaced points on $[-1, 1]$

x	$y = \frac{1}{1+25x^2}$
-1.0	0.03846
-0.809	0.05759
-0.309	0.29520
0.309	0.29520
0.809	0.5759
1.0	0.03846

On plotting the fifth-order interpolating polynomial and the given function in Fig. 1.1, we immediately conclude that the two sets of results do not match well. Except at the six given points, the interpolating polynomial is far away from the true values. Even the variation trends of the two curves are different in ranges [-1, -0.25] and [0.25, 1].

We may consider choosing more points in the interval [-1, 1] to get a better match, but it diverges even more. In fact, Runge found that as the order of the polynomial becomes infinite, the polynomial diverges in the interval of $-1 < x < 0.726$ and $0.726 < x < 1$.

How much can we improve the situation if Chebyshev nodes are used? Reconsider this problem, but take six unequidistantly spaced points in [-1, 1] calculated from Eq. (1.22) and find y at these points as given in Table 1.2.

Now through these six points, we can pass a fifth-order polynomial

$$f_5(x) = 0.355 - 0.716x^2 + 0.399x^4, -1 \le x \le 1 \qquad (1.24)$$

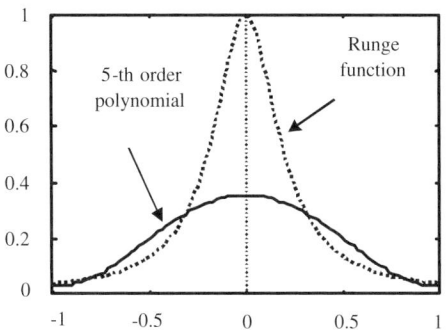

FIGURE 1.2:   The 5th-order polynomial on Chebyshev nodes.

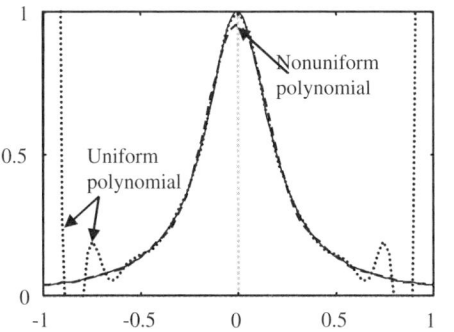

FIGURE 1.3:   Comparison of various interpolation methods for 20 points. The solid line is the Runge function.

On plotting the fifth-order polynomial and the original function in Fig. 1.2, we can see that the two sets of results do not match well this time around the central part. So Chebyshev nodes remove instability at the cost of loosing accuracy around the central part. This is natural if we look at Eq. (1.17) that puts more points near the two ends.

Chebyshev nodes are nearly optimal. In other words, even if we employ an alternative better than Chebyshev interpolation, we would not gain much.

As the number of points is increased to 20, the difference between the Runge function and nonuniform polynomial is not significant, but it remains distinguishable. The uniform polynomial behaves badly around the two ends, as shown in Fig. 1.3.

1.2.2 Fourier expansion basis

The polynomial approximation is suitable for most engineering problems. But for some problems, especially for those with periodic behaviors, Fourier series expansion could be a better choice for the true solution instead of polynomial approximation.

For a continuous function $f(x)$ on the interval $[0, 2\pi]$, the Fourier series expansion can be given by

$$f(x) = a_0 + \sum_{k=1}^{\infty} a_k \cos kx + \sum_{k=1}^{\infty} b_k \sin kx \qquad (1.25)$$

where the coefficients a_0, a_k and b_k are expressed as

$$a_0 = \frac{1}{2\pi} \int_0^{2\pi} f(x) dx \qquad (1.26a)$$

$$a_k = \frac{1}{\pi} \int_0^{2\pi} f(x) \cos kx dx \qquad (1.26b)$$

$$b_k = \frac{1}{\pi} \int_0^{2\pi} f(x) \sin kx dx \qquad (1.26c)$$

In practice, the truncated Fourier series expansion is usually used. Thus,

$$f(x) \approx F_{N+1}(x) = a_0 + \sum_{k=1}^{N/2} a_k \cos kx + \sum_{k=1}^{N/2} b_k \sin kx \qquad (1.27)$$

The convergence of the above expansion to $f(x)$ as N tends to infinity is guaranteed by Weierstrass's second theorem.

Theorem 1.4 *Let $f(x)$ be a continuous function on the interval $[0, 2\pi]$. Then for any $\varepsilon > 0$, there exists an integer n and a trigonometric S_n such that the inequality*

$$\max_{x \in [0, 2\pi]} |f(x) - S_n(x)| < \varepsilon \tag{1.28}$$

is satisfied, where $S_N(x) = a_0 + \sum_{k=1}^{n} a_k \cos kx + \sum_{k=1}^{n} b_k \sin kx$.

The proof of this theorem can be found in the book of Achieser (1992). It is shown that the approximation in Eq.(1.28) satisfies the operations of vector addition and scalar multiplication. So, $F_{N+1}(x)$ consists of a linear vector space $\mathbf{V}_{N+1}$ in which there exists a linearly independent set of base vectors

$$1, \cos x, \sin x, \ldots, \cos(Nx/2), \sin(Nx/2) \tag{1.29}$$

It is of great importance to determine the function values at discrete nodes for the numerical solution of a partial differential equation. Therefore, the Fourier expansion should be expressed in discrete form. Supposing that x_i, $i = 0, 1, \ldots, N$ are the coordinates of $(N+1)$ nodes on the interval $[0, 2\pi]$, and $f(x_i)$ is the function values at the i-th point, the Fourier expansion $F_{N+1}(x)$ can be expressed as follows:

$$F_{N+1}(x) = \sum_{i=0}^{N} f(x_i) \cdot g_i(x) \tag{1.30}$$

where

$$g_i(x) = \frac{G(x)}{G^{(1)}(x_i) \cdot \sin[(x - x_i)/2]} \tag{1.31a}$$

$$G(x) = \prod_{k=0}^{N} \sin \frac{x - x_k}{2} \tag{1.31b}$$

$$G^{(1)}(x) = \prod_{k=0, k \neq i}^{N} \sin \frac{x_i - x_k}{2} \tag{1.31c}$$

It is clear that the coefficients $g_i(x)$, $i = 0, 1, \ldots, N$, form a set of linearly independent vectors in $\mathbf{V}_{N+1}$. Thus it is also a basis of $\mathbf{V}_{N+1}$.

Besides polynomials and trigonometric functions, other types of functions may also be used as bases. For example, Radial Basis Functions are such an option (see Shu et al., 2004).

1.3   Differential quadrature

Gauss quadrature is a numerical integration method. Its basic idea is to approximate a definite integral with a weighted sum of integrand values at a group of nodes in the form of

$$\int_a^b f(x)dx \approx \sum_{j=1}^N w_j f(x_j) \tag{1.32}$$

where $x_j(j = 1, 2, \cdots, N)$ are nodes and w_j are weighting coefficients. They are determined by solving a system of linear equations.

Extending Gauss quadrature to finding the derivatives of various orders of a differentiable function gives rise to differential quadrature (Bellman and Casti, 1971; Bellman et al., 1972). In other words, the derivatives of a function are approximated by weighted sums of the function values at a group of nodes.

Suppose function $f(x)$ is sufficiently smooth on the interval $[x_1, x_N]$. Being sufficiently smooth means that the function is differentiable up to the order we want. On the interval, N distinct nodes are defined:

$$x_1 < x_2 < \cdots < x_N \tag{1.33}$$

The function values on these nodes are assumed to be

$$f_1, f_2 \cdots, f_N \tag{1.34}$$

Based on DQ, the first and second order derivatives on each of these nodes are given by

$$\frac{df(x_i)}{dx} \approx \sum_{j=1}^N a_{ij} f_j, \qquad i = 1,\, 2,\, \cdots,\, N \tag{1.35a}$$

$$\frac{d^2 f(x_i)}{dx^2} \approx \sum_{j=1}^N b_{ij} f_j, \qquad i = 1,\, 2,\, \cdots,\, N \tag{1.35b}$$

The coefficients a_{ij} and b_{ij} are the weighting coefficients (or simply weights) of the first- and second-order derivatives with respect to x, respectively.

To find the weights a_{ij} and b_{ij}, we need approximation theory introduced in the previous section. Our purpose is to construct a polynomial of x, which is of the form

$$f(x) \approx \sum_{k=1}^{N} \lambda_k(x) f(x_k) \tag{1.36}$$

satisfying

$$\lambda_k(x) = \begin{cases} 1 & x = x_k \\ 0 & x = x_j \neq x_k \end{cases} \quad k = 1, \cdots, N, j = 1, \cdots, N \tag{1.37}$$

Here $\lambda_k(x)$ has the delta function property that guarantees the interpolating polynomial $f(x)$ equal to the nodal value f_k at each node x_k. Direct use of Lagrange polynomial (1.10) yields immediately

$$\lambda_k(x) = \frac{L_k(x)}{L_k(x_k)} \quad L_k(x) = \prod_{\substack{j=1 \\ j \neq k}}^{N} (x - x_j) \tag{1.38}$$

Then the first and second derivatives are obtained by differentiating the terms behind the product sign $\prod$ with respect to x. They are

$$\frac{df(x)}{dx} = \sum_{k=1}^{N} a_k(x) f_k \quad and \quad \frac{d^2 f(x)}{dx^2} = \sum_{k=1}^{N} b_k(x) f_k \tag{1.39}$$

where $a_k(x)$ are the coefficients for the first derivative, obtained by differentiating $L_k(x)$ in Eq. (1.38). After some manipulations, we obtain the explicit formulas for $a_k(x)$ in the form of

$$a_k(x) = \frac{d\lambda_k(x)}{dx} = \frac{1}{x - x_k} \sum_{\substack{\ell=1 \\ \ell \neq k}}^{N} \frac{L_\ell(x_\ell)}{L_k(x_k)} \quad \text{if} \quad x \neq x_k \quad k = 1, \cdots, N \tag{1.40}$$

if x is not coincident with any node and

$$a_k(x) = \sum_{\substack{\ell=1 \\ \ell \neq k}}^{N} a_\ell(x) \frac{L_\ell(x_\ell)}{L_k(x_k)} \quad \text{if} \quad x = x_k, k = 1, \cdots, N \tag{1.41}$$

if x is coincident with one of the nodes. The coefficients for the second-order derivative are obtained in the similar way as

$$b_k(x) = \frac{da_k(x)}{dx} = \frac{d^2 \lambda_k(x)}{dx^2} = \frac{1}{x - x_k} \sum_{\substack{\ell=1 \\ \ell \neq k}}^{N} a_\ell(x) \frac{L_\ell(x_\ell)}{L_k(x_k)} - \frac{a_k(x)}{x - x_k} \tag{1.42}$$

if $\quad x \neq x_k, k = 1, \cdots, N$

and

$$b_k(x) = \frac{1}{2} \sum_{\substack{\ell=1 \\ \ell \neq k}}^{N} b_\ell(x) \frac{L_\ell(x_\ell)}{L_k(x_k)}, \quad \text{if } x = x_k, k = 1, \cdots, N \qquad (1.43)$$

It is obvious that if N nodes are used, the interpolating polynomial degree is $(N-1)$. Using this idea, it can be shown that the approximation errors for the first-order derivative (R^1) and the second-order derivative (R^2) are

$$R^1(x_k) = \frac{f^{(N)}(\xi) W^{(1)}(x_k)}{N!} \quad k = 1, \ldots, N \qquad (1.44a)$$

$$R^2(x_k) = \frac{f^{(N)}(\xi) W^{(2)}(x_k)}{N!} \quad k = 1, \ldots, N \qquad (1.44b)$$

where $W(x) = \prod_{j=1}^{N} (x - x_j)$. These residual estimates show that very high accuracy can be achieved if the number of nodes N is large. Accuracy is proportional to N or its powers. By "its powers" we mean here that accuracy may also be proportional to squared or cubic N, or even higher order term of N.

Therefore, explicit formulas for these coefficients are found to be (Quan and Chang, 1989a,b)

$$a_{ij} = \frac{1}{(x_j - x_i)} \prod_{\substack{k=1 \\ k \neq i,j}}^{N} \frac{x_i - x_k}{x_j - x_k}, \quad i, j = 1, 2, \cdots, N, \quad i \neq j \qquad (1.45a)$$

$$a_{ii} = -\sum_{\substack{j=1 \\ j \neq i}}^{N} a_{ij}, \quad i = 1, 2, \cdots, N \qquad (1.45b)$$

$$b_{ij} = 2[a_{ij} a_{ii} - \frac{a_{ij}}{x_j - x_i}], \quad i, j = 1, 2, \cdots, N, \quad i \neq j \qquad (1.45c)$$

$$b_{ii} = -\sum_{\substack{j=1 \\ j \neq i}}^{N} b_{ij}, \quad i = 1, 2, \cdots, N \qquad (1.45d)$$

Equations (1.40) to (1.45) are the key formulas in DQ. Similarly we may obtain formulas for higher order derivatives by using the higher order weighting coefficients, which are expressed as $e_{ij}^{(m)}$ to avoid confusion. They are characterized by recurrence

$$e_{ij}^{(m)} = m(a_{ij}e_{ii}^{(m-1)} - \frac{e_{ij}^{(m-1)}}{x_j - x_i}), i, j = 1, 2, \cdots, N, i \neq j, m = 2, 3, \ldots, N-1$$

$$(1.46a)$$

$$e_{ii}^{(m)} = -\sum_{\substack{j=1 \\ j \neq i}}^{N} e_{ij}^{(m)}, i = 1, 2, \cdots, N \qquad (1.46b)$$

Here we assume that $a_{ij} = e_{ij}^{(1)}$ and $b_{ij} = e_{ij}^{(2)}$. The error estimated for Eqs. (1.44) is determined by

$$Error \leq \frac{1}{N!} \left| \prod_{j=1}^{N} (x - x_j) \right| \max \left| f^{(N)}(x) \right| \qquad (1.47)$$

If the N-th order derivative is finite, then the error decreases to zero very fast because of the presence of the N-th factorial in the denominator.

Shu (2000a) has proven that Eqs. (1.35) is $(N-1)$-th order accurate, and thus it can yield very accurate results using a significantly small number of nodes.

But Eqs. (1.40) ~ (1.43) are sensitive to how nodes are distributed. If nodes are uniformly distributed, it does not converge to the true function as $N \to \infty$. This has been demonstrated in Example 1.3 through Runge example. So, uniform spacing of nodes may lead to disastrous consequences. Among non-uniform spacing of nodes which ensure the convergence of Eqs. (1.40) ~ (1.43), the Chebyshev nodes defined by Eq. (1.17) are nearly optimal. For convenience, it is rewritten here in the form of

$$x_i = x_1 + \frac{1}{2}(1 - \cos \frac{i-1}{N-1}\pi)(x_N - x_1), \quad i = 1, 2, \cdots, N \qquad (1.48)$$

Equation (1.20) ensures that Eq. (1.36) defined on Chebyshev nodes converges to the true function. Nodal distribution defined by Eq. (1.48) is most frequently employed in various DQ formulations.

Example 1.4 *Derivative evaluation using DQ*

Consider triangular function $\cos(\theta)$ on the interval $[0, 2]$. Its first and second derivatives are $-\sin(\theta)$ and $-\cos(\theta)$, respectively. These two derivatives are given in Figs. 1.4 and 1.5, denoted by dots. Differential Quadrature is applied to numerically evaluate the two derivatives.

If uniformly distributed nodes are employed, the numerical values of these derivatives are shown in Fig. 1.4, where 41 nodes are used. Both derivatives

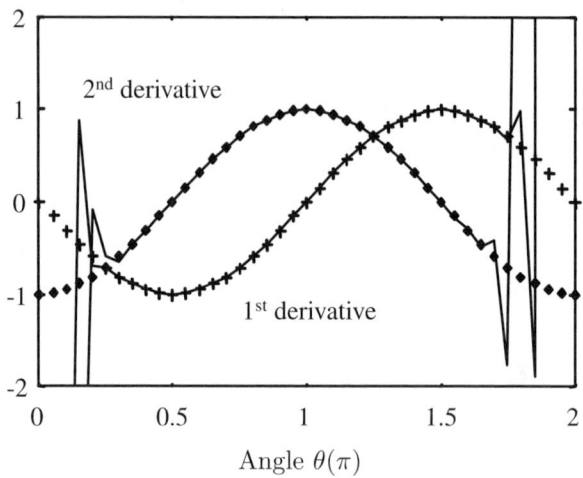

FIGURE 1.4: 1st and 2nd derivatives of function obtained from uniform-node DQ (solid lines) and analytical method (dots).

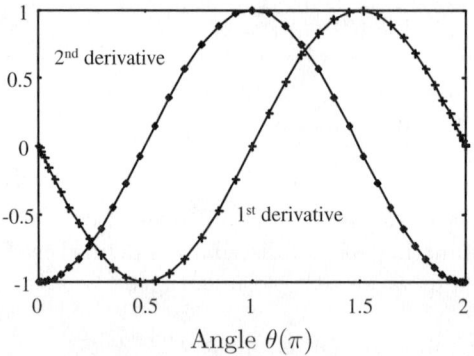

FIGURE 1.5: 1st and 2nd derivatives of function obtained from Chebyshev node DQ (solid lines) and analysis (dots).

evaluated by use of DQ are in good agreement with the true values in the middle portion of the interval $\theta \in [0, 2\pi]$, but are not in agreement with the true values near the left- and right-ends exhibiting significant divergence. This numerically demonstrates the fallacy of uniform node distribution in DQ. Thus, it is always recommended that non-uniform node distribution like Chebyshev nodes be employed.

Figure 1.5 shows the results obtained from DQ evaluated on Chebyshev node distribution given by Eq. (1.48). The results are significantly improved, in perfect accord with the exact results. Thus, it is clearly illustrated from Figs. 1.4 and 1.5 that node distributions play a crucial role in the application of DQ. This is a research topic of interest. The reader is referred to monographs on approximation and DQ methods, for example Zong (2006) and Shu (2000a).

1.4 Direct differential quadrature method

Solving differential equations (ordinary and partial) is one of the most important ways engineers, physicists and applied mathematicians use to tackle a practical problem. Although analytical techniques typical of which is separation of variable, their applications are restricted to over-simplified problems. Numerical methods, the most famous of which is the Finite Element Method (FEM), are widely applied today to solve more practical and complicated problems.

The Finite Element Method is a low-order numerical tool. The displacements predicted by the FEM are continuous, but the stresses, which are essentially first order derivatives of displacements, are discontinuous in the ordinary FEM. We therefore say that the accuracy of FEM is low, thus requiring fine meshes. To improve the accuracy, we may increase mesh density using more elements in a domain. This is not an efficient way because the rate for accuracy improvement is very slow. An alternative which exhibits high accuracy at the cost of low mesh density is to use elements of high order accuracy. This demands an extensive study of high-order numerical methods. Direct differential quadrature method is such a technique, which also provides a cost-effective tool for solving many nonlinear partial differential equations. In the following, however, a linear example is used to illustrate the direct DQ method.

The key procedure in the direct DQ method is to approximate the derivatives in a differential equation by Eq. (1.39). Substituting Eq. (1.39) into the governing equations and equating both sides of the governing equations, we obtain simultaneous equations which can be solved by use of Gauss

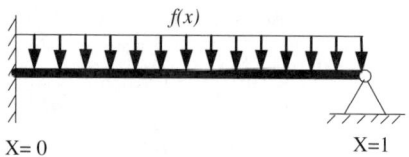

FIGURE 1.6:   Euler beam under uniform loading.

elimination or other methods. We will elaborate this point through the following example.

Example 1.5 *Bending of Euler beam*
   A uniform Euler beam under pure bending shown in Fig. 1.6 is governed by the following fourth-order Euler-beam equation:

$$EI\frac{d^4w}{dx^4} + f(x) = 0 \qquad 0 < x < L \tag{1.49}$$

where EI is the flexural rigidity of the beam, $f(x)$ the external distributed load, and L the length of the beam. Equation (1.49) may be further transformed to a dimensionless form for the convenience of calculation. With non-dimensionalisation procedures neglected, we obtain

$$\frac{d^4W}{dX^4} + F(x) = 0 \qquad 0 < X < 1 \tag{1.50}$$

where $X = x/L$, $W = w/a$, $a = f_0L/EI$, $F(x) = f(x)/f_0$, f_0 is a constant for non-dimensionalisation.
   As shown in Fig. 1.6 the beam is clamped at the left end and simply supported at the right end. The boundary conditions are then

$$W = \frac{dW}{dX} = 0 \qquad \text{at } X = 0 \tag{1.51a}$$

$$W = \frac{d^2W}{dX^2} = 0 \qquad \text{at } X = 1 \tag{1.51b}$$

The exact solution to this problem is $W(X) = \frac{1}{48}X^2(5X - 2X^2 - 3)$.
   Divide the beam domain $0 \le X \le 1$ into $N = 21$ nodes distributed in the form of Eq. (1.48). Writing the deflection $W(x)$ in the form of Eq. (1.32) and substituting it in Eq. (1.50) we obtain

$$\sum_{j=1}^{N} e_{ij}^{(4)} W_j = -F(X_i) \quad \text{for } i = 1, 2, \ldots N \tag{1.52}$$

where $e_{ij}^{(4)}$ are given in Eq. (1.46).

Inserting Eq. (1.35) into Eq. (1.51) we obtain the boundary conditions in the discrete form of

$$W_1 = 0, \quad \sum_{j=1}^{N} e_{1j}^{(1)} W_j = 0 \tag{1.53a}$$

$$W_N = 0, \quad \sum_{j=1}^{N} e_{Nj}^{(2)} W_j = 0 \tag{1.53b}$$

Consider a uniform load with the value $f(x) = f_0$. Then $F(X) = 1$. The deflections of the beam at the nodes is then governed by the system of linear equations

$$\sum_{j=3}^{N-2} C_{ij} W_j = -1 \quad for \quad i = 3, 4, \ldots, N - 2 \tag{1.54}$$

where

$$C_{ij} = e_{ij}^{(4)} + \frac{e_{i,2}^{(4)}(a_{1,j}b_{N,N-1} - a_{1,N-1}b_{N,j}) + e_{i,N-1}^{(4)}(a_{1,2}b_{N,j} - a_{1,j}b_{N,2})}{b_{N,2} \cdot a_{1,N-1} - a_{1,2}b_{N,N-1}} \tag{1.55}$$

Solving Eq.(1.54) using Gauss elimination technique yields the numerical result of the problem. In the mathematical compendium (Chapter 8), a short introduction to Gauss elimination is given for the convenience of the reader.

The numerical and exact solutions are listed in Table 1.3 for comparison. It can be seen from the table that the results obtained from direct DQ method are very close to the exact solutions and up to 11 digits accuracy can be achieved by using only 21 nodes. The numerical results obtained from direct DQ method are also compared with the exact solutions in Fig. 1.7. No difference is observed between the DQ results denoted by the solid line and the exact solutions denoted by open circles. The computational effort is small due to the small number of nodes used.

In summary, direct DQ method is composed of the following procedures:

(1) The function to be determined is replaced by a group of function values at a group of selected nodes. Chebyshev nodes are strongly recommended for numerical stability.

(2) Approximate derivatives in a differential equation by these N unknown function values.

(3) Form a system of linear equations and

(4) Solving the system of linear equation yields the desired unknowns.

It should be noted that these are basic procedures, thus allowing for suitable adaptations for a particular problem.

TABLE 1.3:   Deflection of a beam under uniformly distributed load

X	W (Exact)	W (DQ,N=21)	Error (%)
0	0.0000000000E+0	0.000000000E+0	0.0
0.00616	-2.34415033E-6	-2.34415033E-6	-3.45620757E-11
0.02447	-3.59174775E-5	-3.59174775E-5	-3.35252396E-11
0.0545	-1.69126555E-4	-1.69126555E-4	-3.28543093E-11
0.09549	-4.82675398E-4	-4.82675398E-4	-3.32891966E-11
0.14645	-1.03241288E-3	-1.03241288E-3	-3.43403416E-11
0.20611	-1.81817534E-3	-1.81817534E-3	-3.57907255E-11
0.273	-2.77015180E-3	-2.77015180E-3	-3.73696574E-11
0.34549	-3.75816531E-3	-3.75816531E-3	-3.90387931E-11
0.42178	-4.62129499E-3	-4.62129499E-3	-4.10473715E-11
0.5	-5.20833314E-3	-5.20833314E-3	-4.38982201E-11
0.57822	-5.41611948E-3	-5.41611948E-3	-4.73707426E-11
0.65451	-5.21393933E-3	-5.21393933E-3	-5.09543520E-11
0.727	-4.64733857E-3	-4.64733857E-3	-5.42551513E-11
0.79389	-3.82187643E-3	-3.82187643E-3	-5.74855657E-11
0.85355	-2.87383765E-3	-2.87383765E-3	-5.94420824E-11
0.90451	-1.93844959E-3	-1.93844959E-3	-6.07303242E-11
0.9455	-1.12560121E-3	-1.12560121E-3	-5.93151190E-11
0.97553	-5.08927378E-4	-5.08927378E-4	-4.78693501E-11
0.99384	-1.28232166E-4	-1.28232166E-4	1.05687423E-11
1	0.00000000E+0	0.00000000E+0	0

1.5   Block marching in time with DQ discretization

In the previous section DQ is meant solely for spatial differentiation. Mathematically speaking, however, there should be no difficulty if DQ is applied

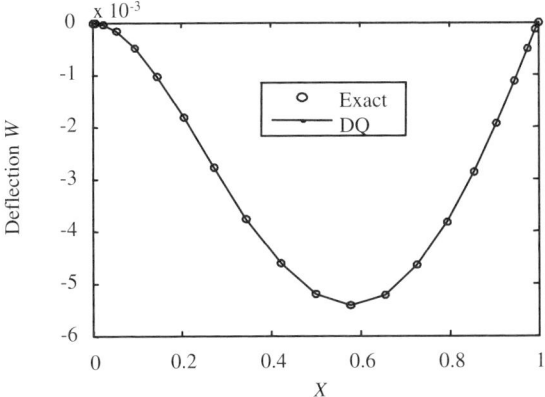

FIGURE 1.7: Deflection of the beam under bending.

to find time-dependent derivatives. That is, the derivatives of a function with respect to spatial variable x and with respect to temporal variable t can be found by use of the same procedures. Based on such a guess, Shu et al (2002b) made an attempt to use DQ for temporal discretization.

A temporal discretization scheme is either implicit or explicit. The Euler forward scheme, Runge–Kutta schemes, Adams–Bashforth schemes and Adams–Moulton schemes are examples of explicit schemes while the Euler backward scheme and Crank–Nicolson scheme are examples of implicit schemes. An explicit scheme can update the solution by a simple algebraic formula with the use of a very small time step size due to the limitation of stability condition. In contrast, implicit scheme allows for a larger time step size but they involve the solution of a coupled set of algebraic equations. The common feature of conventional explicit and implicit schemes is that the numerical solution in the time direction is obtained layer by layer (or called level by level). In other words, the computation is marched in time direction layer by layer. Using this way, the numerical solution at a time level $(n + 1)$ depends on the solutions at its previous levels. Due to accumulation of numerical errors, it can be expected that the accuracy of the numerical solution at the time level $(n+1)$ is less than that at the time level n. This tendency of decreasing the accuracy of numerical solution is undesirable for many time-dependent problems. To keep accurate numerical results for very long time, conventional explicit or implicit schemes need to use very small time step size. To remove the limitation on the time step while keeping high order accuracy, a more efficient temporal discretization method is demanded. Although some high order time difference schemes have been proposed by various authors, they are all the layer-marching method. In other words, they also obtain the solution in the time direction layer by layer. Hence, although the accuracy of numerical solution in the time direction is improved, the large accumulation of numerical errors for a long time remains unresolved.

In this section, an efficient approach developed by Shu (2002a) is introduced. The approach is based on the block-marching in time and DQ discretization in both the spatial and temporal directions.

The block-marching technique yields the solution in the time direction block by block. In each block, there are several time levels (layers). The solutions at these levels are obtained simultaneously and they have the same order of accuracy. Therefore, the accumulation of numerical errors is block by block instead of layer by layer, which can be very small since the number of blocks can be much less than that of time levels.

We will use a two-dimensional, time-dependent problem to illustrate this technique. Consider the solution of the following equation

$$\frac{\partial f}{\partial t} + \frac{\partial f}{\partial x} + \frac{\partial f}{\partial y} = c \left(\frac{\partial^2 f}{\partial x^2} + \frac{\partial^2 f}{\partial y^2} \right) \qquad (1.56)$$

on the domain $0 \leq x \leq L_x, 0 \leq y \leq L_y, 0 \leq t < \infty$. This is a mixed initial/boundary value problem. On the physical domain defined by $0 \leq x \leq L_x, 0 \leq y \leq L_y$, it is a boundary value problem. As such, the boundary conditions on the four boundaries of $x = 0, L_x$ and $y = 0, L_y$ must be imposed. In the time direction, it is an initial value problem, thus only one initial condition at $t = 0$ is provided. To use direct DQ method, a block-marching technique is employed by decomposing the semi-infinite domain in the time direction into many intervals, $\delta t_1, \delta t_2, \cdots$. Each time interval δt and the spatial domain, $0 \leq x \leq L_x, 0 \leq y \leq L_y$, form a block. The configuration of the block is depicted in Fig. 1.8. In each block, we distribute the mesh points respectively in the x, y and t direction, and discretize all the spatial and temporal derivatives by a numerical scheme. Note that each block may include many time levels (layers). Numerical solutions at all time levels in each block are obtained simultaneously, and have the same order of accuracy. It should be mentioned that each block is considered as an open box in the numerical computation. We proceed from block 1 (BLK1). For this block, the function values at the bottom boundary ($t = 0$) are given from the initial condition. The function values at interior points and the top boundary ($t = \delta t_1$) are considered as unknowns, which are given from the solution to Eq. (1.56). After obtaining the solution in BLK1, we march to BLK2, where the bottom boundary ($t = \delta t_1$) is exactly the same as the top boundary of BLK1. In other words, the solution at the top boundary of BLK1 is taken as the initial condition in BLK2. Numerical solution in BLK2 is obtained using the same procedures as in BLK1. We carry on this process until the specified time is reached.

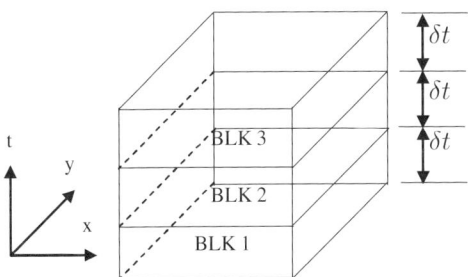

FIGURE 1.8: Configuration of block marching technique.

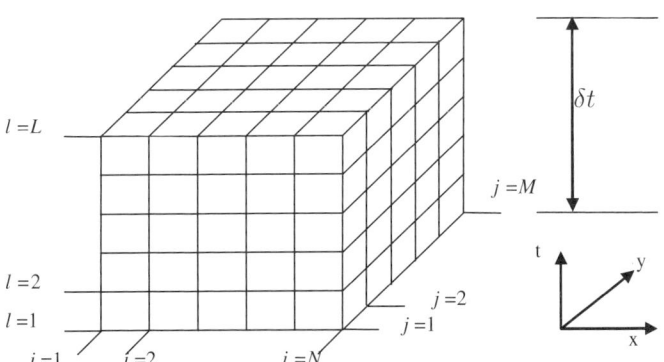

FIGURE 1.9: Discretization in three directions.

In the block-marching technique, the length scales in the x and y directions are fixed as L_x and L_y for different blocks. In this way, we can use the same DQ weighting coefficients in the L_x and L_y directions at different blocks. However, in the time direction, the length scale may be different at different blocks. This implies that different DQ weighting coefficients in the time direction may be used at different blocks. This may lead to complexity of numerical computation. To simplify the problem, we use the same length scale in the time direction for different blocks. Alternatively, we set $\delta t = \delta t_1 = \delta t_2 = \cdots$. Thereby, the mesh point distribution is exactly the same in each block, and the same DQ weighting coefficients can be applied to different blocks. In the K-th block, the non-uniform mesh point distribution is used in the x, y and t direction respectively, which is given in Eq. (1.48).

In the above, N and M are the number of nodes in the x and y direction respectively; L is the number of time levels in the block. The above mesh point distribution is shown in Fig. 1.9. The accuracy of DQ discretization depends on the number of mesh points. For the problem considered, only the first-order temporal derivative is involved. So, the DQ discretization of the temporal derivative has $(L\text{-}1)$-th order of accuracy (see Eq.(1.44)). Obviously, by adjusting the value of L, we can easily generate the high order discretization scheme to reduce the truncation error.

Applying DQ to each block, Eq. (1.56) can be discretized as

$$\sum_{k=1}^{N} \hat{a}_{lk} f_{ijk} + u_{ijl} \sum_{k=1}^{N} a_{ik} f_{kjl} + v_{ijl} \sum_{k=1}^{M} \bar{a}_{jk} f_{ikl} = \sum_{k=1}^{N} b_{ik} f_{kjl} + \sum_{k=1}^{M} \bar{b}_{jk} f_{ikl}$$

$$(1.57)$$

where $\mathbf{a}, \bar{\mathbf{a}}, \hat{\mathbf{a}}$ are the DQ weighting coefficients of the first-order derivatives with respect to x, y and t, respectively, and $\mathbf{b}, \bar{\mathbf{b}}$ are the DQ weighting coefficients of the second-order derivatives with respect to x and y, respectively. Equation (1.57) can be solved by SOR iterative method or other suitable methods. A short introduction to SOR is given in chapter 8.

Example 1.6 *Slow viscous flow*
Consider a two-dimensional slow viscous flow governed by time-dependent Navier Stokes. The vorticity-stream function formulation is taken as the governing equation for the problem, which can be written as $(Re = 1)$

$$\frac{\partial^2 \psi}{\partial x^2} + \frac{\partial^2 \psi}{\partial y^2} = \omega \tag{1.58}$$

$$\frac{\partial \omega}{\partial t} + u \frac{\partial \omega}{\partial x} + v \frac{\partial \omega}{\partial y} = \frac{\partial^2 \omega}{\partial x^2} + \frac{\partial^2 \omega}{\partial y^2} \tag{1.59}$$

where

$$u = \frac{\partial \psi}{\partial y}, v = -\frac{\partial \psi}{\partial x} \text{ and } \omega = \frac{\partial u}{\partial y} - \frac{\partial u}{\partial x} \qquad (1.60)$$

The computational domain for the problem is $0 \le x \le \pi/2$, $0 \le y \le \pi/2$, $0 \le t < \infty$. The boundary conditions for ψ are given by

$$\psi(0, y, t) = \cos(y)e^{-2t}, \quad 0 \le y \le \pi/2 \text{ and } t \ge 0 \qquad (1.61a)$$
$$\psi(\pi/2, y, t) = 0, 0 \le y \le \pi/2 \text{ and } t \ge 0 \qquad (1.61b)$$
$$\psi(x, 0, t) = \cos(x)e^{-2t}, \quad 0 \le x \le \pi/2 \text{ and } t \ge 0 \qquad (1.61c)$$
$$\psi(x, \pi/2, t) = 0, \quad 0 \le x \le \pi/2 \text{ and } t \ge 0 \qquad (1.61d)$$

and the initial condition for ψ is

$$\psi(x, y, 0) = \cos(x)\cos(y) \quad 0 \le x \le \pi/2, 0 \le y \le \pi/2 \qquad (1.62)$$

The exact solution to the problem is

$$\psi(x, y, t) = \cos(x)\cos(y)e^{-2t} 0 \le x \le \pi/2, 0 \le y \le \pi/2 \qquad (1.63)$$

from which we are able to obtain the expressions for u, v and ω of the form

$$u(x, y, t) = -\sin(x)\cos(y)e^{-2t} 0 \le x \le \pi/2, 0 \le y \le \pi/2 \qquad (1.64)$$

To apply the block-marching technique with DQ discretization, the semi-infinite domain in the time direction is decomposed into many intervals. For simplicity, the uniform time interval is used, which is noted as δt. Each δt and the physical domain of $0 \le x \le \pi/2$, $0 \le y \le \pi/2$ form a block. Since the length scales in every block are the same, the DQ weighting coefficients can be uniformly used at different blocks. The DQ weighting coefficients are first computed in BLK1, and then stored for the application in the following blocks. At each block, applying DQ to Eqs. (1.58) and (1.59) leads to

$$\sum_{k=1}^{N} b_{ik}\psi_{ikl} + \sum_{k=1}^{N} \bar{b}_{jk}\psi_{ikl} = \omega_{ijl} \qquad (1.65)$$

$$\sum_{k=1}^{N} \hat{a}_{lk}\omega_{ijk} + u_{ijl}\sum_{k=1}^{N} a_{ik}\omega_{kjl} + v_{ijl}\sum_{k=1}^{M} \bar{a}_{jk}\omega_{ikl} = \sum_{k=1}^{N} b_{ik}\omega_{kjl} + \sum_{k=1}^{M} \bar{b}_{jk}\omega_{ikl}$$
$$(1.66)$$

for $i = 2, 3, ..., N - 1; j = 2, 3, ..., M - 1; l = 2, 3, ..., L$,

where N and M are the numbers of mesh points in the x and y direction, respectively, L is the number of time levels in the block, $\mathbf{a}, \bar{\mathbf{a}}$, and $\hat{\mathbf{a}}$ are the DQ weighting coefficients of the first-order derivatives with respect to x, y and t, and $\mathbf{b}, \bar{\mathbf{b}}$ are the DQ weighting coefficients of the second-order derivatives with respect to x and y. The velocity components u, v in the block can also be approximated through DQ as

$$u_{ijl} = \sum_{k=1}^{M} \bar{a}_{jk} \psi_{ikl} \tag{1.67a}$$

$$v_{ijl} = -\sum_{k=1}^{N} \bar{a}_{ik} \psi_{kjl} \tag{1.67b}$$

The DQ-resulting Eqs. (1.65) and (1.66) in each block are algebraic equations, which can be solved by some iterative methods. In the present computation, SOR iterative method is used. To validate the accuracy of numerical results, the following relative errors are computed,

$$Er_{\psi} = \sum |\psi_{\text{numerical}} - \psi_{\text{exact}}| / \sum |\psi_{\text{exact}}| \tag{1.68a}$$

$$Er_{\omega} = \sum |\omega_{\text{numerical}} - \omega_{\text{exact}}| / \sum |\omega_{\text{exact}}| \tag{1.68b}$$

$$Er_{u} = \sum |u_{\text{numerical}} - u_{\text{exact}}| / \sum |u_{\text{exact}}| \tag{1.68c}$$

$$Er_{v} = \sum |v_{\text{numerical}} - v_{\text{exact}}| / \sum |v_{\text{exact}}| \tag{1.68d}$$

The convergence criteria in each block are set as

$$|R_{\psi}|_{\text{max}}^{(i)} / |R_{\psi}|_{\text{max}}^{(0)} < \varepsilon_1 \tag{1.69a}$$

$$|R_{\omega}|_{\text{max}}^{(i)} / |R_{\omega}|_{\text{max}}^{(0)} < \varepsilon_2 \tag{1.69b}$$

where $|R_{\psi}|_{\text{max}}^{(i)}$ is the maximum absolute value of residual for Eq. (1.65) at i-th iteration in a block, while $|R_{\psi}|_{\text{max}}^{(0)}$ is the maximum absolute value of residual for Eq. (1.65) at the beginning of iteration in the block. $|R_{\omega}|_{\text{max}}^{(i)}$, $|R_{\omega}|_{\text{max}}^{(0)}$ have the similar meaning to $|R_{\psi}|_{\text{max}}^{(i)}$, $|R_{\psi}|_{\text{max}}^{(0)}$, but they are the maximum absolute values of residuals for Eq. (1.66). Here ε_1 and ε_2 are taken the same and fixed at 10^{-8}. In other words, for the computation in each block, the numerical results are considered to be convergent when the maximum absolute value of residuals is reduced by eight orders.

For the numerical computation, the time domain is limited to $0 \leq t \leq 20$. The efficiency of block-marching technique is tested in terms of the total iteration number, NTI, and the CPU time required when the computation reaches $t = 20$. Note that the number of total iterations (NTI) is the sum of all iteration numbers required at all blocks. Since DQ is a global method with high order of accuracy, it was found from numerical experiments that the use of seven nodes in the x and y directions respectively can provide accurate numerical results for this example. Thus, in the following studies, N and M are fixed at 7. On the other hand, the length scale in the time direction in each block, δt, is adjustable. Thus, the number of time levels in each block, L, is not fixed, which depends on the choice of δt. The criterion for the choice of δt and L is set in such a way that approximately the same order of accuracy can be achieved in both the time and spatial directions. For example, when $\delta t = 1$, the mesh size is chosen as $N \times M \times L = 7 \times 7 \times 7$. There is no improvement in accuracy if more than seven nodes are used in the time direction.

TABLE 1.4: Relative errors, NTI and CPU times for the case of $\delta t = 0.25 (N = M=7, L=5)$

$t(s)$	NTI	Er_ϕ	Er_ω	Er_w	Er_φ	CPU(s)
0.25	70	5.56×10^{-7}	8.63×10^{-7}	7.52×10^{-6}	7.53×10^{-6}	0.397
0.50	2×70	4.64×10^{-7}	9.46×10^{-7}	7.75×10^{-6}	7.75×10^{-6}	0.854
0.75	3×70	4.70×10^{-7}	9.66×10^{-7}	7.75×10^{-6}	7.75×10^{-6}	1.322
1.00	4×70	4.71×10^{-7}	9.71×10^{-7}	7.75×10^{-6}	7.75×10^{-6}	1.787
2.00	8×70	4.72×10^{-7}	9.72×10^{-7}	7.75×10^{-6}	7.75×10^{-6}	3.647
3.00	12×70	4.72×10^{-7}	9.73×10^{-7}	7.75×10^{-6}	7.75×10^{-6}	5.523
5.00	20×70	4.72×10^{-7}	9.73×10^{-7}	7.75×10^{-6}	7.75×10^{-6}	9.182
10.00	40×70	4.72×10^{-7}	9.73×10^{-7}	7.75×10^{-6}	7.75×10^{-6}	18.34
20.00	80×70	4.72×10^{-7}	9.73×10^{-7}	7.75×10^{-6}	7.75×10^{-6}	36.73

The effect of δt on accuracy and efficiency of numerical results is investigated in detail. At first, δt is chosen as 0.25. For this case, L is taken as 5. In other words, five time levels are involved in each block. For the selected time interval $0 \leq t \leq 20$, 80 blocks will be used when $\delta t = 0.25$. The relative four errors defined by Eqs. (1.68a) to (1.68d), NTI, and CPU time required on a computer of the type LEONIS are listed in Table 1.4 for the case of $\delta t = 0.25$. It can be seen clearly from Table 1.4 that the accuracy of present results is very high. From the fourth block, the relative error for ψ is kept around 4.7×10^{-7}, the relative error for ω remains around 9.7×10^{-7}, while the relative errors for u and v are the same and kept around 7.75×10^{-6}. Obviously, the accuracy for ω and ψ is higher than that for u and v. This is because ω and ψ are directly computed from the governing Eqs.(1.65) and (1.66) while u and v are indirectly computed from the stream function ψ by using Eqs. (1.67a) and (1.67b).

It was found that the iteration number to satisfy Eqs. (1.69a) and (1.69b) in each block is the same and equals to 70 for the case of $\delta t = 0.25$. Therefore, when t reaches 20, NTI is $80 \times 70 = 5600$, and the CPU time required is 36.73 s. Next, we increase δt from 0.25 to 0.50, and the results of this case are listed in Table 1.5.

TABLE 1.5: Relative errors, NTI and CPU times the case of $\delta t = 0.50(N = M = L=7)$

$t(s)$	NTI	Er_ϕ	Er_ω	Er_w	Er_φ	CPU(s)
0.5	76	4.51×10^{-7}	4.86×10^{-7}	7.64×10^{-6}	7.64×10^{-6}	0.597
1.0	76+72	4.72×10^{-7}	4.08×10^{-7}	7.70×10^{-6}	7.70×10^{-6}	1.297
1.5	76+2×70	4.74×10^{-7}	4.11×10^{-7}	7.70×10^{-6}	7.70×10^{-6}	1.980
2.0	76+3×70	4.74×10^{-7}	4.11×10^{-7}	7.70×10^{-6}	7.70×10^{-6}	2.665
3.0	76+5×70	4.74×10^{-7}	4.11×10^{-7}	7.70×10^{-6}	7.70×10^{-6}	4.014
5.0	76+9×70	4.74×10^{-7}	4.11×10^{-7}	7.70×10^{-6}	7.70×10^{-6}	6.754
10.0	76+19×70	4.74×10^{-7}	4.11×10^{-7}	7.70×10^{-6}	7.70×10^{-6}	13.595
20.0	76+39×70	4.74×10^{-7}	4.11×10^{-7}	7.70×10^{-6}	7.70×10^{-6}	27.277

Since δt is increased, the number of time levels is also increased from 5 to 7. It was found that for this case, the accuracy for ψ, u and v are kept the same as the previous case, but the accuracy for ω is improved. The number of iterations needed is 76 for the first block, and 72 for the following 39 blocks. So, the number of total iterations for this case is NTI $= 76 + 39 \times 72 = 2884$ when $t = 20$, which is almost the half of the previous case. However, the CPU time required for this case is slightly reduced from 36.73 to 27.28 s. The reduction of CPU time is not so promising. The reason is that since L is changed from 5 to 7, the number of unknowns for this case is much larger than the previous case. So, each iteration requires much more CPU time. Furthermore, when the number of time levels is fixed at 7, and δt is increased to 1, it was found that the accuracy of numerical results can be kept the same as the case of $\delta t = 0.5$, but the efficiency is greatly improved. The results of this case are shown in Table 1.6. From Table 1.6, it can be observed that for all the 20 blocks, the number of iterations to satisfy convergence criteria Eqs. (1.6a) and (1.6b) is the same and equals to 70, which is slightly less than that for the case of $\delta t = 0.5$. So, NTI for this case is $20 \times 70 = 1400$. Since the number of unknowns for this case is the same as the case of $\delta t = 0.5$, the CPU time required to reach $t=20$ is just half (13.26 s) of the previous case.

TABLE 1.6: Relative errors, NTI and CPU times given by present method for the case of $\delta t = 1.0 (N = M = L=7)$

$t(s)$	NTI	Er_ϕ	Er_ω	Er_w	Er_φ	CPU(s)
1.00	70	8.31×10^{-7}	3.21×10^{-6}	7.25×10^{-6}	7.26×10^{-6}	0.557
2.00	2×70	7.85×10^{-7}	3.13×10^{-6}	8.47×10^{-6}	8.47×10^{-6}	1.226
3.00	3×70	7.85×10^{-7}	3.13×10^{-6}	8.48×10^{-6}	8.48×10^{-6}	1.907
4.00	4×70	7.85×10^{-7}	3.13×10^{-6}	8.48×10^{-6}	8.48×10^{-6}	2.586
5.00	5×70	7.85×10^{-7}	3.13×10^{-6}	8.48×10^{-6}	8.48×10^{-6}	3.254
6.00	6×70	7.85×10^{-7}	3.13×10^{-6}	8.48×10^{-6}	8.48×10^{-6}	3.933
7.00	7×70	7.85×10^{-7}	3.13×10^{-6}	8.48×10^{-6}	8.48×10^{-6}	4.601
8.00	8×70	7.85×10^{-7}	3.13×10^{-6}	8.48×10^{-6}	8.48×10^{-6}	5.280
9.00	9×70	7.85×10^{-7}	3.13×10^{-6}	8.48×10^{-6}	8.48×10^{-6}	5.948
10.00	10×70	7.85×10^{-7}	3.13×10^{-6}	8.48×10^{-6}	8.48×10^{-6}	6.627
20.00	20×70	7.85×10^{-7}	3.13×10^{-6}	8.48×10^{-6}	8.48×10^{-6}	13.258

TABLE 1.7: Relative errors, NTI and CPU times given by present method for the case of $\delta t = 2.0 (N = M=7, L=11)$

$t(s)$	NTI	Er_ϕ	Er_ω	Er_w	Er_φ	CPU(s)
2.00	81	4.05×10^{-7}	5.53×10^{-7}	7.67×10^{-6}	7.67×10^{-6}	1.183
4.00	2×81	1.38×10^{-7}	4.89×10^{-7}	7.56×10^{-6}	7.56×10^{-6}	2.366
6.00	3×81	1.38×10^{-7}	4.94×10^{-7}	7.56×10^{-6}	7.56×10^{-6}	3.545
8.00	4×81	1.38×10^{-7}	4.93×10^{-7}	7.56×10^{-6}	7.56×10^{-6}	4.720
10.00	5×81	1.38×10^{-7}	4.93×10^{-7}	7.56×10^{-6}	7.56×10^{-6}	5.897
20.00	10×81	1.38×10^{-7}	4.93×10^{-7}	7.56×10^{-6}	7.56×10^{-6}	11.802

From numerical experiments, it was found that the efficiency of block-marching technique could be further improved when δt is increased from 1 to 2. For this case, to keep the high order of accuracy in the time direction, L should be chosen to be 11. This means that the number of unknowns for each block is increased. There are 10 blocks for this case. Table 1.7 gives the results of this case.

Clearly, the accuracy of numerical results is kept as compared with previous cases, but the number of total iterations is cut to 810 and the CPU time is reduced to 11.802 s. The above test cases show that the increase of δt and the proper choice of L can enhance the efficiency of block-marching technique while keeping the high order of accuracy for numerical results. However, it should be noted that there is a limitation on the choice of δt. If δt is too large, the accuracy of numerical results can be greatly reduced if L is not large enough.

TABLE 1.8: Relative errors, NTI and CPU times given by present method for the case of $\delta t = 5.0(N = M{=}7)$

$t(s)$	L	NTI	Er_ϕ	Er_ω	Er_w	Er_φ	CPU(s)
5.0	13	81	4.56×10^{-4}	2.87×10^{-3}	9.56×10^{-4}	9.56×10^{-4}	1.183
	21	97	9.71×10^{-6}	5.23×10^{-6}	7.32×10^{-6}	7.32×10^{-6}	2.355
10.0	13	81+82	6.12×10^{-3}	7.69×10^{-2}	1.85×10^{-2}	1.84×10^{-2}	2.377
	21	97+87	1.43×10^{-5}	5.92×10^{-5}	4.10×10^{-4}	4.10×10^{-4}	4.775
15.0	13	81+2×82	1.74×10^{-1}	2.52×10^{0}	5.75×10^{-1}	5.72×10^{-1}	3.577
	21	97+2×87	1.54×10^{-4}	8.29×10^{-4}	4.02×10^{-4}	4.02×10^{-4}	7.214
20.0	13	81+3×82	5.33×10^{0}	9.41×10^{1}	1.95×10^{1}	1.93×10^{1}	4.760
	21	97+2×87	3.08×10^{-4}	1.15×10^{-2}	1.19×10^{-3}	1.18×10^{-3}	9.661

TABLE 1.9: Relative errors, time steps and CPU times given by 4-stage Runge–Kutta method for the case of $h = 0.005(N = M{=}7)$

$t(s)$	NTI	Er_ϕ	Er_ω	Er_w	Er_φ	CPU(s)
0.5	100	1.62×10^{-3}	1.17×10^{-3}	3.22×10^{-3}	3.22×10^{-3}	2.82
1.0	200	1.59×10^{-3}	1.23×10^{-3}	3.19×10^{-3}	3.19×10^{-3}	5.58
1.5	300	1.59×10^{-3}	1.23×10^{-3}	3.19×10^{-3}	3.19×10^{-3}	8.35
2.0	400	1.59×10^{-3}	1.23×10^{-3}	3.19×10^{-3}	3.19×10^{-3}	11.16
2.5	500	1.59×10^{-3}	1.23×10^{-3}	3.19×10^{-3}	3.19×10^{-3}	13.79
5.0	1000	1.59×10^{-3}	1.23×10^{-3}	3.19×10^{-3}	3.19×10^{-3}	27.87
10.0	2000	1.59×10^{-3}	1.23×10^{-3}	3.19×10^{-3}	3.19×10^{-3}	56.20
20.0	4000	1.59×10^{-3}	1.23×10^{-3}	3.19×10^{-3}	3.19×10^{-3}	111.78

This can be seen in Table 1.8, where t is taken as 5 and L is chosen as 13 and 21 respectively. Only four blocks are involved in this case. From Table 1.8, we can see that the accuracy of numerical results for the first block is still very high, but as marching block by block in the time direction, it declines very fast. Clearly, the accuracy of numerical results of $L{=}13$ is much less than that of $L{=}21$. In fact, the numerical results of $L{=}13$ are incorrect after $t{=}15.00$. To obtain accurate numerical results in the whole time domain, a large value of L has to be used. Thereby, the number of unknowns in each block is very large. This may reduce the efficiency of the method. To show the superiority of the block-marching method, the conventional fourth order Runge–Kutta method, which is a layer-marching approach, is also applied to solving the same problem. For this application, the Runge–Kutta method is only applied to the vorticity equation (1.59). The stream function equation (1.58) is still

TABLE 1.10: Relative errors, time steps and CPU times given by 4-stage Runge–Kutta method for the case of $h = 0.01 (N = M = 7)$

$t(s)$	NTI	Er_ϕ	Er_ω	Er_w	Er_φ	CPU(s)
0.5	50	2.32×10^{-3}	2.34×10^{-3}	6.48×10^{-3}	6.46×10^{-3}	1.41
1.0	100	2.20×10^{-3}	2.44×10^{-3}	6.42×10^{-3}	6.41×10^{-3}	2.82
1.5	150	2.20×10^{-3}	2.45×10^{-3}	6.41×10^{-3}	6.41×10^{-3}	4.12
2.0	200	2.20×10^{-3}	2.45×10^{-3}	6.41×10^{-3}	6.41×10^{-3}	5.61
2.5	150	2.20×10^{-3}	2.45×10^{-3}	6.41×10^{-3}	6.41×10^{-3}	6.95
5.0	500	2.20×10^{-3}	2.45×10^{-3}	6.41×10^{-3}	6.41×10^{-3}	13.95
10.0	1000	2.20×10^{-3}	2.45×10^{-3}	6.41×10^{-3}	6.41×10^{-3}	27.99
20.0	2000	2.20×10^{-3}	2.45×10^{-3}	6.41×10^{-3}	6.41×10^{-3}	56.06

solved by SOR method, and the relaxation factor is also chosen as 1.8. Two time step sizes of 0.005 and 0.01 are used for the layer-marching in the time direction. The numerical results of these two cases are listed respectively in Tables 1.9 and 1.10. From these two tables, we can see that the accuracy of numerical results obtained by 4-stage Runge–Kutta method is just of the order of 10^{-3}, which is much less than that given by the blocking-marching method with DQ discretization. On the other hand, the CPU time required to reach $t=20$ is much larger. This example shows that the proposed method can give better accuracy and efficiency over conventional layer-marching methods for a time-dependent problem.

1.6 Implementation of boundary conditions

A differential equation is underdetermined if boundary conditions are not provided. For a one-dimensional second order partial differential equation, for example, boundary conditions are required at both ends. For such cases, direct DQ method can be applied without difficulty. What should be done is to let the function values or the first derivatives at both ends to equal the prescribed values; see Shu (2000a). Difficulty, however, arises for applying multi-boundary conditions, for example, to solve forth-order differential equations where two boundary conditions are present at each end. Bert and his coworkers (1993, 1994) introduced a delta-point apart from the boundary point by a small distance as an additional boundary point and apply the other boundary condition at that point. It is found that, however, the solution accuracy may not be assured since δ is problem-dependent. Moreover there are some difficulties in applying the multi-boundary conditions accurately at the corner points for two-dimensional problems.

There are several approaches available in the literature for implementing multi-boundary conditions. Some DQ equations at inner nodes are replaced by the additional boundary conditions. It is found that, however, the solution accuracy may vary depending on which DQ equations at inner grids are replaced by the boundary conditions. A complete different approach introduced by Wang and Bert (1993) is that the boundary conditions are built during formulation of the weighting coefficients for higher order derivatives. Later Malik and Bert (1996) tried to extend this idea to all boundary conditions. However, the δ approach has still to be used for some combinations of boundary conditions. Wang et al. (1996, 1997) proposed a method (Differential Quadrature Element Method, or DQEM) to applying the multi-boundary conditions by assigning two degrees of freedom to each end point for a fourth-order differential equation. Later Wu and his coworkers (2000, 2001) proposed a generalized differential quadrature rule (GDQR) also introducing multiple degrees of freedom at boundary points. All boundary conditions can be easily applied by the DQEM or GDQR and accurate solutions can be obtained. Actually, both methods are exactly the same for one-dimensional problems.

Recently, Karami and Malekzadeh (2002) proposed a method to applying the multi-boundary conditions. In formulations of the weighting coefficients of third- and fourth-order derivatives, the second derivatives at boundary points are viewed as an additional independent variable. Wang extended the method (Wang and Bert, 1993) to all boundary conditions by viewing the first derivatives at boundary points as an additional independent variable. Both methods introduce an additional degree of freedom at boundary points, but compute the weighting coefficients of the first-order derivative by using explicit formulae (Shu and Richards, 1992; Wang, 2001). Since the multiple degrees of freedoms are introduced at boundary points merely for the application of boundary conditions, the additional degree of freedom for the beam case is not necessary for the rotation. It could be also the curvature (Karami and Malekzadeh, 2002) or even a displacement at a point not in the solution domain. Similar to the δ approach, Wang et al (2005) proposed a virtual point method in which the additional boundary points are not necessarily close to the end points. In the following, all these methods are summarized for complete reference.

For simplicity, beam problems are considered herein. For the case of a Bernoulli-Euler beam under general loadings, the governing differential equation can be expressed as

$$EI\frac{d^4 w}{dx^4} = q(x); EI\frac{d^3 w}{dx^3} = Q(x); EI\frac{d^2 w}{dx^2} = M(x); (x \in [0, L]) \qquad (1.70)$$

where E, I, L, q, P, ρ, A and w are the Young's modulus, principal moment of inertia about the y-axis, beam length, distributed load, axial load, mass density, cross-sectional area, and deflections, respectively. The shear force Q and bending moment M are

$$EI\frac{d^3 w}{dx^3} = EIw^{(3)} = Q; (x \in [0, L]) \tag{1.71a}$$

$$EI\frac{d^2 w}{dx^2} = EIw^{(2)} = M; (x \in [0, L]) \tag{1.71b}$$

There are four boundary conditions considered herein, namely,

1. *Simply supported*(SS) : $w = M = 0$; $\tag{1.72a}$
2. *Clamped*(C) : $w = w^{(1)} = 0$; $\tag{1.72b}$
3. *Free*(F) : $Q = M = 0$; $\tag{1.72c}$
4. *Guided*(G) : $w^{(1)} = Q = 0$. $\tag{1.72d}$

Based on Eq. (1.46), the k-th order derivative of the solution function at node i is rewritten in the form of

$$w_i^{(k)} = \sum_{j=1}^{N} e_{ij}^k w_j \qquad (i = 1, 2, ..., N) \tag{1.73}$$

where e_{ij}^k are the weighting coefficients of the k-th order derivative, N and w_j are the total number of nodes in the entire domain including the end points, and the solution values at node j, respectively. Here e_{ij}^k are used to avoid potential confusions. Denote $a_{ij}, b_{ij}, c_{ij}, d_{ij}$ the weighting coefficients of the first-, second-, third-, and fourth-order derivatives. In terms of direct DQ, the governing differential equations at an inner node is approximated by

$$\sum_{j=1}^{N} \{EId_{ij}w_j - Pb_{ij}w_j\} - q(x_i) - \rho A\omega^2 w_i = 0, \ (i = M, M+1, ..., N-M+1)$$
$$\tag{1.74}$$

where $w(x, t) = w(x)e^{i\omega t}$ is assumed for the purpose of free vibration analyses, and M is the starting number of inner point, depending on the method of applying boundary conditions. When appropriate boundary conditions are applied, a unique solution can be obtained by the direct DQ method.

(a) *δ approach* (Bert el al., 1988; Wang and Bert, 1993)

Node 1 (or $N-1$) and node 2 (or N) are viewed as the end points, separated by a very small distance δ. Then the two boundary conditions, Eq. (1.72), are applied to these points. Consider an SS-C beam, for example, one has

$$w_1 = 0 \tag{1.75a}$$

$$\sum_{j=1}^{N} b_{1j} w_j = 0 \tag{1.75b}$$

$$w_N = 0 \tag{1.75c}$$

$$\sum_{j=1}^{N} a_{Nj} w_j = 0 \tag{1.75d}$$

In this case, the starting number of inner node in Eq. (1.74) is 3, namely, $M = 3$. It should be pointed out that no much difference is experienced if b_{2j} and $b_{N-1\ j}$ are used in Eq. (1.75) instead of b_{1j} and b_{Nj}, since δ is very small and could be adjusted.

(b) Equation replaced approach (Wang, 2001)

Instead of using δ separation, two DQ equations at inner nodes are replaced by the second boundary conditions, similar to the mixed collocation method. It is found that, however, the solution accuracy may vary depending on which DQ equations at inner grids are replaced by the boundary conditions. Usually equations at grid 2 and grid $N-1$ are replaced by the boundary conditions to achieve the best accuracy, since the discrete errors at these two points are the largest. In this case, the method is the similar to the δ approach and $M = 3$.

(c) Build-in method(1) (Stritz, 1994)

One of the two boundary conditions is built during formulation of the weighting coefficients for higher order derivatives, since one can only apply one condition at the boundary condition directly. Take the SS-C beam as an example. The weighting coefficients for higher order derivatives are computed by

$$b_{ij} = \sum_{k=1}^{N-1} a_{ik} a_{kj} \tag{1.76a}$$

$$c_{ij} = \sum_{k=2}^{N} a_{ik} b_{kj} \tag{1.76b}$$

$$d_{ij} = \sum_{k=2}^{N} b_{ik} b_{kj} \tag{1.76c}$$

Parts of the boundary conditions have been built by simply changing the summation range. The N in Eq. (1.76a) is changed to $N-1$ to build in the condition $w_N^{(1)} = 0$. The 1 in Eq. (1.76b-c) is changed to 2 to build in the condition $w_1^{(2)} = 0$. Thus, only one condition at each end is left. One has, namely, $w_1 = 0$ and $w_N = 0$. In this case, the starting number of inner grid in Eq. (1.74) is 2, namely, $M = 2$. The accuracy is higher than the previous two methods if the number of nodes are small. However, the method cannot be used for all boundary conditions.

(d) DQEM (Malik and Bert, 1996) *and GDQR* (Wang and Gu, 1997)

The essence of DQEM and GDQR are that two degrees of freedoms, namely, $w_1, w_1^{(1)}, w_N, w_N^{(1)}$, are used at the end points to handle the two boundary conditions. Since Hermit type polynomials, instead of Lagrangian polynomials, are used in determination of the weighting coefficients, the method differs from the conventional DQ method. Let $a_{ij}, b_{ij}, c_{ij}, d_{ij}$ be the weighting coefficients of the first-, second-, third-, and fourth-order derivatives of the DQEM or GDQR, and N the number of nodes. Consider the SS-C beam again, one has

$$w_1 = 0 \tag{1.77a}$$

$$\sum_{j=1}^{N+2} b_{1j} u_j = 0 \tag{1.77b}$$

$$w_N = 0 \tag{1.77c}$$

$$w_N^{(1)} = 0 \tag{1.77d}$$

where $\{u\}^T = \{w_1, w_1^{(1)}, w_2, ..., w_{N-1}, w_N, w_N^{(1)}\}$. And Eq. (1.74) becomes

$$\sum_{j=1}^{N+2} \{EI \, d_{ij} u_j - P \, b_{ij} u_j\} - q(x_i) - \rho A \omega^2 w_i = 0, \ (i = 2, 3, ..., N-1) \tag{1.78}$$

It can be seen that the sum is from 1 to $N+2$, different from the DQ method. The multi-boundary conditions can be applied without any difficulty for all cases.

(e) Build-in method (2) (Liu and Wu, 2001a, 2001b)

As is noticed, the build-in method (1) is simple and accurate but cannot be used for all boundary conditions, while the multi-degree methods (DQEM or GDQR) can be conveniently used for all boundary conditions. A new way to apply the boundary condition that combines these two ideas is proposed by Karami and Malekzadeh (2002). The essence is that $w_1^{(2)}$ and $w_N^{(2)}$, the end curvature, are viewed as independent variables in computing the weighting coefficients of the third- and fourth-order derivatives, namely,

$$w_i^{(3)} = \sum_{j=1}^{N} \sum_{k=2}^{N-1} a_{ik} b_{kj} w_j + a_{i1} w_1^{(2)} + a_{iN} w_N^{(2)} = \sum_{j=1}^{N+2} c_{ij} u_j \qquad (1.79a)$$

$$w_i^{(4)} = \sum_{j=1}^{N} \sum_{k=2}^{N-1} b_{ik} b_{kj} w_j + b_{i1} w_1^{(2)} + b_{iN} w_N^{(2)} = \sum_{j=1}^{N+2} d_{ij} u_j \qquad (1.79b)$$

where $\{u\}^T = \{w_1, w_1^{(2)}, w_2, ..., w_{N-1}, w_N, w_N^{(2)}\}$. The DQ equations at inner grids are similar to Eq. (1.78). As can be seen that the symbolic expressions are similar to the ones of DQEM or GDQR, and all boundary conditions can be conveniently applied.

(f) Build-in method (3) (Wu and Liu, 2000)

The build-in method (3) is also an extension of the build-in method (1), similar to the build-in method (2) described above. The essential difference from the build-in method (2) is that $w_1^{(1)}$ and $w_N^{(1)}$, instead of $w_1^{(2)}$ and $w_N^{(2)}$, are used as independent variables in computing the weighting coefficients of the third- and fourth-order derivatives. The weighting coefficients of the second derivative at the end points are computed slightly different from Eq. (1.76a), namely,

$$w_i^{(2)} = \sum_{j=1}^{N} \sum_{k=2}^{N-1} a_{ik} a_{kj} w_j + a_{i1} w_1^{(1)} + a_{iN} w_N^{(1)} = \sum_{j=1}^{N+2} \bar{b}_{ij} u_j \ (i = 1, N) \quad (1.80a)$$

$$w_i^{(2)} = \sum_{j=1}^{N} \sum_{k=1}^{N} a_{ik} a_{kj} w_j = \sum_{j=1}^{N} b_{ij} w_j = \sum_{j=1}^{N+2} \bar{b}_{ij} u_j \ (i = 2, 3, ..., N-1)$$

$$(1.80b)$$

The weighting coefficients of the third- and fourth-order derivatives are computed by

$$w_i^{(3)} = \sum_{j=1}^{N+2} \sum_{k=2}^{N-1} a_{ik} \bar{b}_{kj} u_j = \sum_{j=1}^{N+2} c_{ij} u_j \qquad (1.81a)$$

$$w_i^{(4)} = \sum_{j=1}^{N+2} \sum_{k=2}^{N-1} b_{ik} \bar{b}_{kj} u_j = \sum_{j=1}^{N+2} d_{ij} u_j \qquad (1.81b)$$

where $\{u\}^T = \{w_1, w_1^{(1)}, w_2, ..., w_{N-1}, w_N, w_N^{(1)}\}$ and b_{ik} is computed from Eq. (1.76a). As can be seen that the symbolic expressions are exactly the same as the ones of DQEM or GDQR, thus all boundary conditions can be conveniently applied. It is noted that this method is readily extended to DQEM by using $w_1^{(1)}$ and $w_N^{(1)}$ as independent variables. The only difference from the existing DQEM is the way to compute the weighting coefficients.

(g) Build-in method (4)

A new way to apply the multi-boundary conditions is proposed here without employing the multi-degree of freedoms at the end points. The method is an extension of the method (C) but different from method (G). Take an F-C beam as an example to describe the procedures. Let $NN = N \div 2$. Then, one has

$$\bar{b}_{ij} = \sum_{k=1}^{N-1} a_{ik}a_{kj} \ (i = 1, 2, ..., NN), \tag{1.82a}$$

$$\bar{c}_{ij} = \sum_{k=2}^{N} a_{ik}\bar{b}_{kj} \ (i = 1, 2, ..., NN), \tag{1.82b}$$

$$\bar{d}_{ij} = \sum_{k=2}^{N} b_{ik}\bar{b}_{kj} \ (i = 1, 2, ..., NN) \tag{1.82c}$$

$$\bar{c}_{ij} = \sum_{k=2}^{N} a_{ik}b_{kj} \ (i = NN + 1, NN + 2, ..., N), \tag{1.82d}$$

$$\bar{d}_{ij} = \sum_{k=2}^{N} b_{ik}b_{kj} \ (i = NN + 1, NN + 2, ..., N) \tag{1.82e}$$

where a_{ij} and b_{ij} are computed from Eqs. (1.41) and (1.76). Equation (1.74) then becomes

$$\sum_{j=1}^{N} \{EI \ \bar{d}_{ij}w_j - P \ \bar{b}_{ij}w_j\} - q(x_i) - \rho A \ \omega^2 w_i = 0 \ (i = 2, 3, ..., N - 1) \tag{1.83}$$

Since $w_1^{(2)} = 0$ and $w_N^{(1)} = 0$ have been built, the remaining boundary conditions are

$$\sum_{j=1}^{N} \bar{c}_{1j}w_j = 0, \ w_N = 0 \tag{1.84}$$

In this simple way, the limitations existing in the build-in method (1) have been completely removed without employing either the multi-degree of freedom at end points or δ approach. This will be demonstrated by numerical examples presented in the next section.

(h) Virtual boundary point method

Similar to the δ approach or finite difference method, two virtual points can be placed outside the beam, one on each side. It is not necessary for the two virtual points to be δ apart from the boundary point. Then Eq. (1.41) and Eq. (1.42) are used to compute the weighting coefficients, exactly the same as in the direct ordinary DQ method. The non-uniform nodes could be computed from Chebyshev nodes expressed by Eq.(1.48).

Take an F-C beam as an example to describe the procedures for applying the multi-boundary conditions. Note that the boundary points are 2 and $N-1$. Thus the boundary conditions in terms of direct DQ method become

$$\sum_{j=1}^{N} c_{2j}w_j = 0, \ \sum_{j=1}^{N} b_{2j}w_j = 0, \ \sum_{j=1}^{N} a_{N-1j}w_j = 0, \ w_{N-1} = 0 \qquad (1.85)$$

The DQ equations at inner nodes are

$$\sum_{j=1}^{N} \{EI \, d_{ij}w_j - P \, b_{ij}w_j\} - q(x_i) - \rho A \, \omega^2 w_i = 0 \ (i = 3, 4, ..., N-2) \ (1.86)$$

Eight methods in applying the multi-boundary conditions are summarized. Although all methods could be used in two-dimensional problems, such as thin plate problems, some limitations exist in some of the methods. Methods C and G cannot be used for all boundary conditions in two dimensions. There are some difficulties at the corner points for methods A and B. For method H, the weighting coefficients of mixed derivatives should be computed by using deflections at grids on the plate only, namely, w_{ij}, $(i, j = 2, 3, ..., N-1)$. If domain decomposition is to be used, only methods D and F can be conveniently used, and the formulations of the system of DQ equations are similar to the finite element method.

Example 1.7 *Flexural vibration of a prismatic beam*

As the first example, consider the flexural vibration of prismatic beams with ten combinations of boundary conditions, i.e., SS-SS, SS-C, SS-F, SS-G, C-C, C-F, C-G, F-F, F-G, and G-G, where SS, C, F and G denote the simply supported, clamped, free and guided boundaries, respectively. Use the DQ method with method of build-in (4) to solve the problems. For the general cases, the DQ equations for free vibration analyses can be written in the following partition form, namely,

$$\begin{bmatrix} K_{ee} \ K_{ei} \\ K_{ie} \ K_{ii} \end{bmatrix} \begin{Bmatrix} \delta_e \\ \delta_i \end{Bmatrix} = \begin{Bmatrix} 0 \\ \rho A \varpi^2 \{w_i\} \end{Bmatrix} \qquad (1.87)$$

where subscripts e, i denote the quantities at the end points and at the inner points. The $(N-2)$ equations in Eq. (1.87) for obtaining the frequency is

$$[K_{ii} - K_{ie} K_{ee}^{-1} K_{ei}] \{w_i\} - \rho A \omega^2 \{w_i\} = \{0\} \qquad (1.88)$$

Table 1.11 lists the first five non-zero frequency parameters ($\bar{\omega} = L \sqrt[4]{\rho A \omega^2 / EI}$) of beams with ten boundary conditions. As expected, the accuracy of numerical data is very high. The fundamental frequency for all cases is exactly the same as the analytical solution, which is 18. All five frequencies are accurate to five significant figures, proved by changing the number of grids N, and exactly the same as the analytical solutions for the SS-SS beam. It is also noted that the non-zero frequencies for the F-F beam are the same as the C-C beam. Numerical data demonstrate that the proposed method can handle all boundary conditions.

TABLE 1.11: First five non-zero frequency parameters of beams with various boundary conditions (DQM, N=17)

Mode Bc's	First	Second	Third	Fourth	Fifth
SS-SS	3.1415927	6.2831853	9.4247780	12.566371	15.707963
C-C	4.7300407	7.8532046	10.995608	14.137166	17.278741
C-F	1.8751041	4.6940911	7.8547574	10.995541	14.137169
C-SS	3.9266023	7.0685827	10.210176	13.351769	16.493356
F-F	4.7300407	7.8532046	10.995608	14.137166	17.278747
G-G	3.1415927	6.2831853	9.4247780	12.566371	15.707965
G-SS	1.5707963	4.7123890	7.8539816	10.995574	14.137167
C-G	2.3650204	5.4978039	8.6393798	11.780972	14.922566
F-G	2.3650204	5.4978039	8.6393798	11.780972	14.922566
F-SS	3.9266023	7.0685827	10.210176	13.351769	16.493357

Example 1.8 *Flexural vibration of rectangular plates*

As the second example, consider the flexural vibration of rectangular plates with all boundary clamped (C-C-C-C). This example could not be solved by DQ method with build-in method (1) previously. The first six frequency parameters, $\bar{\omega} = \omega a^2 \sqrt{\rho h / D}$, are listed in Table 1.12, where a, h, D are the length, thickness, and bending rigidity of the square plate, respectively. Methods (G) and (H) are used to apply the boundary conditions. As is expected, the same accurate results as the DQEM are obtained. It should be pointed out that Leissa's analytical solutions are upper-bound solutions.

Numerical experience shows that the difference among various methods in applying boundary conditions exists only when N is small. With the increase of the number of grids, the difference becomes smaller among various methods. More grids are needed to get accurate high order frequencies.

TABLE 1.12:   First six frequency parameters ($\overline{\omega}^2 = \rho\, a^4\, \varpi^2\, /D$) of the square plate with all boundary clamped (C-C-C-C)

Method \ Mode	1	2(=3)	4	5	6
Present(G)	35.986	73.396	108.22	131.58	132.20
Present(H)	35.985	73.393	108.22	131.58	132.20
DQEM	35.985	73.393	108.22	131.57	132.19
DQEM	35.985	73.399	108.22	130.95	131.56
Leissa (1973)	35.992	73.413	108.27	131.64	132.24

1.7   Conclusions

Differential Quadrature is an attractive idea for numerical evaluating derivatives of a given function. Applying it to solving differential equations yields the direct Differential Quadrature Method. Differential Quadrature formulas can be obtained from approximation theory.

Direct DQ method is a global numerical scheme. Just like Gauss quadrature, direct DQ method is able to yield high-accurate solutions by use of a small number of nodes. It is simple and straightforward, thus making it an attractive numerical method.

Differential Quadrature also applies to temporal discretization, resulting in the so-called block-marching technique with DQ discretization. It is implicit and flexible. The numerical results are obtained block by block in the time direction. Compared to the conventional 4-stage Runge–Kutta method, the accuracy of numerical results obtained from this method is about four orders higher, and the computational effort required is much less.

Various methods to apply the multi-boundary conditions in the applications of Differential Quadrature method are summarized and discussed. Numerical examples are given to demonstrate on how to implement the boundary conditions. Boundary conditions have significant influences on the accuracy of numerical solutions, and thus deserve special cautions.

It was recognized soon after direct DQ method was proposed, however, that direct DQ method suffers from several shortcomings. To remove these shortcomings, a variety of numerical schemes have been proposed to improve its accuracy and enhance its capability. In the following six chapters, we will focus on these newly developed methods based on DQ.

Chapter 2

Complex Differential Quadrature Method

Theory of complex variables is the branch of mathematics investigating functions of complex variables and is extremely important in complex analysis. It is hard to find another mathematical branch in which complex variable has not been mentioned. This is a very special mathematical phenomenon because most of mathematical branches are irrelevant. It is often an important mathematical event if two branches of mathematics are linked through a subtle way, but complex analysis comes into other mathematical branches in a natural way. Without the aid of complex analysis, we are even unable to prove the fundamental theorem of algebra.

Theory of complex variable is very important not only as a tool for mathematical analysis, but also as an effective technique for solving some tough problems in areas as diverse as electromagnetism, thermodynamics, fluid mechanics, thermal mechanics, elastic mechanics and acoustics. It is even able to provide a surprisingly concise solution to singular problems, which are hard to obtain by using an alternative method. In mechanics of elasticity, the stress field of a crack tip in a plate is singular, indicating that the stress there becomes larger and larger as we get closer and closer to the crack tip. The solution obtained from complex variable theory reveals that the singular behavior of such a tip is described by the reciprocal of distance r from the crack tip. Another striking application of complex variable theory is that it provides a complete solution to the lifting problem of a two-dimensional wing, which singularity at the trailing edge is removed through smart introduction of Kutta condition. Therefore, theory of complex variables remains important in the future. In view of this, it is necessary to generalize DQ method from the real domain to the complex domain, termed as complex DQ method.

In this chapter, complex DQ method (Zong, 2003a; Zhang et al., 2007) is introduced. First, extension of Lagrange interpolation to the complex plane is presented. It turns out to be composed of two parts: polynomial and rational interpolations of an analytic function. Based on Lagrange interpolation in the complex plane, complex differential quadrature method is constructed suited to solving the potential problem (Laplace's equation) and the plane linear elastic problem (the so-called bi-harmonic equation). Three important issues in the method are:

(1) Extension of Lagrange interpolation to the complex plane;

(2) Use of analytic function or Airy's stress function in the form of complex variable; and

(3) Application of boundary conditions to determination of the unknowns.

The last step involves solving a system of linear equations, which can be neglected if conformal mapping, a powerful technique able to transform a complicated physical problem into a readily analyzable one, is employed. On the contrary, conformal mapping can be readily performed if Lagrange interpolation in the complex plane is used together. We will thus in section 2.4 develop a methodology which combines conformal mapping technique and complex DQ method for easily solving some two-dimensional potential problems.

2.1 Differential quadrature in the complex plane

Lagrange interpolation on the real line is one of the simplest interpolation methods and has been widely investigated over the past two centuries (Buchanan and Turner, 1997). It is possible to generalize Lagrange interpolation to the complex plane. Some changes are, however, necessary to account for more complicated cases special in the complex plane. In what follows, extension of Lagrange interpolation to the complex plane and differentiation of a function of a complex variable (or more exactly analytical function) by use of complex Lagrange polynomial are presented. Because books on complex variable theory are plenty, the reader is assumed to be familiar with basic knowledge of complex variables and analytic functions. Several books are listed in the reference section and the reader is referred to Chapter 8.

It is assumed that in the domain S in the two-dimensional complex plane encircled by its boundary Γ, an analytic function $\varphi(z)$ is defined. On the boundary Γ, N nodes are specified as shown in Fig. 2.1. These nodes are assumed to be a sequence in the form of

$$z_1, z_2, \ldots, z_N \tag{2.1}$$

on which the function values of the analytic function $\varphi(z)$ are known

$$\varphi_1, \varphi_2, \cdots, \varphi_N \tag{2.2}$$

Our purpose is to construct a polynomial of $z = x + iy$, where $i = \sqrt{-1}$ is the imaginary unit, in the form of

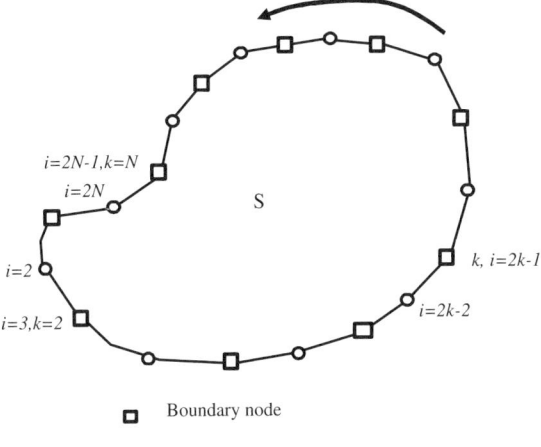

FIGURE 2.1: Boundary nodes and collocation points.

$$\varphi(z) = \sum_{k=1}^{N} \lambda_k(z)\varphi_k \tag{2.3}$$

satisfying

$$\lambda_k(z) = \begin{cases} 1 & z = z_k \\ 0 & z = z_j \neq z_k \end{cases} \qquad k = 1, \cdots, N, \quad j = 1, \cdots, N. \tag{2.4}$$

This is the Kronecker delta function. It is its property that guarantees the interpolating polynomial $\varphi(z)$ equal to the function value φ_k at each node z_k. Direct extension of Lagrange polynomial on the real line to the complex plane leads to the following equations (Buchanan and Turner, 1997):

$$\lambda_k(z) = \frac{L_k(z)}{L_k(z_k)}, \quad L_k(z) = \prod_{\substack{j=1 \\ j \neq k}}^{N} (z - z_j), \tag{2.5}$$

where the symbol $\prod$ represents multiplication of the terms following the symbol.

It is straightforward to find the first- and second-order derivatives by differentiating Eq. (2.5) with respect to z

$$\varphi'(z) = \frac{d\varphi(z)}{dz} = \sum_{k=1}^{N} a_k(z)\varphi_k \text{ and } \varphi''(z) = \frac{d^2\varphi}{dz^2} = \sum_{k=1}^{N} b_k(z)\varphi_k \tag{2.6}$$

where $a_k(z)$ are the coefficients for the first-order derivative. After some manipulations, we obtain the coefficients for $z \neq z_k$ as

$$a_k(z) = \frac{d\lambda_k(z)}{dz} = \frac{1}{z - z_k} \sum_{\substack{\ell=1 \\ \ell \neq k}}^{N} \frac{L_\ell(z_\ell)}{L_k(z_k)}, \quad k = 1, \cdots, N. \qquad (2.7a)$$

And for $z = z_k$

$$a_k(z) = \sum_{\substack{\ell=1 \\ \ell \neq k}}^{N} a_\ell(z) \frac{L_\ell(z_\ell)}{L_k(z_k)}, \quad k = 1, \cdots, N. \qquad (2.7b)$$

The coefficients for the second-order derivative can be obtained in the similar manner as

$$b_k(z) = \frac{da_k(z)}{dz} = \frac{d^2\lambda_k(z)}{dz^2} = \frac{z - z_k}{1} \sum_{\substack{\ell=1 \\ \ell \neq k}}^{N} a_\ell(z) \frac{L_\ell(z_\ell)}{L_k(z_k)} - \frac{z - z_k}{a_k(z)} \qquad (2.8a)$$

$k = 1, \cdots, N,$ if $z \neq z_k$, and

$$b_k(z) = \frac{1}{2} \sum_{\substack{\ell=1 \\ \ell \neq k}}^{N} b_\ell(z) \frac{L_\ell(z_\ell)}{L_k(z_k)}, \quad k = 1, \cdots, N, \quad \text{if} \quad z = z_k. \qquad (2.8b)$$

Equations (2.7) and (2.8) are the most important formulas in this chapter. It is clearly shown that both first- and second-order derivatives are reduced to the weighted sum of the function values at the nodes. This greatly simplifies the numerical differentiation process without deteriorating numerical accuracy.

It is obvious that if N nodes are used, the interpolating polynomial degree is $(N-1)$. Using this idea, it can be shown that the approximation errors for the first-order derivative (R^1) and the second-order one (R^2) are

$$R^1(z_k) = \frac{\varphi^{(N)}(\xi)W^{(1)}(z_k)}{N!}, \quad k = 1, \ldots, N \qquad (2.9a)$$

$$R^2(z_k) = \frac{\varphi^{(N)}(\xi)W^{(2)}(z_k)}{N!}, \quad k = 1, \ldots, N \qquad (2.9b)$$

where $W(z) = \prod_{j=1}^{N} (z - z_j)$. These residual estimates show that it is possible to achieve high accuracy if the number of grid points N is large. Moreover accuracy is proportional to N or even to its powers.

The aforementioned formulas are identical to those on the real line if complex variable z is replaced by real variable x. But a complex polynomial is really different from a real polynomial in the sense that the former also allows for the appearance of the terms of negative powers. This is because a Laurent's series, in which terms of negative powers appear, in the complex plane replaces the Taylor's series on the real line. Terms of negative powers in a Laurent's series bring in convenience to treat infinite domain as these terms vanish or are constants at infinity. In "domain" type methods such as Finite Element Method (FEM), an infinite domain is approximated by a very large domain. This is not cost-effective. On the other hand, infinite domain problems can be easily handled by boundary methods such as Boundary Element Method (BEM). In the following, we will construct a Lagrange polynomial which contains terms of negative powers, and thus allows for effective treatment of infinite domain.

It is proposed to construct the Lagrange interpolation for infinite domain in the following way:

$$\lambda_k(z) = \frac{\Lambda_k(z)}{\Lambda_k(z_k)}, \quad \Lambda_k(z) = \prod_{\substack{j=1 \\ j \neq k}}^{N} \left(\frac{1}{z} - \frac{1}{z_j} \right). \tag{2.10}$$

It is clearly indicated that $\lambda_k(z)$ possesses the Kronecker delta function property defined by Eq. (2.4). After some algebraic operations, Eq. (2.10) turns out to be a rational interpolation. Hence, Lagrange interpolation in the complex plane is composed of two parts: polynomial and rational interpolations. Note that polynomial and rational approximations are two individual techniques on the real line, but each of them is a part of Lagrange interpolation in the complex plane.

This above procedure is justified by the fact that there are two types of analytical functions: holomorphic and meromorphic (Chapter 8). The former is differentiable in an open subset, and thus can be approximated by Lagrange polynomial in which all the terms have positive powers. The latter is differentiable except at finitely many points and thus can be approximated by a complex ration number.

To find the first- and second-order derivatives of $\Lambda_k(z)$, multiply both sides of Eq. (2.10) by $(1/z - 1/z_k)$ and then differentiate both sides with respect to z. The results after simple manipulations are

$$\varphi'(z) = \frac{d\varphi(z)}{dz} = \sum_{k=1}^{N} a_k(z)\varphi_k \tag{2.11}$$

for the first derivative, and

$$\varphi''(z) = \frac{d^2\varphi}{dz^2} = \sum_{k=1}^{N} b_k(z)\varphi_k \tag{2.12}$$

for the second derivative. The coefficients for the first-order derivative are

$$a_k(z) = \frac{d\lambda_k(z)}{dz} = -\frac{1}{z - z^2/z_k} \sum_{\substack{\ell=1 \\ \ell \neq k}}^{N} \frac{\Lambda_\ell(z_\ell)}{\Lambda_k(z_k)}, \quad k = 1, \cdots, N, \text{if} \quad z \neq z_k \tag{2.13}$$

and

$$a_k(z) = \sum_{\substack{\ell=1 \\ \ell \neq k}}^{N} a_l(z) \frac{\Lambda_\ell(z_\ell)}{\Lambda_k(z_k)}, \quad k = 1, \cdots, N, \text{if} \quad z = z_k. \tag{2.14}$$

The coefficients for the second-order derivative are

$$b_k(z) = \frac{da_k(z)}{dz} = \frac{d^2\lambda(z)}{dz^2} = -\frac{1}{z - z^2/z_k}[\sum_{\substack{\ell=1 \\ \ell \neq k}}^{n} a_\ell(z) \frac{\Lambda_\ell(z_\ell)}{\Lambda_k(z_k)} - (1 - \frac{2z}{z_k})a_k(z)] \tag{2.15}$$

if $z \neq z_k, k = 1, \cdots, N$ and

$$b_k(z) = \frac{z_k}{2 + z_k} \sum_{\substack{\ell=1 \\ \ell \neq k}}^{N} b_\ell(z) \frac{\Lambda_\ell(z_\ell)}{\Lambda_k(z_k)}, \text{if} \quad z = z_k, k = 1, \cdots, N. \tag{2.16}$$

Equations (2.6) to (2.8) are suitable for finite domain and Eqs. (2.11) to (2.16) are suitable for infinite domain.

Example 2.1 *Lagrange interpolation in the complex plane*
  Lagrange interpolation in the complex plane is an extension of Lagrange interpolation on the real line. It plays a crucial role in complex DQ method, and the accuracy related with interpolation and numerical differentiation is of major concern. This example is designed to check its accuracy.
  Suppose an analytic function $\varphi(z)$ is given in a domain encircled by its boundary Γ. Because the analytic function is given, its values on the boundary Γ are also known. Based on the function values on N nodes on the boundary Γ and Lagrange interpolation in Eqs. (2.5) to (2.8), the function value and its first- and second-order derivatives at any point within the domain can be obtained. Comparison of numerical and analytical results will reveal the accuracy of Lagrange interpolation in the complex plane.

Take $\varphi(z) = z^3$ in a unit circle bounded by Γ: $|z| = 1$ for example. N nodes on Γ are specified. The relative errors are defined

$$e_1 = \frac{|\varphi(z) - \varphi_0(z)|}{|\varphi_0(z)|}, \quad e_2 = \frac{|\varphi'(z) - \varphi_0'(z)|}{|\varphi_0'(z)|}, \quad e_3 = \frac{|\varphi''(z) - \varphi_0''(z)|}{|\varphi_0''(z)|} \quad (2.17)$$

where the subscript "$_0$" denotes the known function values, and thus $\varphi_0 = z^3$, $\varphi_0' = 3z^2$ and $\varphi_0'' = 6z$. The letters in Eq. (2.17) without subscripts denote the numerical results.

The relative errors defined by Eq. (2.17) for the point $z = \sqrt{2}(1 + i)/4$ are shown in Fig. 2.2. When $N = 5$ the relative errors are up to 3%. As N increases to 10, the relative errors are dramatically reduced to the order of $10^{-5} \sim 10^{-4}$ %. As N increases to larger numbers, the relative errors increase rather than decrease. The reason is that as N increases, the order of Lagrange polynomial $L(z)$ in Eq. (2.5) is correspondingly raised and more higher-order terms appear in Lagrange polynomial. From numerical analysis, it is well known that higher-order monomials (of the type z^n) might introduce local oscillations to the numerical results, as a consequence, deteriorating numerical accuracy. However, the numerical accuracies as shown in Fig. 2.2 remain below 10^{-2}% for the second-order derivatives, 10^{-3}% for the first-order one and 10^{-4}% for the function values over the whole range. Such accuracy is encouraging.

If more similar analyses are performed at other points, which is omitted here for brevity, we are able to obtain the same conclusions (Zong, 2003a). Thus, as on the real line, Lagrange interpolation in the complex plane may also yield highly accurate results by use of a small number of nodes.

2.2 Complex DQ method for potential problems

In this and the next sections, the applicability and validity of the complex DQ method are demonstrated through the two-dimensional potential problems governed by Laplace's equation and plane linear elastic problems governed by bi-harmonic equation.

Consider a domain S encircled by its boundary Γ. An analytic function defined on S is of the form

$$\varphi(z) = u(x, y) + iv(x, y) \quad (2.18)$$

where $i = \sqrt{-1}$ is the imaginary unit, u, v are real and imaginary parts, respectively, and $z = x + iy$.

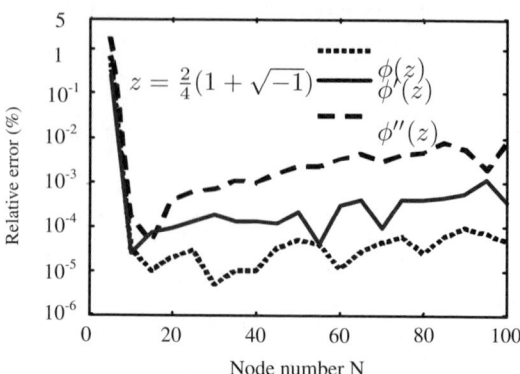

FIGURE 2.2:   Influence of node number on differentiation accuracy by use of Lagrange interpolation in the complex domain.

The real and imaginary parts of $\varphi(z)$ are harmonic functions, satisfying the following Laplace's equation

$$\nabla^2 u = \nabla^2 v = 0 \qquad (2.19)$$

where ∇^2 is the Laplace operator. A harmonic function is fully determined by its values on the boundary. According to the different boundary conditions, the potential problems governed by Eq. (2.19) can be classified into three categories, that is,

(a)*Dirichlet problem: The boundary conditions for this type of problem are*

$$u = Re(\varphi) = m \quad on \, z \in \Gamma \qquad (2.20)$$

where Re denotes the real part.

(b) *Neumann problem: The corresponding boundary conditions are*

$$\frac{\partial u}{\partial n} = Re(n \cdot \frac{d\varphi}{dz}) = m \quad on \, z \in \Gamma \qquad (2.21)$$

(c) *Mixed (or Robin) problem: The boundary conditions are*

$$\alpha u + \beta \frac{\partial u}{\partial n} = \alpha Re(\varphi) + \beta Re(n \cdot \frac{d\varphi}{dz}) = m \quad on \, z \in \Gamma \qquad (2.22)$$

where m, α and β are given functions on the boundary Γ.

Note that Eq. (2.22) is the general form of Eqs. (2.20) and (2.21). If $\alpha = 1$ and $\beta = 0$, Eq. (2.22) is identical to Eq. (2.20); if $\alpha = 0$ and $\beta = 1$, it turns out to be same as Eq. (2.21). Thus the three boundary value problems can be written in the form of Eq. (2.22) with different coefficients.

A harmonic function is fully determined by its values on the boundary. Assume that the number of nodes on the boundary Γ is N and the function values on these N nodes are denoted by $\varphi_1, \varphi_2, \cdots \varphi_N$. The unknown number is $2N$, that is, a pair of φ^R and φ^I, the real and imaginary parts of $\varphi(z)$ at each node. Herein, it is worth noting that according to complex variable theory, the Cauchy-Riemann equations holds for any analytic complex function:

$$\frac{\partial \varphi^R}{\partial x} = \frac{\partial \varphi^I}{\partial y}, \quad \frac{\partial \varphi^R}{\partial y} = -\frac{\partial \varphi^I}{\partial x}. \tag{2.23}$$

Assume n and s are the normal and tangential unit vectors at the boundary, respectively. Then the Cauchy-Riemann equations along the boundary are

$$\frac{\partial \varphi^R}{\partial n} = \frac{\partial \varphi^I}{\partial s}, \quad \frac{\partial \varphi^R}{\partial s} = -\frac{\partial \varphi^I}{\partial n}. \tag{2.24}$$

The real and imaginary parts are associated with each other by the Cauchy-Riemann equation. An attempt has been made by Zhang (2003) to solve the potential $\varphi(z)$ with N boundary conditions in combination of Cauchy-Riemann conditions, leading the unknown number to N, either the real part φ^R or the imaginary part φ^I as unknowns. When one of them is determined, the other one is fixed up to an arbitrary constant. The results are excellent, but it requires more coding efforts. As stated in the previous section, there exist three types of boundary value potential problems, i.e. Dirichlet, Neumann and Robin problems. If Cauchy-Riemann conditions are employed, different problems need different computer codes. Efforts have also been made to solve plane linear elasticity problem by employing Cauchy-Riemann conditions (Zhang, 2003). The results turned out to be unsatisfactory.

In view of the aforementioned challenges, an alternative technique is often adopted. The real and imaginary parts of the analytic function $\varphi(z)$ are treated as independent variables, resulting in $2N$ unknowns instead of N (Zong, 2003a). Then for potential problems, $2N$ boundary conditions are imposed on the boundary nodes denoted by

$$\xi_1, \xi_2, \cdots, \xi_{2N}. \tag{2.25}$$

The two sets of nodes on the boundary Γ are related in the following way

$$\xi_i = z_{2i-1}, \quad i = 1, 2, \cdots N. \tag{2.26}$$

We are thus able to express the function values of $\varphi(z)$ on the nodes of $\xi_1, \xi_2, \cdots, \xi_{2N}$ in the form of Lagrange interpolating polynomial using Eq. (2.3). Mathematically, we have

$$\varphi(\xi_i) = \sum_{k=1}^{N} \lambda_k(\xi_i) \cdot \varphi_k, \quad i = 1, 2, \cdots 2N. \tag{2.27}$$

Straightforward application of Eq. (2.6) or Eq. (2.10) to Eq. (2.27) yields the formula for evaluating the first derivative of $\varphi(\xi_i)$. For convenience, we rewrite the formula again here as follows

$$\varphi'(\xi_i) = \sum_{k=1}^{N} a_{ik}\varphi_k, \quad i = 1, 2, \cdots 2N. \tag{2.28}$$

By substituting Eqs. (2.27) and (2.28) into Eq. (2.22), one can obtain the following governing equation for the potential problems

$$\alpha_i Re(\sum_{k=1}^{N} \lambda_k(\xi_i)\varphi_k) + \beta_i Re(n_i \sum_{k=1}^{N} a_{ik}\varphi_k) = g_i, \quad i = 1, 2, \cdots 2N. \tag{2.29}$$

It is worth noting that both sides of Eq. (2.29) are real numbers, so it can be rewritten as

$$(\mathbf{A}^R + \mathbf{B}^R)\varphi^R - (\mathbf{A}^I + \mathbf{B}^I)\varphi^I = \mathbf{G} \tag{2.30}$$

where $\mathbf{A} = \{\alpha_i\lambda_k(\xi_i)\}$, $\mathbf{B} = \{\beta_i n_i \alpha_{ik}\}$, $\mathbf{G} = \{g_i\}^T$ $i = 1, 2, \cdots 2N$, $k = 1, 2, \cdots N$, the superscripts R and I indicate the real and imaginary parts, respectively.

Equation (2.30) can be expressed finally in the following matrix form

$$\left[\mathbf{A}^R + \mathbf{B}^R \; -(\mathbf{A}^I + \mathbf{B}^I) \right] \begin{bmatrix} \varphi^R \\ \varphi^I \end{bmatrix} = \mathbf{G}. \tag{2.31}$$

This system of linear algebraic equations can be solved by use of Gauss elimination method.

Example 2.2 *Two dimensional potential problem*
 Numerical results obtained from complex DQ method for Dirichlet, Neumann and mixed problems, are solved in the following and compared with exact solutions. In two dimension, the Laplace's equation

$$\nabla^2 u = 0 \tag{2.32}$$

is solved based on appropriate boundary conditions on the domains with piece-wise straight or curved boundaries, where u represents the real part of the analytic function φ. The imaginary part of the analytic function φ also satisfies the Laplace's equation. The geometry and the boundary conditions for these examples are shown in Fig. 2.3.

(a) *Dirichlet problem on an ellipse*

The first example is a Dirichlet problem on an ellipse centered at the origin with the major axis $a = 1$ and the minor axis $b = 0.5$ as shown in Fig. 2.3(a). The prescribed Dirichlet boundary condition for the problem is

$$u(x, y) = y \tag{2.33}$$

The exact solution for the normal derivative q, on the ellipse is found to be

$$q = \frac{\partial u}{\partial n} = \frac{a \sin \theta}{\sqrt{a^2 \sin^2 \theta + b^2 \cos^2 \theta}}. \tag{2.34}$$

The numerical and exact solutions for q around the ellipse are shown in Fig. 2.4 with 60 nodes used on the boundary. From Fig. 2.4, it is observed that the two solutions are in excellent agreement.

In complex DQ method, only the nodes are specified on the boundary to model the problem, thus the number of nodes exerts significant influence on the accuracy of the numerical results. A convergence study is then carried out on the present Dirichlet problem to examine the effect of the number of the nodes on the accuracy of the solutions.

The accuracy of the numerical results is assessed using a global error measure

$$\varepsilon = \frac{1}{|q^{(e)}|_{\max}} \sqrt{\frac{1}{N} \sum_{i=1}^{N} \left[q_i^{(e)} - q_i^{(c)} \right]^2} \tag{2.35}$$

where ε is the error in the solution, N is the number of nodes on the boundary, and the superscripts (e) and (c) refer to the exact and numerical solutions, respectively.

The relationship between the error and node number is plotted in Fig. 2.5. From the figure, it is readily seen that the error decreases with increasing node number, indicating that more nodes used in complex DQ method lead to more accurate numerical results. In most cases, a certain number of nodes

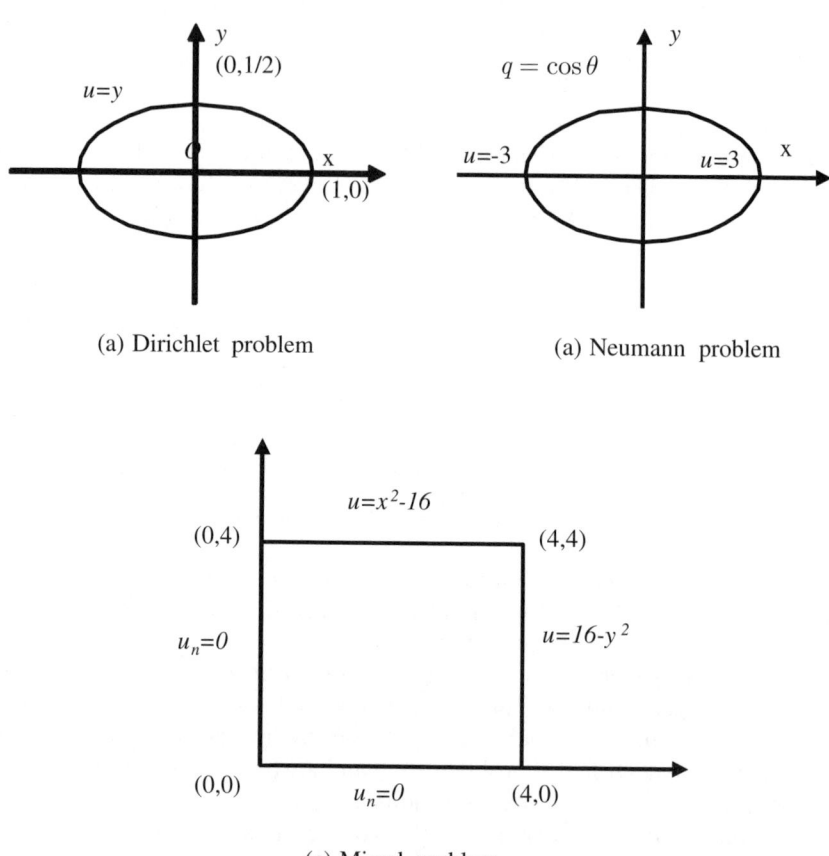

(a) Dirichlet problem

(a) Neumann problem

(c) Mixed problem

FIGURE 2.3: Geometry and boundary conditions for potential examples.

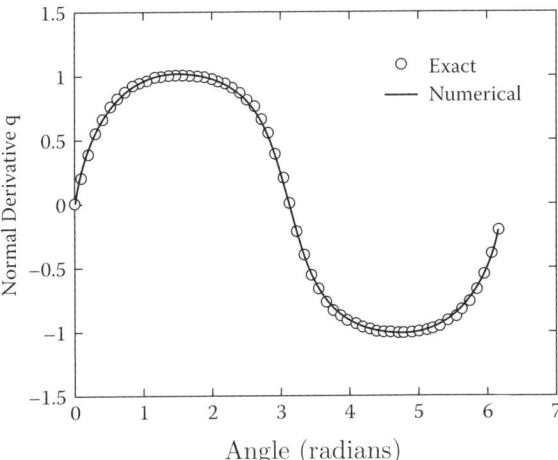

FIGURE 2.4: Dirichlet problem on an ellipse with $u(x, y) = y$ prescribed on the ellipse with 60 nodes on the boundary.

are enough to obtain results with high accuracy. In this respect, the present method is an effective method for potential problems with high computational efficiency.

(b) Neumann problem on a circle

The second example is a Neumann problem on a circle of radius 3 units with the center at the origin as shown in Fig. 2.3(b). The Neumann problem is solved based on the boundary condition

$$q = \frac{\partial u}{\partial n} = \cos(\theta) \tag{2.36}$$

imposed on the circle. In order to obtain a unique solution, the potential is specified at the two points (3, 0) and (-3, 0). The numerical results, together with the exact solution

$$u = x = 3\cos(\theta) \tag{2.37}$$

are compared in Fig. 2.6. The agreement, with 40 nodes uniformly distributed on the circle, are excellent.

(c) Mixed problem on a square

Consider a mixed boundary problem governed by Laplace's equation on a square of side 4 units with corners at (0,0), (4,0), (4,4) and (0,4). The geometry and the boundary conditions are shown in Fig. 2.3(c). The prescribed boundary conditions are

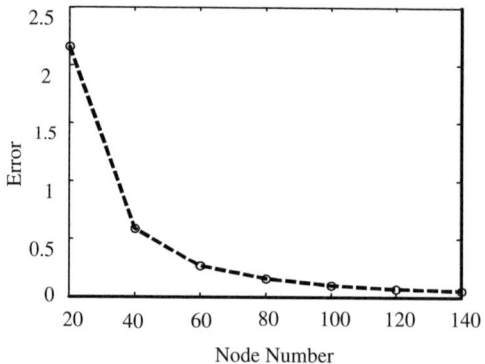

FIGURE 2.5: Convergence on an ellipse for a Dirichlet problem with $u(x, y) = y$ prescribed on the boundary.

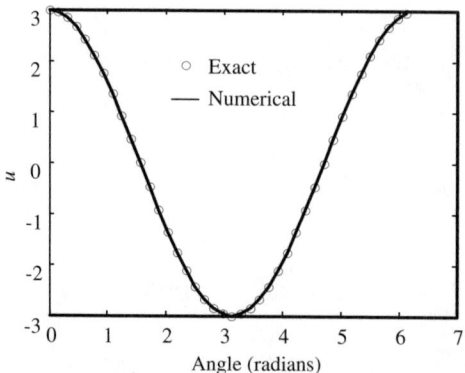

FIGURE 2.6: Neumann problem on a circle of 3 units with $q = 3 \cos \theta$ prescribed on the circle. Forty nodes are used on the circle.

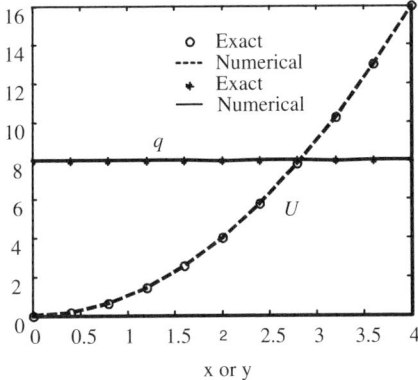

FIGURE 2.7:   Mixed problem on a square, u on $y = 0$(circles), q on $x = 4$(stars).

$$q = \frac{\partial u}{\partial n} = 0 \quad \text{on the lines } x = 0 \text{ and } y = 0$$
$$u = 16 - y^2 \quad \text{on the line } x = 4 \tag{2.38}$$
$$u = x^2 - 16 \quad \text{on the line } y = 4.$$

The exact solution for this problem is given by

$$u = x^2 - y^2. \tag{2.39}$$

The numerical and exact solutions are illustrated in Fig. 2.7, where the dashed line denotes the values of u on the line $y = 0$ while the solid line represents the normal derivatives q on the line $x = 4$. The stars and circles are the exact solutions. 40 nodes are used on the square, i.e., 10 nodes on each side. Also, it is observed that the numerical results agree well with the exact one. The foregoing three examples for potential problems subject to different boundary conditions are successfully solved by the present method. Few nodes are used but the results are of high accuracy. The convergence of the numerical results is excellent since only one analytic function needs to be found.

2.3   Complex DQ method for plane linear elastic problems

2.3.1   Revisit of two-dimensional problems in the theory of linear elasticity in an isotropic medium

Most of this part can be found elsewhere. A more detailed description of the theory of linear elasticity can be found in the books by Muskhelishvili (1953) and Timoshenko and Goodier (1970). Thus, only basic theory of linear elasticity, which is useful for the development of complex DQ method, is briefed here.

Two-dimensional problems in the theory of linear elasticity in the absence of body forces require the integration of a system of equations consisting of the equilibrium equations

$$\frac{\partial \sigma_y}{\partial x} + \frac{\partial \tau_{xy}}{\partial y} = 0, \quad \frac{\partial \tau_{xy}}{\partial x} + \frac{\partial \sigma_y}{\partial y} = 0 \tag{2.40}$$

and the compatibility equation

$$\left(\frac{\partial^2}{\partial x^2} + \frac{\partial^2}{\partial y^2} \right) (\sigma_x + \sigma_y) = 0 \tag{2.41}$$

with corresponding boundary conditions. In the above, σ_x and σ_y are stress components in x-, y-axes while τ_{xy} is the shear stress.

In the case of the first fundamental equation, i.e, the external stresses acting on the boundaries Γ of an area S are given, the boundary conditions are expressed as

$$\begin{cases} \sigma_x \cos(n, x) + \tau_{xy} \cos(n, y) = X_n \\ \tau_{xy} \cos(n, x) + \sigma_y \cos(n, y) = Y_n \end{cases} \tag{2.42}$$

where X_n and Y_n are the given components of the external stress vector acting on the boundary Γ and n is the exterior normal.

In the case of the second fundamental equation, that is, when the displacements along the boundary Γ of the area S are given, the boundary conditions are read as:

$$u = \tilde{u}, \quad v = \tilde{v} \tag{2.43}$$

where $\tilde{u}$ and $\tilde{v}$ are known displacements of points on the boundary line. They are given function of the arc s of the boundary line, calculated from an arbitrarily chosen point.

Equations (2.40) and (2.41), as a set of equations can be further simplified if we introduce the following formulas relating the so-called Airy's stress function (or bi-harmonic function) U and the three stress components

$$\sigma_x = \frac{\partial^2 U}{\partial x^2}, \quad \tau_{xy} = -\frac{\partial^2 U}{\partial x \partial y}, \quad \sigma_y = \frac{\partial^2 U}{\partial y^2} = 0. \tag{2.44}$$

Keeping them in mind, we are able to combine the two equations (2.40) and (2.41) into a single biharmonic equation defined by:

$$\frac{\partial^4 U}{\partial x^4} + 2\frac{\partial^4 U}{\partial x^2 \partial y^2} + \frac{\partial^4 U}{\partial y^4} = 0. \tag{2.45}$$

The solution to the above equation is that biharmonic function U must be of the following form

$$U(x, y) = Re\left[\bar{z}\varphi(z) + \chi(z)\right] \tag{2.46}$$

where $\varphi(z)$ and $\chi(z)$ are analytic functions of the complex variable $z = x + iy$ and $\bar{z} = x - iy$. Consequently, the solution to the two-dimensional problem is reduced to the determination of two analytic functions $\varphi(z)$ and $\psi(z) = d\chi(z)/dz$ from the boundary conditions which can be proven to be of the form:

(a)*for the first fundamental problem*

$$\varphi(z) + z\overline{\varphi'(z)} + \overline{\psi(z)} = i\int_{z_0}^{z}(X_n + iY_n)ds \quad \text{for } z \in \Gamma. \tag{2.47}$$

(b)*for the second fundamental problem*

$$\kappa\varphi(z) - z\overline{\varphi'(z)} - \overline{\psi(z)} = 2\mu(\tilde{u} + i\tilde{v}) \quad \text{for } z \in \Gamma, \tag{2.48}$$

where $\kappa = (3 - \nu)/(1 + \nu)$ for the plane stress and $\kappa = 3 - 4\nu$ for plane strain; ν is Poisson's ratio and $\mu = G = E/2(1 + \nu)$ is the shear modulus; E is Young's modulus. When the functions $\varphi(z)$ and $\psi(z)$ are determined, the stress components σ_x, τ_{xy} and σ_y can be directly calculated from the functions $\varphi(z)$ and $\chi(z)$ by using the following well-known formulas:

$$\begin{cases} \sigma_x + \sigma_y = 4Re\left[\varphi'(z)\right] \\ \sigma_y - \sigma_x + 2i\tau_{xy} = 2\left[\bar{z}\varphi''(z) + \psi'(z)\right] \end{cases} \tag{2.49}$$

Equations (2.47) and (2.48) do not contain integrals, and thus allow for the use of element free methods. In the following sections, approximation of analytic functions by means of Lagrange polynomial in the complex plane will be investigated and employed to solve these equations.

2.3.2 Complex DQ method for plane linear elasticity

Solving the bi-harmonic equation is reduced to find the two analytic functions $\varphi(z)$ and $\psi(z)$ which satisfy the boundary conditions in Eqs. (2.47) and (2.48). In fact, by introducing two indicators $\alpha(z)$ and $\beta(z)$, Eqs. (2.47) and (2.48) can be rewritten in the general form:

$$\alpha(z)\varphi(z) + \beta(z)z\overline{\varphi'(z)} + \beta(z)\overline{\psi(z)} = F \ for \ z \in \Gamma \qquad (2.50)$$

where α, β and F are associated with the boundary conditions. For force boundary conditions, we have

$$\alpha = 1, \quad \beta = 1, \quad F = i \int_{z_0}^{z} (X_n + iY_n)ds \qquad (2.51a)$$

whereas for displacement boundary conditions,

$$\alpha = \kappa, \quad \beta = -1, \quad F = 2\mu(\tilde{u} + i\tilde{v}). \qquad (2.51b)$$

The two-dimensional problem is thus reduced to solving Eq. (2.50).

Similar to the foregoing potential problems, there exists $2N$ unknowns (φ_k and ψ_k) for the plane elastic problems if N nodes are selected on the boundary. $2N$ collocation points ξ_m ($m = 1, \cdots, 2N$), on the boundary are chosen. Suppose on these $2N$ collocation points ξ_m, the boundary values of the complex function F in Eq. (2.50) are given, denoted by F_m ($m = 1, \cdots, 2N$). The total number of nodes is half of that of collocations points, and part of collocation points may be coincident with the nodes, say $z_k = \xi_{2k-1}$ as shown in Fig. 2.1. Then the function value at any point in the solution domain by means of Lagrange interpolation in the complex plane is

$$\varphi(z) = \sum_{k=1}^{N} \lambda_k(z)\varphi_k, \quad \psi(z) = \sum_{k=1}^{N} \lambda_k(z)\psi_k. \qquad (2.52)$$

Particularly on the collocation points it holds

$$\varphi(\xi_m) = \sum_{k=1}^{N} \lambda_k(\xi_m)\varphi_k = \sum_{k=1}^{N} \lambda_{mk}\varphi_k, \quad \psi(\xi_m) = \sum_{k=1}^{N} \lambda_{mk}\psi_k \qquad (2.53)$$

By substituting Eq. (2.53) into Eq. (2.50), and then by using Eq. (2.6) or Eq. (2.12), we have

$$\sum_{k=1}^{N} \alpha_m \lambda_{mk} \varphi_k + \sum_{k=1}^{N} \xi_m \beta_m \bar{a}_{mk} \bar{\varphi}_k + \sum_{k=1}^{N} \beta_m \bar{\lambda}_{mk} \bar{\psi}_k = F_m \tag{2.54}$$

$$\text{for } \xi_m \in \Gamma \quad m = 1, \cdots, 2\text{N}.$$

Introducing the following matrices

$$\mathbf{C}_1 = \{\alpha_m \lambda_{mk}\}_{2N \times N}, \mathbf{C}_2 = \{\xi_m \beta_m \bar{a}_{mk}\}_{2N \times N}, \mathbf{C}_3 = \{\beta_m \bar{\lambda}_{mk}\} \tag{2.55a}$$

$$m = 1, \ldots, 2N$$

and vectors

$$\mathbf{\Phi} = \{\varphi_k\}_N, \quad \mathbf{\Psi} = \{\psi_k\}_N, \quad \mathbf{F} = \{F_k\}_N, \quad k = 1, \cdots, N \tag{2.55b}$$

we rewrite Eq. (2.54) in the following matrix form

$$\mathbf{C}_1 \mathbf{\Phi} + \mathbf{C}_2 \bar{\mathbf{\Phi}} + \mathbf{C}_3 \mathbf{\Psi} = \mathbf{F}. \tag{2.56}$$

This is a system of linear equations of complex coefficients. It may be recast in the form of real-coefficient system of linear equations. Separating the real and imaginary parts of both sides of the above equation yields

$$\begin{pmatrix} \mathbf{C}_1^R + \mathbf{C}_2^R & \mathbf{C}_2^I - \mathbf{C}_1^I & \mathbf{C}_3^R & -\mathbf{C}_3^I \\ \mathbf{C}_1^I + \mathbf{C}_2^I & \mathbf{C}_1^R - \mathbf{C}_2^R & \mathbf{C}_3^I & \mathbf{C}_3^R \end{pmatrix} \begin{pmatrix} \mathbf{\Phi}^R \\ \mathbf{\Phi}^I \\ \mathbf{\Psi}^R \\ \mathbf{\Psi}^I \end{pmatrix} = \begin{pmatrix} \mathbf{F}^R \\ \mathbf{F}^I \end{pmatrix} \tag{2.57}$$

where the superscript ^{R} denotes the real part and superscript ^{I} denotes imaginary part. Equation (2.57) is a system of linear algebraic equations of size $4N \times 4N$, which can be solved by using Gauss elimination method. The $2N$ complex unknowns (or $4N$ real unknowns), i.e., $\mathbf{\Phi} = \mathbf{\Phi}^R + i\mathbf{\Phi}^I$ and $\mathbf{\Psi} = \mathbf{\Psi}^R + i\mathbf{\Psi}^I$ can be obtained by solving Eq. (2.57). The solution procedures for the plane elastic problems are as follows

(1) Determine finite or infinite domain problem, that is, determine to use Eq. (2.5) or Eq. (2.11)

(2) Find the coefficients λ_{mk}, a_{mk} and b_{mk} either from Eqs. (2.5) to (2.9) for finite domain problem or from Eqs. (2.11) to (2.16) for infinite domain problem

(3) Form matrices $\mathbf{C}_1$, $\mathbf{C}_2$, $\mathbf{C}_3$ and vector $\mathbf{F}$ calculated from Eqs. (2.55a) and (2.55b)

(4) Form the coefficient matrix on the right-hand side of Eq. (2.57)

(5) 4N unknowns obtained by solving Eq. (2.57)

(6) Find stresses from Eq. (2.49) and

(7) Find displacements from Eq. (2.48).

Example 2.3 *Test against given analytic functions*
Consider a unit circle. The two analytic functions $\varphi(z)$ and $\psi(z)$ are specified as

$$\varphi(z) = z^3, \quad \psi(z) = \exp(z) \tag{2.58}$$

By substituting Eq. (2.58) into Eq. (2.50) and differentiating with respect to z we boundary conditions

$$F = \alpha z^3 + 3\beta z \overline{z^2} + \beta \overline{\exp(z)} \quad \text{on the boundary } |z| = 1 \tag{2.59}$$

Suppose the first fundamental equation is considered, that is, $\alpha=1$ and $\beta=1$, and uniformly distributed N nodes on the boundary Γ: $|z| = 1$. By substituting Eq. (2.59) into Eq. (2.57) and solving it by use of Gauss elimination method, we obtain the $2N$ function values φ_k and ψ_k, $(k=1,\ldots, N)$. Then the stress states inside the circle can be explored by using Eq. (2.49). On the other hand, the exact solutions of the stress components, denoted by σ_x^0, σ_y^0 and τ_{xy}^0, can be obtained from Eq. (2.49), too. Defining relative errors similar to those in Eq. (2.17), we obtain accuracy estimates at the particular point $z = \frac{\sqrt{2}}{4}(1 + i)$ as shown in Fig. 2.8.

When N (=5) is small, the relative errors for three stress components are up to 5%. As N increases to 10, errors are dramatically reduced to 10^{-6} to 10^{-3} %. Further increase of N results in loss of accuracy. In general, the errors for the three stress components are below 0.1%, showing similar tendency to the curves shown in Fig. 2.2. As N is beyond 60, relative errors are kept stable at 10^{-3}%. This example clearly demonstrates that the present complex DQ method has very good accuracy. Investigations of other points lead to the same conclusions.

Example 2.4 *Two Concentrated Forces Acting on a Disk*
Concentrated forces acting on an elastic body can simplify theoretical analysis significantly, but present great challenge for numerical analysis because infinitely large stresses are usually created at the force-application point. A circular 2-D elastic body subjected to two concentrated forces shown in Fig.

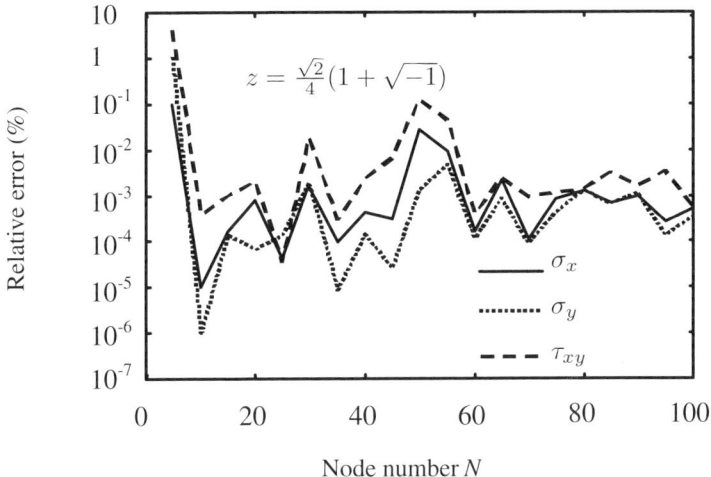

FIGURE 2.8: Influence of node number on accuracy of stress components.

2.9 is one of these examples. As shown in Fig. 2.9, the circular disk is compressed by two concentrated forces of same magnitude P acting on two points $z_1 = 1$ and $z_2 = -1$. The boundary conditions are then

$$F = \left\{ \begin{array}{ll} -iP & 0 \leq \theta < \pi \\ 0 & \pi \leq \theta < 2\pi \end{array} \right\}. \tag{2.60}$$

The exact solutions to this problem are given by (Timenshenko and Goodier, 1970)

$$\sigma_x^0 = \frac{2P}{\pi} \left(\frac{\cos^3 \theta_1}{r_1} + \frac{\cos^3 \theta_2}{r_2} \right) - \frac{P}{\pi} \tag{2.61a}$$

$$\sigma_y^0 = \frac{2P}{\pi} \left(\frac{\sin^2 \theta_1 \cos \theta_1}{r_1} + \frac{\sin^2 \theta_2 \cos \theta_2}{r_2} \right) - \frac{P}{\pi} \tag{2.61b}$$

$$\tau_{xy}^0 = \frac{2P}{\pi} \left(\frac{\cos^2 \theta_1 \sin \theta_1}{r_1} - \frac{\cos^2 \theta_2 \sin \theta_2}{r_2} \right) \tag{2.61c}$$

where r_1 and r_2 are the distances from any point z in the domain to the two force-application points as shown in Fig. 2.9, that is,

$$z_1 - z = r_1 \exp(i\theta_1) \text{ and } z_2 - z = r_2 \exp(i\theta_2) \tag{2.61d}$$

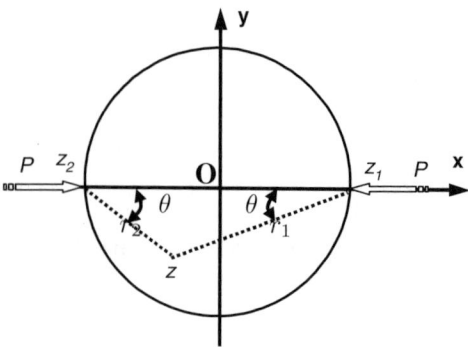

FIGURE 2.9: A disk subjected to two concentrated forces.

By substituting Eq. (2.60) into Eq. (2.57) and solving it with $N = 61$ equidistance nodes, the numerical results can be furnished. The numerical and exact results in the form of contour are illustrated in Figs. 2.10(a) to 2.10(d). In these figures, the upper half part represents the numerical results and the lower half part denotes the exact results. In Fig. 2.10(a), which displays the distribution of stress component σ_x, the contours start from -4.5 to -0.5. The lowest stress is located at the force-application points, which is infinite. In Fig. 2.10(b) which shows the distribution of stress σ_y, the contours start from -0.2 up to 0.3. The lowest stresses are negative, occurring at the two force-application points. The stress changes from compression (negative) to tension when one moves from boundary to the centre. It reaches its maximum at the centre. Figure 2.10 (c) shows the shear stress distribution within the circle. The contours start from -0.8 up to 0.8 with an increment of 0.2. The upper and lower parts are symmetric about center.

Example 2.5 *Uniform force acting on a cantilever beam*

A cantilever beam of length a and depth b subjected to uniform force on the upper boundary is considered in this section as shown in Fig. 2.11. A plane stress state is assumed. The elastic constants for the beam is taken as $E = 2.11 \times 10^{11}$ N/m^2 and Poisson ratio $v = 0.3$. This is a problem of mixed boundary conditions. The rightmost end is fixed with zero displacements, and the rest are stress boundary conditions. The size of the beam is assumed to be $a=2$ and $b=1$.

This problem is solved using the present method with each side discretized by 5 nodes. The stresses at the section $x = -1$ are shown in Figs. 2.12. Figure 2.12 (a), (b) and (c) are results for stress components σ_x, σ_y and τ_{xy}, respectively. The ordinates in the three figures are these three stress components divided by applied force p. The accuracy of the numerical results is checked by comparing with the FEM results provided by NASTRAN (Caffrey

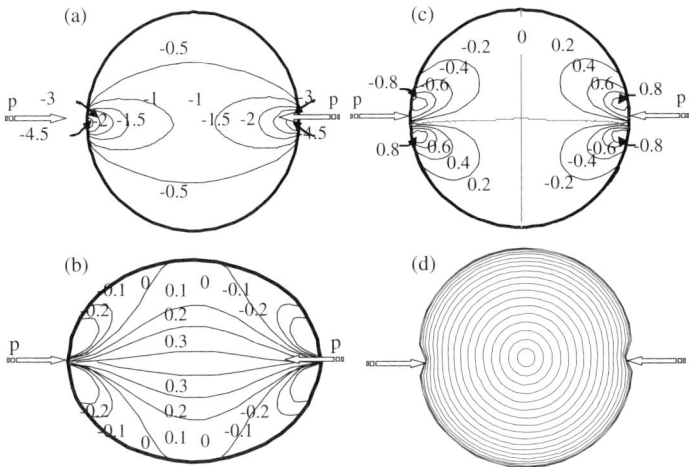

FIGURE 2.10: (a) Distribution of stress component σ_x. The contours in the upper part are drawn from the results given by the present method and the contours in lower part are drawn from analytical results. The rest are same. (b) Distribution of stress component σ_y; (c) Distribution of stress component τ_{xy}; (d) Distribution of magnified deformations.

and Lee, 1994) and the analytical results (Timenshenko and Goodier, 1970). In the FEM calculation in NASTRAN, $20 \times 20 = 400$ elements were adopted. The analytical results are expressed as

$$\sigma_x^0 = -\frac{6}{b^3}(\frac{b}{2} - y)(x + a)^2 + \frac{1}{b^3}(\frac{b}{2} - y)[4(\frac{b}{2} - y)^2 - \frac{3}{5}b^2] \tag{2.62a}$$

$$\sigma_y^0 = -\frac{1}{2}[1 + (\frac{1}{2} - \frac{y}{b})(\frac{2y}{b} - 1)^2] \tag{2.62b}$$

$$\tau_{xy}^0 = -\frac{6}{b^3}(x + a)[\frac{b^2}{4} - (\frac{b}{2} - y)^2]. \tag{2.62c}$$

The three stress components obtained from the present method agree well with analytical results while the agreement between FEM results and the analytical results is much worse than the present method. In view of the computation efforts, it is confirmed that the present method yields results with high accuracy by using a small number of nodes. It is more cost effective than FEM.

Figure 2.13 shows the convergence rate of this problem. The errors in the figure are defined as the squared root of the squared error integral divided by the node number

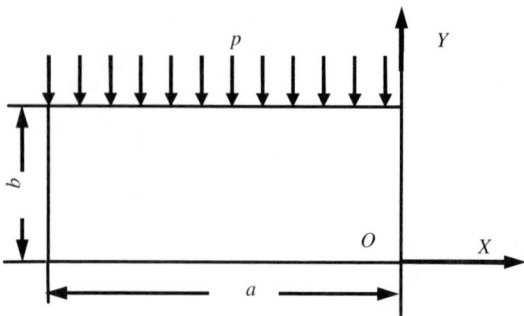

FIGURE 2.11: Cantilever beam subjected to uniform force.

$$e_x = \frac{1}{N}\sqrt{\iint (\sigma_x - \sigma_x^0)^2 \, dx dy}$$

with similar formulas for e_y and e_τ. The figure indicates fast and good rate of convergence.

Example 2.6 *Hole in an infinite medium*

Consider an infinite plate with a central circular hole subjected to unidirectional tension of magnitude 1; see Fig. 2.14. Suppose the radius of the hole is 1. This is an infinite domain problem and the Eqs. (2.11) to (2.16) will be used for Lagrange interpolation. The exact solutions in polar coordinate system to the problem are (Timoshenko and Goodier, 1970)

$$\sigma_r^0 = \frac{1}{2}(1 - \frac{1}{r^2}) + \frac{1}{2}\cos(2\theta)(1 - \frac{1}{r^2})(1 - \frac{3}{r^2})^2 \qquad (2.63a)$$

$$\sigma_\theta^0 = \frac{1}{2}(1 + \frac{1}{r^2}) - \frac{1}{2}\cos(2\theta)(1 + \frac{3}{r^4}) \qquad (2.63b)$$

$$\tau_{r\theta}^0 = -\frac{1}{2}\sin(2\theta)(1 - \frac{1}{r^2})(1 + \frac{3}{r^2}) \qquad (2.63c)$$

The numerical calculations can be simplified if we notice that the solution to the problem is a superposition of the following two parts: one is a uniform infinite plate subjected to uniform tension in x direction and the other is the weakening due to the presence of the hole. The solution to the first problem is well known. Substituting it into Eq. (2.50) yields the boundary condition on the hole

$$F = -iy \qquad (2.64)$$

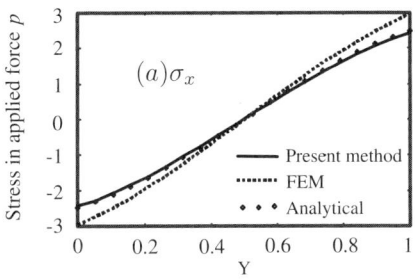

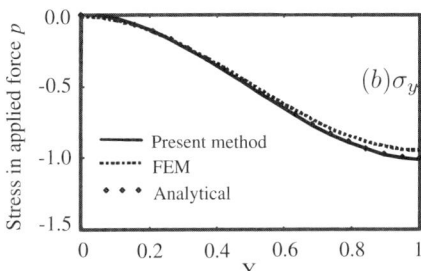

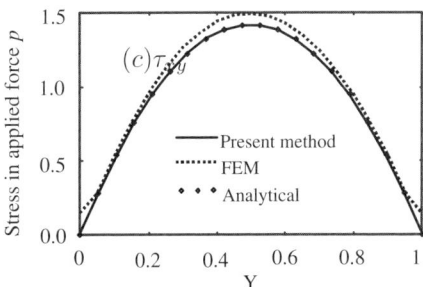

FIGURE 2.12: Comparison of stress components at section $x = -1$ for the cantilever problem.

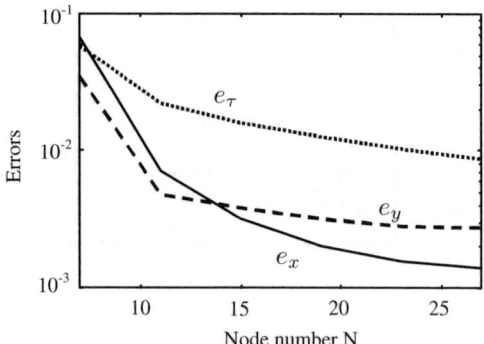

FIGURE 2.13:   Rate of convergence for the three stress components.

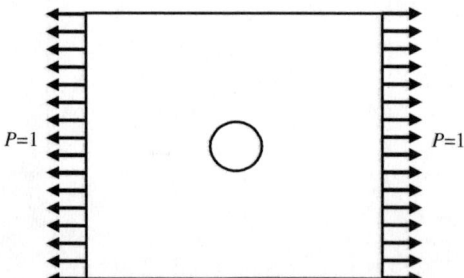

FIGURE 2.14:   An infinite plate with a central circular hole subjected to uniform tension in horizontal direction.

Both exact solution and numerical results are plotted in Figs. 2.15(a) to (c) in the form of contours for the three stress components, respectively. The upper part in each figure denotes the results obtained from the present method using $N = 20$ and the lower part represents the exact results from Eqs. (2.63). Again, excellent agreement between the numerical and analytical results is observed.

One more comment on necessary extension of DQ onto complex domain may be justified by the following discussions. Differentiation, integration and solution of differential equation, etc., are usually an operation on functions of continuous variables. When we carry out actual computation for such operations in a computer we must first approximately replace the continuous operand with finite and discrete numerical parameters. Then we carry out computation for these numerical values and get information about the consequence of differentiation, integration, etc. from the result of computation (which again consists of finite number of parameters). For such finite number of parameters we use, in many cases, function values at discrete values of the independent variable (nodes). Sometimes we also use coefficients of an expansion in terms of some suitable set of functions (power series or the Fourier series, for example). However, in any case, in order to express a function in sufficiently high precision with a relatively small number of parameters it is required that the function be moderately smooth. Although in many books the requirement on the smoothness is not so explicitly mentioned, it is implicitly included in the error representation of the numerical formula. In fact, in many cases the error representation of numerical formulas includes the maximum absolute value of the nth order derivative of the function in question (n is determined from the formula), and hence it is implicitly assumed that the function is n-times differentiable and that its nth order derivative is not very large. However, in practical applications, it is quite rare that, for example, we know that the function is three times differentiable but do not know whether it is four times differentiable or not. In actual applications, on the other hand, it is usually assumed implicitly that the function under consideration is infinite times differentiable, or more strongly, is analytic in a region which includes the real axis. In addition, there are many cases in which the analyticity can actually be proved in some explicitly given region. This is obviously the case if the function is defined explicitly in an analytical form. As another example from the solid state physics, there is a strong reason to believe that the function which describes the current-voltage characteristics of a certain electronic device, a diode for example, is regular in the strip region about the real axis with half width $2\pi kT$ (where k is the Boltzmann constant and T is the absolute temperature). Also, to investigate what kind of analyticity the state function of a material has, or in other words, where and what kind of singularities exist in the complex plane, is an important subject in the field of recent statistical mechanics. As seen from the discussion given above we

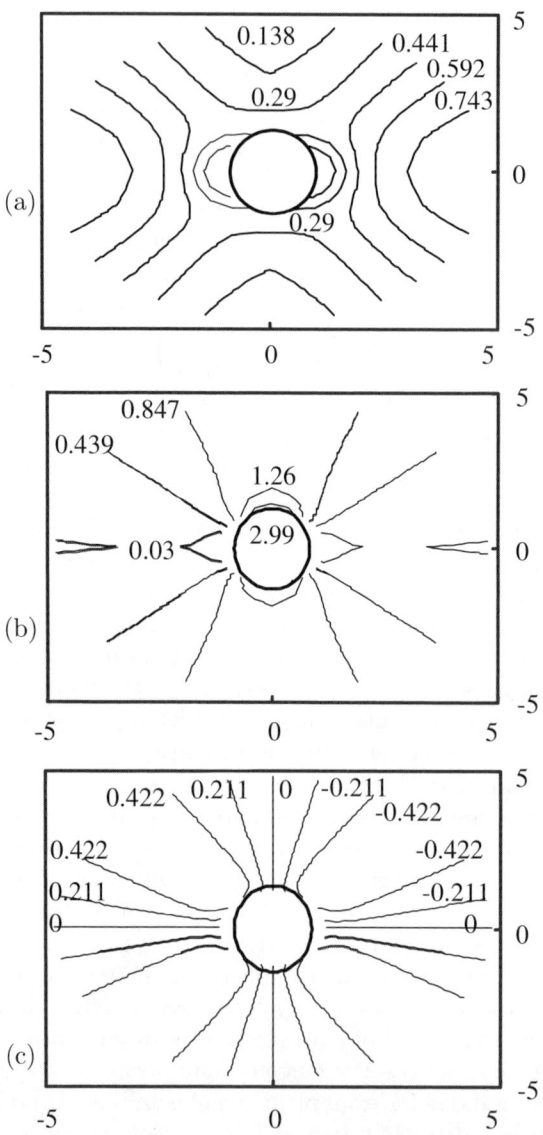

FIGURE 2.15: (a) Distribution of stress component σ_r. The contours in the upper part are drawn from the results obtained by the present method and the contours in the lower part are drawn from analytical results. The rest are the same. (b) Distribution of stress component σ_θ; (c) Distribution of stress component $\tau_{r\theta}$.

should recognize that it is by far more realistic to study in what region the function is regular rather than to establish the n times differentiability. If we base our argument on the analyticity of the function in this way, the complex function theory will play a fundamental role as a matter of course (Takahashi, 2005).

2.4 Conformal mapping aided complex differential quadrature

The key steps of the method introduced in the previous sections are

(1) Approximation of the solution by use of Lagrange interpolating polynomial with function values on nodes as unknowns

(2) Substitution of the interpolating polynomial into the boundary conditions and formation of a system of linear equations

(3) Application of Gauss Elimination method to solving the system of linear equations

The procedures can be further simplified without forming and solving the system of linear equations if conformal mapping technique is applied. Conformal mapping starts from a known solution to a hypothetical profile in, say, the ζ-plane. Through a mapping function which is analytic, this simple solution is mapped onto the complicated geometry in the physical z-plane under consideration. The key step in applying conformal mapping thus lies in finding the mapping function.

Conformal mapping, also called a conformal map, conformal transformation, angle-preserving transformation, or biholomorphic mapping, is extremely important in complex analysis, as well as in many areas of physics and engineering. It seems, however, that conformal mapping is becoming old-fashioned compared with the widespread popularity of numerical methods, such as finite element method. This is an erroneous impression once we notice the vast body of the literature on conformal mapping-based applications and new mathematical methods. The true case is two-fold. On one hand, conformal mapping has benefited from the deceasing cost and increasing speed of digital computations. Algorithms are being used for rapidly evaluating many hitherto intractable mappings. On the other hand, capability to provide nearly explicit solution by conformal mapping has made it a valuable partner in raising the efficiency of other numerical techniques. It can transform complicated geometry in a physical plane onto a readily analyzable one and then make subsequent numerical calculations more efficient. It is this two-fold fact that makes the idea of combining conformal mapping and Complex Differential Quadrature (CDQ) method very attractive. Before we go into conformal mapping-aided CDQ, we will develop the methodology for performing conformal mapping numerically using Lagrange interpolation.

2.4.1   Conformal mapping aided by Lagrange interpolation

The task of conformal mapping is to find an analytic function $z = f(\zeta)$ which will map a complicated geometry in the complex z-plane onto a simpler one in the complex ζ-plane. The particular form of the transformation function $f(\zeta)$ will depend on the specific boundaries and the application at hand.

The existence of such mapping was proved by Riemann. An arbitrary closed profile in the complex z-plane can be mapped onto the unit circle in the complex ζ-plane through transformation $z = f(\zeta)$ of the following properties:

(1) The mapping is one-to-one so that to each point in the physical domain, there is one and only one corresponding point in the mapped domain.

(2) Closed curves map to closed curves.

(3) Angles are preserved between the intersections of any two lines in the physical domain and the mapped domain.

The importance of the last property will be utilized in Section 2.4.2. To be precise, we give the following theorem without proof.

Theorem 2.1 (Riemann mapping theorem) *Any simply connected domain D in the complex z-plane which boundary contains more than one point can be mapped onto the interior (or exterior) of a unit circle $|\zeta| < 1$ (or $|\zeta| > 1$) with angle-preserving (conformal map). If there exists a point z_0 inside D such that $f(z_0) = 0$ and $f'(z_0)$ takes positive real value, then the mapping function $\zeta = f(z)$ is unique.*

The theorem applies both to the interior and exterior of a unit circle, and thus we put exterior in the bracket in the above statement of the theorem.

Centuries of efforts have discovered many particular forms of conformal mappings, the best known of which is the Joukowski transformation $z = (\zeta + c^2/\zeta)/2$ mapping a unit circle to the thin wing section. It played an important role in the development of flight theory.

Conformal mapping and analytic function are closely related. An analytic function is conformal at any point where it has nonzero derivatives. Conversely, any conformal mapping of a complex variable which has continuous partial derivatives is analytic. Thus, construction of a conformal mapping is in fact to find an analytic function which meets the specific requirements.

The Riemann mapping theorem enables the solution for any practical two-dimensional body to be obtained by conformal transformation. However, finding the suitable transformation between a unit circle and an arbitrary body in the complex z-plane is a tough job. Numerical technology is promising for providing a general methodology for solving such a problem. Several numerical methods have been proposed in the past with detailed introduction in the book (Schinzinger and Laura, 2003). The widespread procedure is to approximate the mapping function by virtue of a polynomial in ζ, which is analytic by definition. The coefficients of the polynomial are unknowns, which are to be determined by ensuring the one-to-one mapping among the points in the complex z-plane and ζ-plane. The points on the unit circle in the complex ζ-plane are specified *a priori*, but the points in the complex z-plane are yet to be determined. Thus, there are two sets of unknowns: the coefficients and the points in the z-plane. If one set of the two is known, the other can be found. Thus it is an iteration process to find them.

Initial guess is in general very critical for iteration algorithms. Another limitation is that the above method may fail for complicated profile permitting two vertical coordinate values for one horizontal coordinate value. In view of this, the practical importance of the Riemann mapping theorem is often diminished by the numerical complexity of mapping arbitrary body profiles. However, use of complex Lagrange interpolation introduced in Section 2.1 may reduce the complexity significantly, providing a direct and iteration-free technique to find the one-to-one map. The complex Lagrange interpolation method proves to be extremely simple, the essential elements of which are as follows.

(a) Domain transformation is equal to curve transformation

Any conformal mapping given by an analytic function is valid in a domain. However, it can be uniquely determined by its values on the boundary.

Theorem 2.2 *Suppose two simply connected domains D and Δ, which boundaries are Γ and C, are given. If the mapping function $\zeta = f(z)$ is analytic inside C, is continuous on C and one-to-one maps boundary C to boundary Γ with orientation preserved, then $\zeta = f(z)$ is the one-to-one mapping from domain D and Δu*

This theorem states that the conformal mapping can be obtained from one-to-one mapping between the two boundaries in the two complex planes. Finding a conformal mapping in a domain is equal to finding an analytic function whose values on a curve are given.

(b) Discretization of mapping and mapped curves

Theorem 2.2 enables us to focus on finding the analytic function mapping the complicated curve in the z-plane onto the simpler one in the ζ-plane. Consider a unit circle C in the complex ζ-plane as shown in Fig. 2.16(a) and a complicated curve Γ in the complex z-plane as shown in Fig. 2.16(b).

The transformation function mapping C to Γ is denoted by $z = f(\zeta)$ and the inverse transformation mapping Γ to C is $\zeta = f^{-1}(z) = F(z)$. Select N

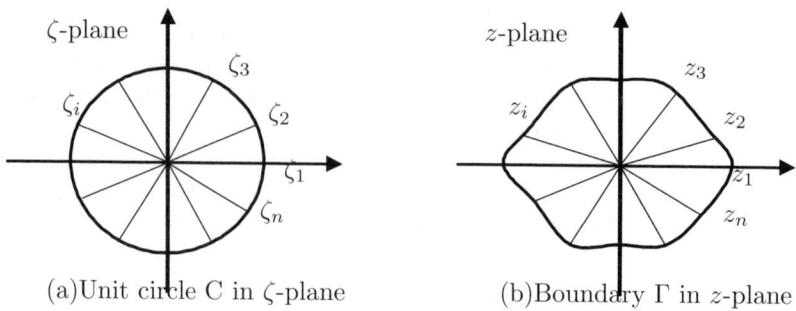

(a)Unit circle C in ζ-plane   (b)Boundary Γ in z-plane

FIGURE 2.16:   Conformal mapping between ζ- and z-planes.

points on the unit circle C. We use $\zeta_j (j = 1, 2, ..., N)$ to denote them. These points are supposed to be arranged counterclockwise as shown in Fig. 2.16(a). The inherent limitation of Lagrange interpolation requires that these points or nodes should be Chebyshev ones. Meanwhile, select N points $z_j (j = 1, 2, ...N)$ on Γ in the z-plane. z_j may be equally spaced or not.

(C) Buildup of the one-to-one mapping

Conformal mapping relates the two complex variables z and ζ through $z = f(\zeta)$. Suppose the conformal mapping to be determined maps point ζ_j to z_j. Then the complex Lagrange interpolation in Section 2.1 gives the desired conformal mapping defined by

$$z_i = f(\zeta_i) = \sum_{j=1}^{N} \lambda_j(\zeta_i) z_j \tag{2.65}$$

The inverse transformation is then

$$\zeta_i = F(z_i) = \sum_{j=1}^{N} \lambda_j(z_i) \zeta_j. \tag{2.66}$$

For points which are not interpolation points or which are not on Γ, we are able to determine them from the Lagrange interpolation

$$z = f(\zeta) = \sum_{j=1}^{N} \lambda_j(\zeta) z_j \tag{2.67}$$

and the inverse transformation is

$$\zeta = F(z) = \sum_{j=1}^{N} \lambda_j(z) \zeta_j \tag{2.68}$$

Equations (2.65) to (2.68), extremely simple in form and calculation, are the essential elements of the complex Lagrange interpolation method for performing conformal mapping. The interpolation function $\lambda_j(\zeta)$ may take different forms depending on whether the interpolation is performed for the interior (Eq. (2.5)) or exterior (Eq. (2.10)) domains.

In the following, several examples are given to show the method.

Example 2.7 *Conformal mapping between a unit circle and an ellipse*

The function to be determined maps a unit circle onto an ellipse. Consider only the upper half plane due to symmetry. Two cases are considered: $N=$ 11 points and $N=$ 21 points. The N points on the unit circle in the ζ-pane are evaluated through the Chebyshev formula

$$\zeta_j = \xi_j + i\eta_j, j = 1, 2, ..., N \tag{2.69}$$

where

$$\xi_j = \cos\theta_j, \eta_j = \sin\theta_j \text{ and } \theta_j = \frac{\pi}{2}[1 - \cos\frac{j-1}{N-1}\pi] \tag{2.70}$$

Here θ_j are Chebyshev nodes. The ellipse in the z-plane is one with major semiaxis $a=$ 2 and minor semi-axis $b=$ 1, the N points on which are evaluated from

$$z_j = x_j + iy_j, j = 1, 2, ..., N \tag{2.71}$$

where

$$x_j = a\cos\theta_j, y_j = b\sin\theta_j \text{ and } \theta_j = \frac{\pi}{2}[1 - \cos\frac{j-1}{N-1}\pi] \tag{2.72}$$

Using Eqs. (2.65) and (2.67) we have found the desired conformal mapping easily with results shown in Fig. 2.17. Figure 2.17(a) shows the results obtained from $N = 17$ points. The curve denoted by line points are results obtained from Eq. (2.67). In the figure the given ellipse is also plotted, denoted by the dotted line. And the unit circle, which should have appeared in the ζ-plane, is plotted in the same figure. It is observed there is large discrepancy between the interpolated and the given ellipse. To improve the accuracy, $N=$ 21 points are used with results shown in Fig. 2.17(b). This time, the two curves obtained from Eq.(2.67) and the given ellipse are coincident completely.

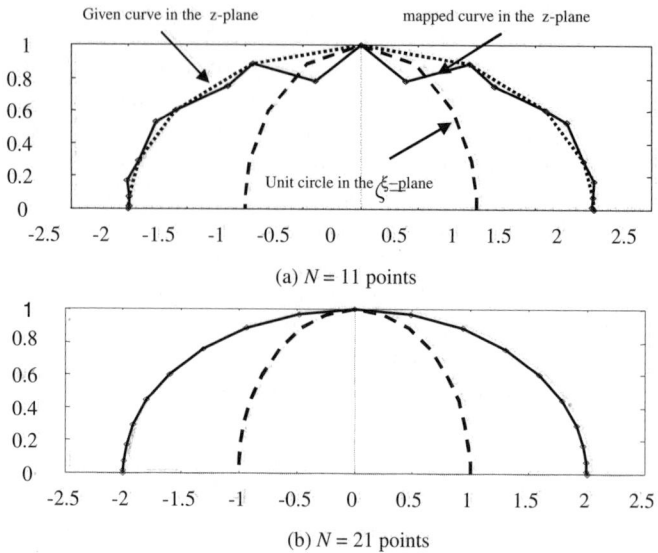

FIGURE 2.17: The conformal mapping between a unit circle and an ellipse found by Eqs. (2.65) and (2.67).

It should be emphasized that node selection (Chebyshev node) defined by Eq. (2.70) is necessary. To see this, we change the points to those given by

$$\xi_j = \cos\theta_j, \eta_j = \sin\theta_j \text{ and } \theta_j = \frac{j-1}{N-1}\pi \tag{2.73}$$

and recalculated the case $N=21$. The results are given in Fig. 2.18(a). Bad results are observed near the two ends. This is due to the instability of Lagrange interpolation on uniformly spaced nodes as explained in Chapter 1. It should be pointed out, however, that the nodes for points z_j are not necessarily Chebyshev nodes. Again, to see this, we use Chebyshev nodes for ζ_j and uniform nodes for z_j. The results are shown in Fig. 2.18(b). From the figure it is concluded that the node selection of ζ_j is very important, but node selection of z_j is not equally important. This is true for the forward transformation $z = f(\zeta)$. For the inverse transformation, we should note that the node selection of z_j will be important while node selection of ζ_j will not be important.

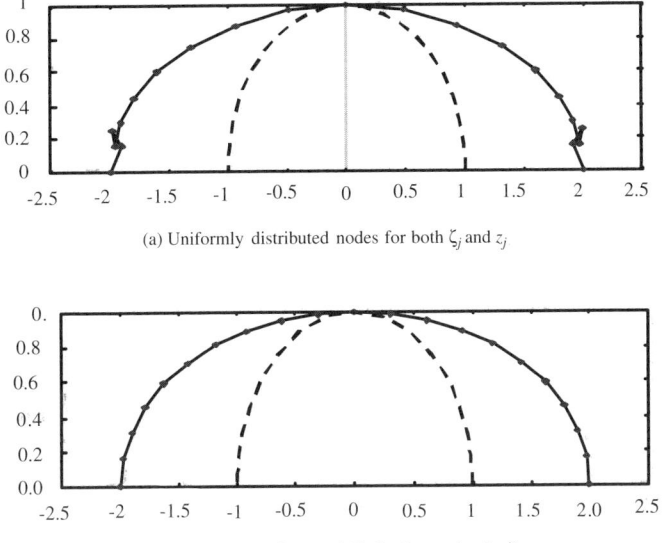

(a) Uniformly distributed nodes for both ζ_j and z_j

(b) Uniform nodes for z_j and Chebyshev nodes for ζ_j

FIGURE 2.18: Influence of node selection on the conformal mapping.

Example 2.8 *Conformal mapping from a unit circle onto a square*

In this example, the mapping function to be determined is one mapping a unit circle onto a square. Again consider the upper half plane only due to symmetry. The square is of unit length. Note that the challenge here arises from the corners of the square under consideration.

Two cases are considered: $N=17$ points and $N=33$ points. The N points on the unit circle in the ζ-plane are determined from Eq. (2.70) in the previous example again, and the N points on the square in the z-plane are uniformly distributed on the sides of the square.

The results obtained from $N=17$ points and Eq. (2.67) are given in Fig. 2.19(a). And the results obtained from $N=34$ points and Eq. (2.67) are plotted in Fig. 2.19(b). In each figure, two curves are plotted. The line points denote the mapped square obtained from Eq. (2.67) and the solid lines denote the given square.

In Fig. 2.19(a), the given square and the mapped square exhibit large difference on the top side, the latter being oscillatory. They show negligible difference on the left and right vertical sides. The reason is that Lagrange polynomials are higher order accurate, but a straight line is zero-th order accurate.

In Fig. 2.19(b), more nodes are used ($N=33$). The accuracy is improved a lot compared with the previous figure. In general, the agreement between the

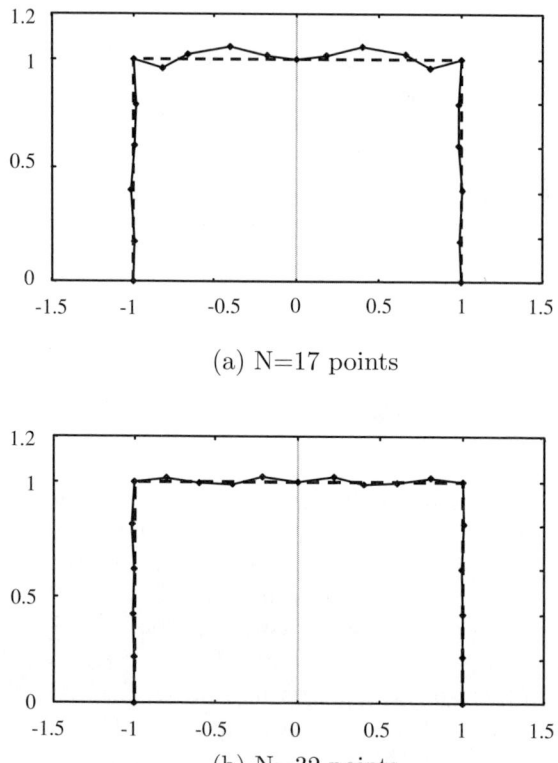

(a) N=17 points

(b) N=32 points

FIGURE 2.19: The square in the z-plane obtained from Eq. (2.67) is denoted by line points and the given square is denoted by solid line.

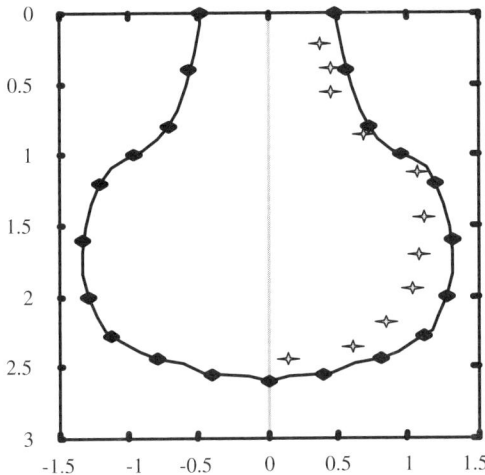

FIGURE 2.20:   Bulbous bow cross section. Points are given offset values of the cross section; the solid line is obtained from Eq. (2.67). Symbol + denotes the results obtained from Smith conformal mapping method. He used ten parameters to give the curve, which however is not sufficient to fully represent the cross section (Smith, 1967).

given square and the mapped square is satisfactory for this difficult problem.

Example 2.9 *Conformal mapping from a unit circle onto a bulbous bow section*

Ships with so-called bulbous bows have smaller resistance in water, thus running faster compared with same ships without bulbous bows. The cross section of a bulbous bow is complicated, as shown in Fig. 2.20. 21 points are given to represent the cross section denoted by dots in Fig. 2.20. Note that the true values should be the data in the figure multiplied by 0.025.

To determine the conformal mapping, one maps a unit circle to this cross section. The mapped cross section obtained from Eq. (2.67) is denoted by solid line in Fig. 2.20. In the figure are also given the results obtained from Smith method (Smith, 1966). They are denoted by the symbol +. Smith method is a numerical conformal mapping method. The transformation function is expressed by a polynomial of ζ. The order of the polynomial determines the number of unknowns (parameters) used. Ten parameter polynomial implies that nineth order polynomial is used.

From the figure there exists large discrepancy between the results obtained from Smith method and the given cross section. It is hard to use Smith method to obtain satisfactory results. The results obtained from the method developed in this section are very close to the true curve with significant improvement in accuracy and capability.

The method presented in this section is straightforward and very easy to employ. It is in essence a complex Lagrange interpolation, and provides an alternative to applying numerical methods to conformal mapping representation of complicated profiles in the complex plane. Meanwhile we should say that the inherent lack of stability of polynomials may limit the applicability of this method in some cases. To fully understand and solve this problem requires further efforts which are, however, rewarding because of wide applications of conformal mapping methodology.

2.4.2 Conformal mapping aided CDQ method

With the aid of conformal mapping technique, the known solution in the ζ-plane may be directly mapped onto the z-plane, thus enabling us to simplify the solution procedures significantly.

Involving Laplace's equation in two independent variables, problems in heat conduction, electrostatic potential and fluid flow, are the three classical instances to which conformal mapping technique is applied. Among them, fluid flow is the most difficult one due to the complicated nature of fluid flow. Thus in the following we will develop the conformal mapping-aided CDQ method, with particular applications to fluid flow. Because complex potential description of two-dimensional fluid flows is a specialized topic requiring a good demand of mathematics and a clear understanding of mechanics, a brief introduction is necessary for the convenience of the reader. More details can be found in Churchill and Brown (1990) and Newman (1977).

We consider only two-dimensional fluid flow, which is further assumed to be incompressible, irrotational and steady. Then there exists a potential function $\varphi(x, y)$ and stream function $\psi(x, y)$, which are related with the velocity vector through

$$u = \frac{\partial \varphi}{\partial x} = -\frac{\partial \psi}{\partial y} \text{ and } v = \frac{\partial \varphi}{\partial y} = \frac{\partial \psi}{\partial x} \tag{2.74}$$

Both the potential and stream functions satisfy the Laplace's equation

$$\nabla^2 \varphi = \nabla^2 \psi = 0 \tag{2.75}$$

The pressure $p(x, y)$ is given by the nonlinear Bernoulli's equation

$$\frac{p}{\rho} + \frac{1}{2}(u^2 + v^2) = \text{constant} \tag{2.76}$$

where ρ is the fluid density, being uniform everywhere in the fluid. Note that the pressure is greatest where the speed $|V| = \sqrt{u^2 + v^2}$ is least.

Introducing complex coordinate $z = x + iy$, we are able to form an analytic function $w(x, y) = \varphi(x, y) + i\psi(x, y)$, called complex potential, whose real and imaginary parts automatically satisfy the Laplace's equation. Moreover, its derivative is the conjugate velocity in the form of

$$u - iv = \frac{dw}{dz} \tag{2.77}$$

Our task is to find the complex potential, which in turn yields velocity through Eq. (2.77) and further gives pressure through Eq. (2.76).

In the following we will go into a class of particular problems in detail: fluid flow past a two-dimensional body of arbitrary shape. This is one of the basic problems in fluid mechanics. On using conformal mapping approach, we transform a profile K in the z-plane onto the unit circle C in the ζ-plane by aid of analytic function. The exterior domain of the unit circle C in the ζ-plane corresponds to the exterior domain of the profile K in the z-plane. Since the solution to the flow past a unit circle is known, the problem of flow past any body can be found. More specifically we have the following theorem.

Theorem 2.3 *Suppose $z = f(\zeta)$, which inverse is $\zeta = F(z)$, is a single-valued analytic function. It maps the exteriors of a unit circle C in the ζ-plane onto the exteriors of any profile K in the z-plane by preserving angle and being single-valued. It is further assumed that the mapping function satisfies*

(1) Point ∞ corresponds to Point ∞; and

(2) $\left(\frac{dz}{d\zeta}\right)_\infty = const.$

Because the complex potential of the flow past a unit circle is given by

$$W(\zeta) = \bar{V}_\infty \zeta + \frac{V_\infty}{\zeta} + \frac{\Gamma}{2\pi i} \ln \zeta \tag{2.78}$$

then the complex potential for the flow past any body in the z-plane is given by

$$w(z) = \bar{V}_\infty F(\zeta) + \frac{V_\infty}{F(\zeta)} + \frac{\Gamma}{2\pi i} \ln F(\zeta) \tag{2.79}$$

where V_∞ is the velocity at infinity and Γ is the circulation.

The theorem simplifies the solution to the flow past any body to looking for the conformal mapping between a unit circle C and any profile K. Once this mapping function is found, the complex potential $w(z)$ can be evaluated from

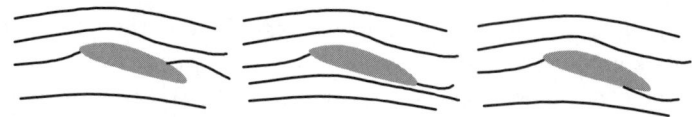

FIGURE 2.21:   Three types of flow patterns: the left indicates that fluid flows makes a U-tern from below at the trailing edge. The right represents that fluid flows makes a U-turn from above at the trailing edge.

Eq. (2.79). Theoretically, the function $z = f(\zeta)$ uniquely exists. In reality, however, it is tough to find the function as explained in the previous section.

Once the complex potential is found, the complex velocity can be evaluated from Eq. (2.79) by use of the chain rule of differentiation

$$\bar{V} = u - iv = \frac{dw}{dz} = \frac{dw}{d\zeta}\frac{d\zeta}{dz} = \left(\bar{V}_\infty - \frac{V_\infty}{\zeta^2} + \frac{\Gamma}{2\pi i}\frac{1}{\zeta}\right)\frac{d\zeta}{dz} \qquad (2.80)$$

where $\bar{V}$ denotes the conjugate velocity and $d\zeta/dz$ can be found from the following equation

$$\frac{d\zeta}{dz} = \left(\frac{dz}{d\zeta}\right)^{-1} = \left(\sum_{j=1}^{n}\lambda_j(\zeta)z_j\right)^{-1} = \left(\sum_{j=1}^{n}a_j(\zeta)z_j\right)^{-1} \qquad (2.81)$$

The coefficient $a_j(\zeta)$ is given by Eq. (2.7a) or (2.14).

The pressure distribution on the surface of a body in the fluid domain is given from the following Bernoullis equation if gravity is neglected

$$p + \frac{1}{2}(u^2 + v^2) = p + \frac{1}{2}\frac{dw}{dz}\overline{\frac{dw}{dz}} = const \qquad (2.82)$$

Integrating the pressure (Eq. (2.82)) on the body surface yields the resultant forces and integrating the pressure about the origin of the coordinate system yields resultant moment. Because complex potential is analytic, finding resultant forces and moments is equal to integrating an analytic function along a closed curve. Up to here, the flow past a body of arbitrary shape has been solved.

It is worth noting that the circulation Γ remains unknown. For a sharp-tailing body as shown in Fig. 2.21, Γ can be determined by using the Joukowski-Kutta condition. There are a variety of theoretically feasible solutions to the flow past a body corresponding to different stagnation points. The same is true of a sharp-tailing body, for which three inherently different types of flow patterns are shown in Fig. 2.21. The left figure represents the flow making a U-turn from below at the trailing edge, and the right figure represents the flow making a U-turn from above at the trailing edge. The middle figure represents flow leaving the trailing edge smoothly. In the left and right figures the U-turn flow is greater than π, resulting in infinitely large velocity at the trailing edge. This is physically impossible. Only the case sketched in the middle is a physically feasible flow because fluid smoothly leaves the trailing edge, on which velocity is finite. In 1909 Joukowski assumed that the velocity at the trailing edge must be finite

$$\left(\frac{dw}{dz}\right)_{trailingedge} = \text{const} \tag{2.83}$$

from which we may determine the value of the circulation Γ. More specifically, suppose the trailing edge B corresponds to the point $E = \exp(i\theta_0)$ in the ζ-plane. If the mapping function $z = f(\zeta)$ is known, then θ_0 is known. It is clear that the analytic function $z = f(\zeta)$ is not angle-preserving at point E because the curve with angle π at point E becomes the curve with angle $2\pi - \tau$ at point B where τ is the angle at the trailing edge. Thus we have at point E

$$\left(\frac{dz}{d\zeta}\right)_E = 0 \tag{2.84}$$

Moreover, from the chain rule of differentiation, the velocities at point E and point B are connected through

$$\left(\frac{dW}{d\zeta}\right)_E = \left(\frac{dw}{dz}\right)_B \left(\frac{dz}{d\zeta}\right)_E \tag{2.85}$$

from which we conclude that

$$\left(\frac{dW}{d\zeta}\right)_E = 0 \tag{2.86}$$

So point E is a stagnation point on the ζ-plane. In the case of flow past a circle, the stagnation point E and the circulation Γ is uniquely connected through

$$\sin(\alpha - \theta_0) = -\frac{\Gamma}{4\pi |V_\infty|} \tag{2.87}$$

Note that the incoming flow in the z-plane is one with amplitude $|V_\infty|$ and angle of attack $\alpha - \theta_0$, so that

$$\Gamma = -4\pi |V_\infty| \sin(\alpha - \theta_0). \tag{2.88}$$

Once Γ is determined, $w(z)$ is fully determined. The problem for determining the flow of arbitrary wing section with sharp trailing edge is simplified to one to find the conformal mapping $z = f(\zeta)$. If the wing section under consideration does not have a sharp trailing edge, however, it is impossible to determine the circulation Γ theoretically. It is alternatively evaluated either *a priori* or through experiment.

Example 2.10 *Irrotational flow past an ellipse without circulation*

Consider the irrotational flow past an ellipse defined by $(x/a)^2 + (y/b)^2 = 1$. Suppose $a = 1$ and $b = 0.5$ in this example and circulation $\Gamma = 0$. The velocity at infinity is V_∞. Example 2.7 gives the transformation and Eqs. (2.80) and (2.81) give the velocity distribution on the ellipse. $N = 30$ nodes are used to approximate the ellipse.

The results are plotted in Fig. 2.22. The two velocity components (horizontal and vertical) are denoted by dashed and dotted lines separately in the figure. The vertical axis is velocity. The incoming flow is schematically shown in the left low corner. Figure 2.22(a) shows the results obtained by setting the angle of attack being zero and Fig. 2.22(b) shows the results for the case of attack angle equal to 0.1πugo the incoming velocity is $V_\infty = 1$ for the case in Fig. 2.22 (a) and $V_\infty = \exp(i \times 0.1\pi) = \cos(0.1\pi) + i\sin(0.1\pi)$ for the case in Fig. 2.22 (b).

The horizontal velocity is symmetric about the vertical axis in Fig. 22(a) and vertical velocity is symmetric about the coordinate origin, representing a flow moving upward in the range [-1,0] and then moving downward in the range [0,1]. The flow is no longer symmetric if an angle of attack 0.1π applied to the flow with numerical results shown in Fig. 2.22(b). The vertical velocity is larger in the range [-1,0] than that in the range [0,1].

Example 2.11 *Flow past an ellipse with circulation*

Consider Example 2.10 by adding circulation $\Gamma = 2\pi\sqrt{a^2 - b^2}\sin\alpha$ where α is the angle of attack to the flow in Fig. 2.22(b). The velocity distribution dramatically changes as shown in Fig. 2.23(b).

Example 2.12 *Flow past a plate with an angle of attack*

Flow past a plate is an analytically tractable problem but numerically tough problem because velocity becomes infinite at the leading edge. This can be readily solved using the method presented in this section. Regard a plate as an ellipse with long axis equal to a (=half plate length) but with a nearly zero short axis $b \to 0$. So the approach in Example 2.11 can be used again with $b = 0.001$, a number obtained through several trials.

The exact horizontal velocity on the upper surface of the plate is

$$u(x, +0) = \cos\alpha + \sin\alpha\sqrt{\frac{a-x}{a+x}}, \quad v = 0 \tag{2.89}$$

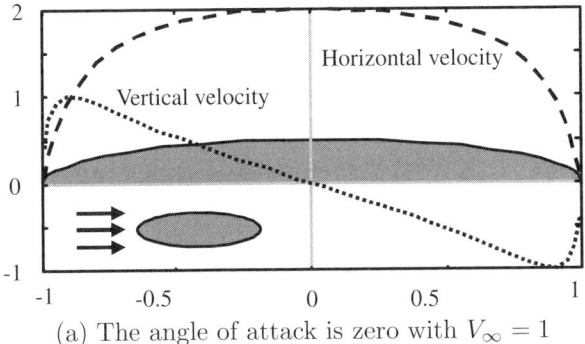

(a) The angle of attack is zero with $V_\infty = 1$

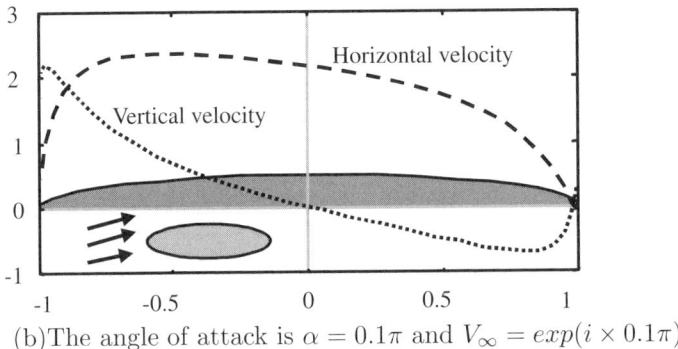

(b)The angle of attack is $\alpha = 0.1\pi$ and $V_\infty = exp(i \times 0.1\pi)$

FIGURE 2.22: Velocity distribution on the surface of an ellipse without circulation.

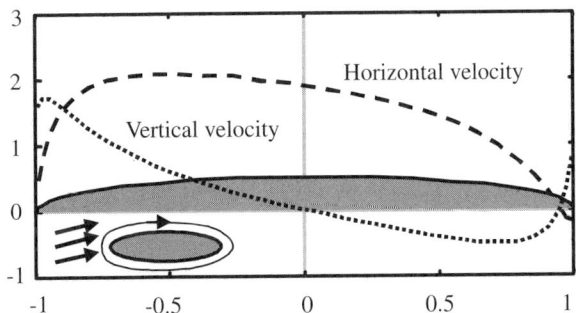

FIGURE 2.23: Velocity distribution on the surface of an ellipse. The angle of attack is $\alpha=0.1$ and the circulation is $\Gamma = 2\pi\sqrt{a^2 - b^2}$.

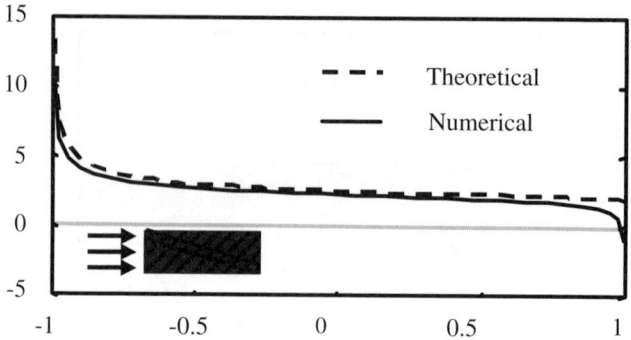

(a) Horizontal velocity distribution on the upper plate surface

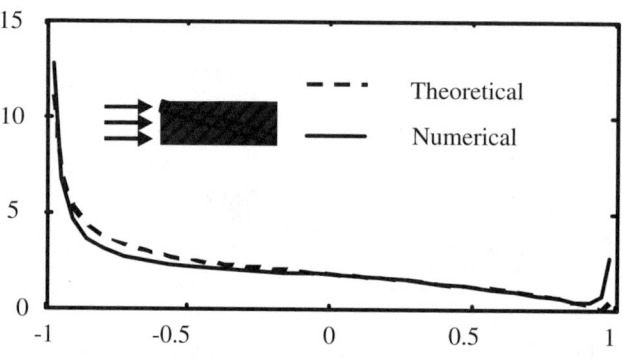

(b) Pressure distribution on the upper plate surface

FIGURE 2.24:   Velocity distribution on the surface of an ellipse. The angle of attack is 0.1π.

The pressure distribution is

$$p(x, +0) = const - \frac{\rho}{2}u^2(x, +0) \tag{2.90}$$

They are plotted in Fig. 2.24 denoted by dashed lines.

Because there exists a corner flow at the trailing edge, a circulation must be added to the flow, the magnitude of which is determined from Eq. (2.81).

$$\Gamma = -4\pi a \left|V_\infty\right| \sin \alpha, \quad \theta_0 = 0 \tag{2.91}$$

The numerical results are given in Fig. 2.24 with solid lines. Note that in the figure, the leading point and the trailing edge are omitted to avoid infinite values. The analytical and numerical results are quite close.

2.5   Conclusions

Lagrange interpolation in the complex plane has two forms, which turn out to be polynomial and rational approximations of an analytical function. The extension of Lagrange interpolation to the complex plane together with collocation method on the boundary result in a new method, termed as complex DQ method. The new method is applicable to the two-dimensional potential problems and plane elastic problems in an isotropic medium. The two kinds of problems can be solved by the present method effectively at low computational cost since both of them can be expressed in a general form with different coefficients determined by different boundary conditions. Moreover, the numerical examples can be solved easily and accurately by using a small number of nodes.

A comparison of the present method and BEM might help to clarify the excellent features of the present method. It is known that BEM is formulated through discretizing an integral equation which uses the weak form of solution (or singular solution) as its kernel. But in the complex plane singular solution such as logarithm of distance is not the sole option. Another large class of possible solutions is polynomial which can also be employed as fundamental solutions to some equations. This is the starting point of the present method. By using polynomials in the complex plane, the method is free of singularity and the numerical solutions can be formulated in the strong form as demonstrated in the former plane elastic problems. The equation to be discretized is a differential equation rather than integral equation. Another advantage of the present method is that the stress states inside the solution domain can be determined easily while it is difficult for BEM. The method is, however, limited to two-dimensional problems due to the use of complex variable theory. On the other hand, BEM can be formulated for both two-dimensional and three-dimensional problems.

Chapter 3

Triangular Differential Quadrature Method

In applying direct DQ method (DQM) to multi-dimensional problems, transformation (Lam, 1993; Bert and Malik, 1996d) is usually needed to map a non-rectangular physical domain into a normalized computational domain. As a result, a simple governing equation may turn into an awkward one, depending on the irregularity of the physical domain. For straight-edged quadrangles and even curvilinear quadrangles, the transformation does not impose significant difficulty to the solution as long as a proper transformation is available. However, for problems on a triangular domain which are often encountered in practice, singularity arises which has to be eliminated in the implementation of DQM, as pointed out by Zhong (1998). In the implementation of DQM on a triangular domain, one edge of the mesh has to degenerate into a single point, resulting in over-dense grid near the point and therefore the unnecessarily high computational cost. Additionally, the DQM is further hindered in this situation if high-order differential equation is to be solved. To overcome the above obstacles, a triangular differential quadrature (TDQ) method (TDQM) was proposed by Zhong (1998, 2000, 2001). In the present method, not only transformation is rendered unnecessary but also singularity does not appear as well.

3.1   TDQ method in standard triangle

Before embarking on how to express the derivatives in a differential equation, the way to discretize a triangular domain is illustrated. In order to resort to the area co-ordinate system L_i, it is first necessary to establish a method of construction and identification of the points in a triangular domain. The three edges are identified by the opposite vertices, respectively, e.g., edge 1 opposite vertex 1. The normal to an edge identifies the corresponding direction.

One may then draw parallel lines dividing the distance between vertex 1 and edge 1 into m equal segments in direction 1. Each line is identified with a digit from 0 to m, the line 0 being coincident with edge 1 and line m passing through vertex 1. A typical line is denoted by p in direction 1. From the

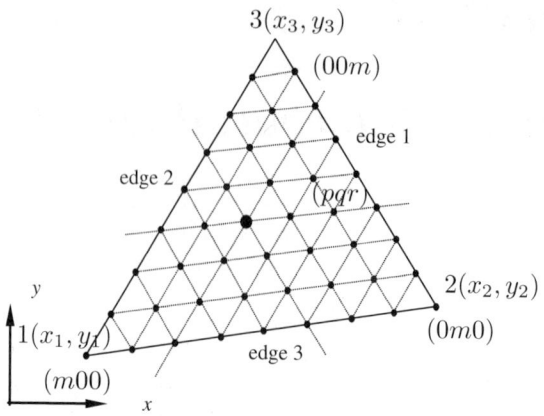

FIGURE 3.1: Grid points in a triangular domain.

intersections of the lines with the other two edges, parallel lines can be drawn with respect to edge 2 and edge 3, respectively. Now, the construction of mesh over the entire domain is completed. Altogether, there are

$$M = (m+1)(m+2)/2 \tag{3.1}$$

grid points in the entire triangular domain. In a similar way, the typical lines in the direction 2 (normal to edge 2) and direction 3 (normal to edge 3) are designated as q and r, respectively. Apparently, a typical point in the mesh is identified by three digits p, q, r, consistent with the designation of typical lines in the three directions. The area coordinates for the typical point are p/m, q/m, r/m. It is noted that

$$p+q+r = m, \quad 0 \le p,q,r \le m \tag{3.2}$$

The system of point designation is illustrated in Fig. 3.1.

The essence of the TDQM is that the partial derivative of a function with respect to a space variable at a given point is approximated by a weighted linear summation of the function values at all discrete points in the triangular domain. Therefore, the approximation of a derivative at a grid point is given as

$$D_n\{f(x,y)\}_{\alpha\beta\gamma} = \sum_{j=0}^{m} \sum_{i=0}^{m-j} e_{\alpha\beta\gamma,pqr}^{(n)} f_{pqr} \tag{3.3}$$

where D_n is a differential operator of order n. The subscript indicates the value of the derivative at grid point (α, β, γ). The summation indices (p, q, r) take the following values in the two summation loops

$$(p, q, r) = (m - i - j, i, j) \tag{3.4}$$

where $e^{(n)}_{\alpha\beta\gamma,pqr}$ are the weighting coefficients related to the function values f_{pqr} at point (p, q, r). Suppose that all the weighting coefficients are known, then the governing differential equation will be rewritten as a set of simultaneous algebraic equations with the function values at all grid points as unknown variables. Analogous to the direct DQM, governing differential equation of a physical problem is expressed in terms of TDQ at interior grid points of the domain. The boundary conditions at the edges of the domain must be invoked. Equation (3.3) needs to be implemented if Neumann boundary conditions appear.

From Eq. (3.1), it is worth noting that the total number of grid points makes it possible to form a complete polynomial through order m. Therefore, one simple approach to determine the weighting coefficients is to require that Eq. (3.3) be exact when f takes the following M base functions

$$f = L_1^p L_2^q L_3^r, \quad 0 \le p, q, r \le m \tag{3.5}$$

where the expressions of the three area coordinates of an arbitrary point (x, y) inside a triangular domain are given as

$$L_i = \frac{1}{2\Delta}(a_i + b_i x + c_i y), \quad i = 1, 2, 3 \tag{3.6}$$

Δ is the area of the triangle which is obtained on the basis of the Cartesian coordinates of the three vertices as

$$2\Delta = \begin{vmatrix} 1 & x_1 & y_1 \\ 1 & x_2 & y_2 \\ 1 & x_3 & y_3 \end{vmatrix} \tag{3.7}$$

and the coefficients in Eq. (3.6) are the values of the determinants of the corresponding cofactor matrices, e.g.

$$a_1 = x_2 y_3 - x_3 y_2, \quad b_1 = y_2 - y_3, \quad c_1 = -(x_2 - x_3) \tag{3.8}$$

The remaining coefficients can be obtained by interchanging the subscripts 1-3. With the chain rule of differentiation, a set of equations can be obtained by substituting Eq. (3.5) into Eq. (3.3), for instance,

$$\left(\frac{\partial f}{\partial x}\right)_{\alpha\beta\gamma} = \sum_{j=0}^{m} \sum_{i=0}^{m-j} e^{(x)}_{\alpha\beta\gamma,pqr} f_{pqr} = \left[\frac{\partial L_1}{\partial x} \frac{\partial L_2}{\partial x} \frac{\partial L_3}{\partial x}\right] \left\{\begin{array}{c} \frac{\partial f}{\partial L_1} \\ \frac{\partial f}{\partial L_2} \\ \frac{\partial f}{\partial L_3} \end{array}\right\}_{\alpha\beta\gamma} \tag{3.9}$$

$$\left(\frac{\partial^2 f}{\partial x^2}\right)_{\alpha\beta\gamma} = \sum_{j=0}^{m} \sum_{i=0}^{m-j} e^{(xx)}_{\alpha\beta\gamma,pqr} f_{pqr}$$

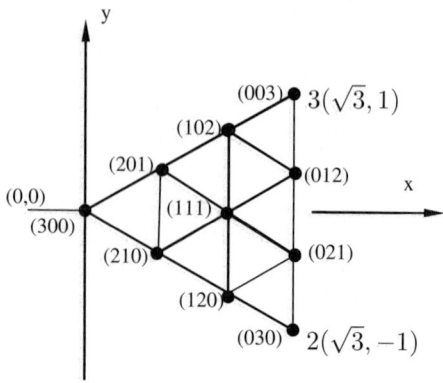

FIGURE 3.2: Mesh of isolateral triangular prismatic shaft.

$$= \begin{bmatrix} \dfrac{\partial L_1}{\partial x} & \dfrac{\partial L_2}{\partial x} & \dfrac{\partial L_3}{\partial x} \end{bmatrix} \begin{bmatrix} \dfrac{\partial^2 f}{\partial L_1^2} & \dfrac{\partial^2 f}{\partial L_1 \partial L_2} & \dfrac{\partial^2 f}{\partial L_1 \partial L_3} \\ \dfrac{\partial^2 f}{\partial L_1 \partial L_2} & \dfrac{\partial^2 f}{\partial L_2^2} & \dfrac{\partial^2 f}{\partial L_2 \partial L_3} \\ \dfrac{\partial^2 f}{\partial L_1 \partial L_3} & \dfrac{\partial^2 f}{\partial L_2 \partial L_3} & \dfrac{\partial^2 f}{\partial L_3^2} \end{bmatrix}_{\alpha\beta\gamma} \begin{Bmatrix} \dfrac{\partial f}{\partial L_1} \\ \dfrac{\partial f}{\partial L_2} \\ \dfrac{\partial f}{\partial L_3} \end{Bmatrix} \qquad (3.10)$$

$e^{(x)}_{\alpha\beta\gamma,pqr}$ and $e^{(xx)}_{\alpha\beta\gamma,pqr}$ can then be determined.

Example 3.1 *Isolateral triangular prismatic shaft under torsion*

To verify the TDQM, the same problem treated in Zhong's work (1998) using DQM is re-investigated in this example for comparison. It is an isolateral triangular prismatic shaft under torsion. The governing equation is given in terms of the Prandtl function, i.e.

$$\frac{\partial^2 \phi}{\partial x^2} + \frac{\partial^2 \phi}{\partial y^2} = -2 \qquad (3.11)$$

The exact solution is a cubic polynomial (Timoshenko and Goodier, 1970)

$$\phi = -\frac{\sqrt{3}}{6}(x - \sqrt{3})(x^2 - 3y^2) \qquad (3.12)$$

The two stress components are

$$\tau_{zx} = \frac{\partial \phi}{\partial y} = \sqrt{3}y(x - \sqrt{3}), \quad \tau_{zy} = \frac{\partial \phi}{\partial x} = \frac{\sqrt{3}}{2}(x^2 - y^2) - x \qquad (3.13)$$

To perform TDQ analysis in the problem, the lowest-order usable uniform grid is constructed (see Fig. 3.2). Since only Dirichlet boundary condition

$$\phi = 0 \tag{3.14}$$

is present around the periphery of the domain, the problem is reduced to solving the following equation

$$(e^{(xx)}_{111,111} + e^{(yy)}_{111,111})\phi_{111} = -2 \tag{3.15}$$

From Eq. (3.14), the two weighting coefficients in the above equation are obtained with ease

$$
\begin{aligned}
e^{(yy)}_{111,111} &= \frac{9}{2\Delta^2}(c_1c_2 + c_2c_3 + c_3c_1) = -\frac{9}{2} \\
e^{(xx)}_{111,111} &= \frac{9}{2\Delta^2}(b_1b_2 + b_2b_3 + b_3b_1) = -\frac{9}{2},
\end{aligned}
\tag{3.16}
$$

ϕ_{111} is found to be 2/9, agreeing exactly with the theoretical value. To furnish the stress components at the centroid (111), two more weighting coefficients are needed, which are

$$e^{(x)}_{111,111} = \frac{3}{2\Delta}(b_1 + b_2 + b_3) = 0, \quad e^{(y)}_{111,111} = \frac{3}{2\Delta}(c_1 + c_2 + c_3) = 0 \tag{3.17}$$

Therefore,

$$\tau_{zx|111} = e^{(y)}_{111,111}\phi_{111} = 0, \quad \tau_{zy|111} = e^{(x)}_{111,111}\phi_{111} = 0 \tag{3.18}$$

Again, they agree exactly with the theoretical solutions. The excellent agreement achieved does not come as a surprise since the base functions in Eq. (3.5) for a 10-point mesh form a complete cubic polynomial with 10 complete terms.

Example 3.2 *Uniformly stretched isotropic square membrane*

To further demonstrate the capability of the TDQM, the fundamental frequency of a uniformly stretched isotropic square membrane is studied. The motion equation is given by (Boyce and Diprima, 1986)

$$\frac{\partial^2 u}{\partial t^2} = \frac{T}{\rho}\left(\frac{\partial^2 u}{\partial x^2} + \frac{\partial^2 u}{\partial y^2}\right) \tag{3.19}$$

where u is the transverse deflection of the membrane, T the uniform tension per unit length of the boundary and ρ the mass per unit area. Due to the symmetry of the problem, half of the square domain is considered as shown in Fig. 3.3.

Consider the free vibration of the membrane. Suppose $u(x, y, t) = Re\left\{U(x, y)e^{i\omega t}\right\}$, where $U(x, y)$ is the complex amplitude, ω is the frequency and Re represents the real part inside the bracket. Therefore, what is meant

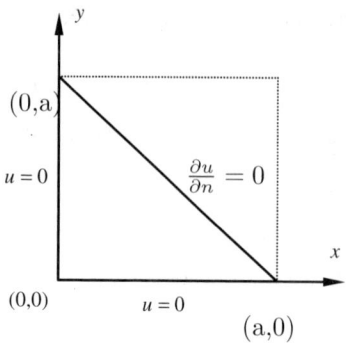

FIGURE 3.3:   Fundamental frequency study of a square membrane.

by the above expression is in fact $u(x, y, t) = U_R \cos(\omega t) - U_I \sin(\omega t)$, where U_R and U_I are real and imaginary parts of $U(x, y)$.

Then the free vibration of the membrane is described by the differential equation

$$-\frac{a^2}{4\pi^2} \left(\frac{\partial^2 U}{\partial x^2} + \frac{\partial^2 U}{\partial y^2} \right) = \lambda^2 U \tag{3.20}$$

subject to the following boundary conditions at the three edges of the isosceles right triangle:

$$U(x, y)|_{x=0} = 0, \quad U(x, y)|_{y=0} = 0, \quad \left.\frac{\partial U}{\partial n}\right|_{x+y=a} = 0 \tag{3.21}$$

By introducing the following non-dimensional term

$$\lambda = \frac{a\omega}{2\pi} \left(\frac{\rho}{T} \right)^{1/2} \tag{3.22}$$

The TDQ analogue of Eq. (3.20) may be written as

$$-\frac{a^2}{4\pi^2} \sum_{j=0}^{m} \sum_{i=0}^{m-j} (e^{(xx)}_{\alpha\beta\gamma,pqr} + e^{(yy)}_{\alpha\beta\gamma,pqr}) U_{pqr} = \lambda^2 U_{\alpha\beta\gamma} \tag{3.23}$$

The assembly of the TDQ analogue Eq. (3.23) at all inner grid points and the boundary condition equations at the three edges of the triangular results in a set of M algebraic equations which can be written in matrix form as

$$\begin{bmatrix} [K_{bb}] & [K_{bi}] \\ [K_{ib}] & [K_{ii}] \end{bmatrix} \begin{Bmatrix} \{U_b\} \\ \{U_i\} \end{Bmatrix} = \begin{Bmatrix} \{0\} \\ \lambda_2\{U_i\} \end{Bmatrix} \tag{3.24}$$

where $\{U_b\}$ is the displacement vector at the three edges of the triangle and $\{U_i\}$ the displacement vector at all inner grid points. By eliminating the $\{U_b\}$ vector, the above equations become a standard eigenvalue problem, i.e.,

$$[K]\{U_i\} = \lambda^2 \{U_i\} \tag{3.25}$$

The eigenvalues of matrix $[K]$ can be obtained by using a double QR algorithm (Press et al, 1986). The theoretical value (Weaver et al., 1990) of the fundamental frequency is

$$\lambda_{min}^{exact} = 1/\sqrt{2} \approx 0.70710678 \tag{3.26}$$

TABLE 3.1: TDQ solution of the fundamental frequency of a square membrane $(a \times a)$

m	M	λ_{min}	$(\lambda_{min} - \lambda_{min}^{exact})/\lambda_{min}^{exact}$ (%)
4	15	0.67718000	4.2
5	21	0.71909966	-1.7
6	28	0.70953492	0.3
7	36	0.70654357	0.1
8	45	0.70702182	0.0
9	55	0.70712921	0.0
10	66	0.70710891	0.0
11	78	0.70710614	0.0
12	91	0.70710673	0.0
13	105	0.70710680	0.0
14	120	0.70710678	0.0
15	136	0.70710678	0.0

The convergence of the non-dimensional fundamental frequency is displayed in Table 3.1. The results are given up to eight significant digits, highlighting the rapid convergence of the present TDQM.

3.2   TDQ method in curvilinear triangle

In this section, the previous method will be extended to the case in a curvilinear triangle.

In general, a point-based transformation is adopted to map a curvilinear triangular domain onto its counterpart—a standard unit isosceles right triangle (see Fig. 3.4) on which the approximations of derivatives are carried out. A uniform grid system is generated first in the standard triangle by following

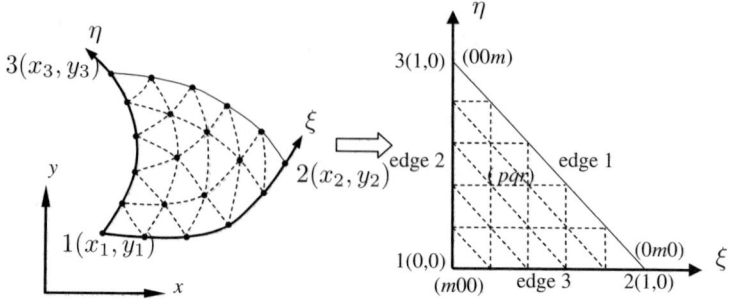

FIGURE 3.4:   A curvilinear triangle and its counterpart-standard triangle.

the usual procedures (Zhong, 2000, 2001). In total, there are grid points in the right triangle, where m is the number of the equal segments on each edge of the triangle.

$$M = (m+1)(m+2)/2 \qquad (3.27)$$

A partial derivative of the function at a grid point (α, β, γ) is approximated as

$$D_n \left\{ f(\xi, \eta) \right\}_{\alpha\beta\gamma} = \sum_{j=0}^{m} \sum_{i=0}^{m-j} C^{(n)}_{\alpha\beta\gamma,pqr} f_{pqr}, \qquad \alpha + \beta + \gamma = m \qquad (3.28)$$

where D_n is a differential operator of order n. The subscript indices (p, q, r) in Eq. (3.28) take the following values in the two summation loops

$$(p, q, r) = (m - i - j, i, j) \qquad (3.29)$$

$C^{(n)}_{\alpha\beta\gamma,pqr}$ are the weighting coefficients related to the function values f_{pqr} at points (p, q, r), which are determined by requiring that Eq. (3.28) be satisfied by the following M trial functions

$$f_{pqr} = \overline{f}_p(L_1)\overline{f}_q(L_2)\overline{f}_r(L_3), \ \ 0 \le p, q, r \le m \qquad (3.30)$$

where the auxiliary functions are given as

$$\overline{f}_p(L_1) = \begin{cases} \prod_{k=1}^{p} \frac{mL_1-k+1}{k}, & 1 \le p \le m; \\ 1, & p = 0; \end{cases} \tag{3.31a}$$

$$\overline{f}_q(L_2) = \begin{cases} \prod_{k=1}^{q} \frac{mL_2-k+1}{k}, & 1 \le q \le m; \\ 1, & q = 0; \end{cases} \tag{3.31b}$$

$$\overline{f}_r(L_3) = \begin{cases} \prod_{k=1}^{r} \frac{mL_3-k+1}{k}, & 1 \le r \le m; \\ 1, & r = 0 \end{cases} \tag{3.31c}$$

with

$$L_1 = 1 - \xi - \eta, \ L_2 = \xi, \ L_3 = \eta \tag{3.32}$$

The explicit formulas for the weighting coefficients of first order derivatives are obtained as

$$C^{(\xi)}_{\alpha\beta\gamma,pqr} = \begin{bmatrix} \frac{\partial L_1}{\partial \xi} & \frac{\partial L_2}{\partial \xi} & \frac{\partial L_3}{\partial \xi} \end{bmatrix} \begin{Bmatrix} \frac{\partial f_{pqr}}{\partial L_1} \\ \frac{\partial f_{pqr}}{\partial L_2} \\ \frac{\partial f_{pqr}}{\partial L_3} \end{Bmatrix}_{\alpha\beta\gamma} = (-\frac{d\overline{f}_p}{dL_1}\overline{f}_q\overline{f}_r + \overline{f}_p\frac{d\overline{f}_q}{dL_2}\overline{f}_r)_{\alpha\beta\gamma}$$

$$\tag{3.33a}$$

$$C^{(\eta)}_{\alpha\beta\gamma,pqr} = \begin{bmatrix} \frac{\partial L_1}{\partial \eta} & \frac{\partial L_2}{\partial \eta} & \frac{\partial L_3}{\partial \eta} \end{bmatrix} \begin{Bmatrix} \frac{\partial f_{pqr}}{\partial L_1} \\ \frac{\partial f_{pqr}}{\partial L_2} \\ \frac{\partial f_{pqr}}{\partial L_3} \end{Bmatrix}_{\alpha\beta\gamma} = (-\frac{d\overline{f}_p}{dL_1}\overline{f}_q\overline{f}_r + \overline{f}_p\overline{f}_q\frac{d\overline{f}_r}{dL_3})_{\alpha\beta\gamma}$$

$$\tag{3.33b}$$

The weighting coefficients for higher order derivatives can be obtained through recurrence relationship as

$$C^{(\xi\xi)}_{\alpha\beta\gamma,pqr} = \sum_{j=0}^{m}\sum_{i=0}^{m-j} C^{(\xi)}_{\alpha\beta\gamma,stu}C^{(\xi)}_{stu,pqr} \tag{3.34a}$$

$$C^{(\xi\eta)}_{\alpha\beta\gamma,pqr} = \sum_{j=0}^{m}\sum_{i=0}^{m-j} C^{(\xi)}_{\alpha\beta\gamma,stu}C^{(\eta)}_{stu,pqr} \tag{3.34b}$$

$$C^{(\eta\eta)}_{\alpha\beta\gamma,pqr} = \sum_{j=0}^{m}\sum_{i=0}^{m-j} C^{(\eta)}_{\alpha\beta\gamma,stu}C^{(\eta)}_{stu,pqr} \tag{3.34c}$$

During the implementation of the TDQ, a localized strategy is frequently employed to overcome the sensitivity of approximation to corner condition formed by two adjacent Neumann boundary conditions and the instability of high order approximation, i.e., a partial derivative of a function with respect to a space variable at a grid point is approximated by a weighted linear sum of the function values at grid points in a local small triangle containing the grid point in question rather than at all the grid points in the entire triangular domain. Analogous to Eq. (3.28), the expression of a partial derivative at a grid point (α, β, γ) is changed in

$$D_n \{f(\xi, \eta)\}_{\alpha\beta\gamma} = \sum_{j=0}^{m'} \sum_{i=0}^{m'-j} C_{\alpha\beta\gamma, pqr}^{(n)} f_{pqr}$$

$$\alpha + \beta + \gamma = m, \quad (p, q, r) = (m' - i - j, i, j) \tag{3.35}$$

where $m'(\mbox{¡}m)$ is the number of the equal segments of an edge in a local triangle containing the grid point (α, β, γ). The weighting coefficients $C_{\alpha\beta\gamma, pqr}^{(n)}$ are also determined in the local triangle.

3.3   Geometric transformation

The geometric transformation of a curvilinear triangle into the standard triangle is usually based on the following point transformation expression

$$x = \sum_{j=0}^{N} \sum_{i=0}^{N-j} f_{\overline{pqr}} x_{\overline{pqr}}, \quad y = \sum_{j=0}^{N} \sum_{i=0}^{N-j} f_{\overline{pqr}} y_{\overline{pqr}} \tag{3.36}$$

where N is the order of transformation; $(x_{\overline{pqr}}, y_{\overline{pqr}})$ are the Cartesian coordinates of the transformation point in the curvilinear triangle, $f_{\overline{pqr}}$ are functions given in the similar form as Eq. (3.30). The three digit subscripts $(\overline{p}, \overline{q}, \overline{r}) = (N - j - i, i, j)$ are introduced to identify the nodes in the transformation. Unlike in finite element analysis where piecewise lower order isoparametric transformations suffice to provide accurate modelling of curved domain boundaries, a global transformation of the curvilinear triangle onto the standard triangle is sought herein. In principle, high order transformations in Eq. (3.36) should be able to model the curved boundary adequately. In practice, however, difficulty may arise since the transformation is rather sensitive to the location of inner transformation points as reported in Mitchell and Wait (1977), Zhong (2002). Ideally, the difficulty can be circumvented by employing transformations leaving out inner transformation points. Fortunately, this can be realized by adopting the "serendipity" idea in finite element method. Thus, a triangular serendipity transformation is developed as

$$x = \sum_{j=1}^{15} f_j x_j, \quad y = \sum_{j=1}^{15} f_j y_j \tag{3.37}$$

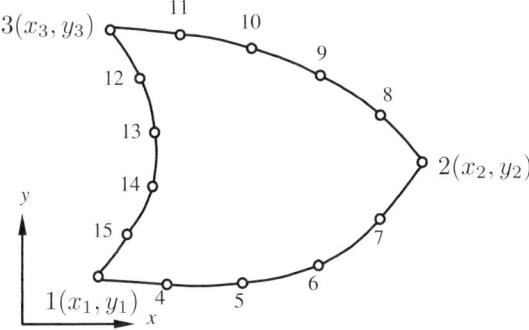

FIGURE 3.5: Points for quintic transformation of a curvilinear triangle.

It is well-known that the quintic order transformation often meets the accuracy requirement of geometric modelling. The shape functions associated with the fifteen nodes on the boundary of a curvilinear triangle (see Fig. 3.5) are given in Appendix A at the end of this chapter. It is worth noting that the high-order serendipity transformation for triangles is developed without the use of inner grid points, different from transformation for quadrilaterals. Thus, the present serendipity transformation enables the TDQM to deal with the problem on complex geometric domains.

The Jacobian matrix of the transformation is

$$[J] = \begin{bmatrix} x_\xi & y_\xi \\ x_\eta & y_\eta \end{bmatrix} \tag{3.38}$$

Its inverse matrix is given as follows:

$$\begin{bmatrix} \xi_x & \eta_x \\ \xi_y & \eta_y \end{bmatrix} = [J]^{-1} = \frac{1}{|J|} \begin{bmatrix} y_\eta & -y_\xi \\ -x_\eta & x_\xi \end{bmatrix}, \ |J| = x_\xi y_\eta - x_\eta y_\xi \tag{3.39}$$

where $|J|$ is the determinant of the Jacobian matrix. With the chain rule of differentiation, the second derivatives of a function in the two coordinate systems are related as

$$\xi_{xx} = |J|^{-3} \left[|J| \left(y_\eta y_{\xi\eta} - y_\xi y_{\eta\eta} \right) - y_\eta \left(y_\eta \left. |J| \right|_\xi - y_\xi \left. |J| \right|_\eta \right) \right]$$
$$\xi_{yy} = |J|^{-3} \left[|J| \left(x_\eta x_{\xi\eta} - x_\xi x_{\eta\eta} \right) - x_\eta \left(x_\eta \left. |J| \right|_\xi - x_\xi \left. |J| \right|_\eta \right) \right]$$

$$\eta_{xx} = |J|^{-3} \left[|J| \left(y_\xi y_{\xi\eta} - y_\eta y_{\xi\xi} \right) - y_\xi \left(y_\eta \left. |J| \right|_\xi - y_\xi \left. |J| \right|_\eta \right) \right]$$

$$\eta_{yy} = |J|^{-3} \left[|J| \left(x_\xi x_{\xi\eta} - x_\eta x_{\xi\xi} \right) - x_\xi \left(x_\eta \left. |J| \right|_\xi - x_\xi \left. |J| \right|_\eta \right) \right] \qquad (3.40)$$

$$\xi_{xy} = |J|^{-3} \left[|J| \left(x_\xi y_{\eta\eta} - x_\eta y_{\xi\eta} \right) - y_\eta \left(x_\eta \left. |J| \right|_\xi - x_\xi \left. |J| \right|_\eta \right) \right]$$

$$\eta_{xy} = |J|^{-3} \left[|J| \left(x_\eta y_{\xi\xi} - x_\xi y_{\xi\eta} \right) - y_\xi \left(x_\eta \left. |J| \right|_\xi - x_\xi \left. |J| \right|_\eta \right) \right]$$

The above relations will be used later to transform the governing equations and boundary conditions from the $x - y$ domain into the $\xi - \eta$ domain.

3.4 Governing equations of Reissner-Mindlin plates on Pasternak foundation

The TMDQ coupled with geometric transformation is employed to deal with elliptic Reissner-Mindlin plates on Pasternak foundation under uniformly distributed load. Due to the double symmetry of the problem, a quadrant of the plate is considered in the analysis and a triangular serendipity transformation is adopted to map the quadrant into the computational domain—the unit isosceles right triangle.

3.4.1 Governing equations in physical coordinate system

For flexural analysis of Reissner plates on a Pasternak foundation, the differential equilibrium equations are given as follows (Kobayashi and Sonoda, 1989):

$$\frac{D}{2} \left((1 - \nu) \nabla^2 \psi_x + (1 + \nu) \frac{\partial \phi}{\partial x} \right) + \kappa G h \left(\frac{\partial w}{\partial x} - \psi_x \right) = 0, \qquad (3.41a)$$

$$\frac{D}{2} \left((1 - \nu) \nabla^2 \psi_y + (1 + \nu) \frac{\partial \phi}{\partial y} \right) + \kappa G h \left(\frac{\partial w}{\partial y} - \psi_y \right) = 0, \qquad (3.41b)$$

$$\kappa G h \left(\nabla^2 w - \phi \right) + q - p = 0. \qquad (3.41c)$$

where

$$\phi = \frac{\partial \psi_x}{\partial x} + \frac{\partial \psi_y}{\partial y}, \qquad (3.42)$$

$$D = \frac{Eh^3}{12(1 - \nu^2)} \tag{3.43}$$

$$p = k_f w - G_f \nabla^2 w \tag{3.44}$$

and $w(x, y)$ is the transverse deflection; $\psi_x(x, y)$ and $\psi_y(x, y)$ the rotations of the normal about y−axis and x−axis; D, E, G and ν the plate flexural rigidity, Young's modulus, shear modulus and Poisson's ratio, respectively; h is the plate thickness; $q(x, y)$ the surface load intensity; and ∇^2 is the Laplacian operator in Cartesian coordinates. The shear correction factor κ is taken to be 5/6, p is the subgrade reaction intensity, k_f and G_f are the subgrade reaction coefficient and shear modulus of foundation, respectively. The force resultants are given by:

$$M_x = -D \left(\frac{\partial \psi_x}{\partial x} + \nu \frac{\partial \psi_y}{\partial y} \right) \tag{3.45}$$

$$M_y = -D \left(\nu \frac{\partial \psi_x}{\partial x} + \frac{\partial \psi_y}{\partial y} \right) \tag{3.46}$$

$$M_{xy} = -\frac{1 - \nu}{2} D \left(\frac{\partial \psi_x}{\partial y} + \frac{\partial \psi_y}{\partial x} \right) \tag{3.47}$$

$$Q_x = \kappa G h \left(\frac{\partial w}{\partial x} - \psi_x \right) \tag{3.48}$$

$$Q_y = \kappa G h \left(\frac{\partial w}{\partial y} - \psi_y \right) \tag{3.49}$$

where M_x, M_y and M_{xy} are the moments and twisting moment; Q_x and Q_y the shear forces.

3.4.2   Governing equations in the standard triangle

By using the geometric coordinate mapping relations, the governing equations in Eqs. (3.41) in the physical domain $x - y$ can be transformed into the computational domain $\xi - \eta$ as follows:

$$[K]_{3 \times 18} \begin{bmatrix} [\partial]_{6 \times 1} & \mathbf{0} & \mathbf{0} \\ \mathbf{0} & [\partial]_{6 \times 1} & \mathbf{0} \\ \mathbf{0} & \mathbf{0} & [\partial]_{6 \times 1} \end{bmatrix} \begin{Bmatrix} \psi_x \\ \psi_y \\ w \end{Bmatrix} = \begin{Bmatrix} 0 \\ 0 \\ q/D \end{Bmatrix} \tag{3.50}$$

where

$$[\partial]_{6 \times 1} = \left\{ \frac{\partial^2}{\partial \xi^2} \ \frac{\partial^2}{\partial \xi \partial \eta} \ \frac{\partial^2}{\partial \eta^2} \ \frac{\partial}{\partial \xi} \ \frac{\partial}{\partial \eta} \ 1 \right\}^T \tag{3.51}$$

is a differential operator vector and the entries in $[K]_{3\times18}$ are given in Appendix B. With the TDQ approximation, Eq. (3.50) will be recast into a set of algebraic equations at any inner grid point in the standard triangular domain.

3.4.3   Boundary conditions

To impose boundary conditions, the following transformations of displacement components and internal forces are used

$$\psi_n = n_1\psi_x + n_2\psi_y; \quad \psi_s = -n_2\psi_x + n_1\psi_y. \tag{3.52}$$

$$
\begin{aligned}
M_n &= n_1^2 M_x + 2n_1 n_2 M_{xy} + n_2^2 M_y; \\
M_{ns} &= n_1 n_2 (M_y - M_x) + (n_1^2 - n_2^2) M_{xy}; \\
Q_n &= n_1 Q_x + n_2 Q_y.
\end{aligned}
\tag{3.53}
$$

where (n_1, n_2) are the directional cosines of a unit normal at an arbitrary point on the boundary of the physical domain (see Fig. 3.6). Two typical boundary conditions are considered in this chapter, the clamped and the simply supported boundary conditions, which are mathematically expressed as follows:

(a) *Clamped (C)*

$$w = 0, \quad \psi_n = 0, \quad \psi_s = 0; \tag{3.54}$$

(b) *Simply supported (S):*

(i) Type I(S) : w = 0, $M_n = 0$, $\psi_s = 0$;

$$\tag{3.55}$$

(ii) Type II(S′) : w = 0, $M_n = 0$, $M_{ns} = 0$.

$$\tag{3.56}$$

In Eqs. (3.53), (3.55) and (3.56), M_n, M_{ns} are the bending and twisting moments on the edge of the plate respectively; ψ_n and ψ_s are the rotations of the mid-plane in the normal plane, nz, and in the tangent plane, sz, to the boundary; the subscripts n and s represent the normal and tangent directions at a point on the boundary, respectively. Equation (3.55) is designated for plates with simply supported boundary conditions.

3.4.4   Symmetry consideration

In the present case, all elliptic plates on Pasternak foundation are assumed subjected to uniformly distributed load. Consequently, the flexural deformation symmetry of plates can be employed to reduce the computational effort

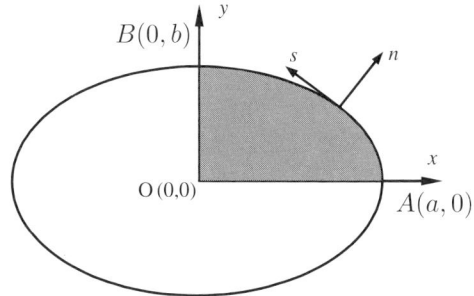

FIGURE 3.6: A quadrant of an elliptic plate.

and increase the computational accuracy. The deformation symmetry conditions of the Reissner/Mindlin plate about a symmetric plane are expressed as

$$\psi_n = 0, \quad \frac{\partial \psi_s}{\partial n} = 0, \quad \frac{\partial w}{\partial n} = 0 \tag{3.57}$$

By combining the boundary conditions and symmetric conditions at the boundary nodes with Eqs. (3.50) at all inner grid points, a set of algebraic equations with at most $3M$ unknowns is established.

Example 3.3 *Plates*

For clarity of presentation, the following normalized quantities are introduced

$$r_0 = \frac{rD}{qb^4}, w_0 = \frac{wD}{qb^4}, M_{r0} = \frac{M_r}{qb^2}, M_{x0} = \frac{M_x}{qb^2}, M_{y0} = \frac{M_y}{qb^2} \tag{3.58}$$

The following parameters are taken in all computations: flexural rigidity $D = 10^6$kNm, Poisson's ratio $\nu = 0.3$, subgrade reaction coefficient $k_f = 2 \times 10^4$kN/m^3. The foundation shear modulus is taken as $G_f = 3 \times 10^4$kN/m unless otherwise stated.

(a) Convergence study

First a circular plate resting on a Pasternak foundation under uniformly distributed load is studied to examine the convergence of the present method. The thickness to width ratio is taken as $h/b = 0.01$. Both simply supported and clamped boundary conditions are considered and the results are compared with the analytical ones (Yu, 1957).

In view of the double symmetry of the plate, a quadrant of the plate is analyzed. The central deflection and the central moment of circular plates are given in Table 3.2. It is found that results with relative error less than

TABLE 3.2: Central deflection and central moment of thin circular plates on a Pasternak foundation with clamped or simply supported edge

m	M	Clamped		Simply supported	
		ω_0	M_{r0}	ω_0	M_{r0}
14	120	0.004651	0.01735	0.005748	0.01088
15	136	0.004637	0.01775	0.005754	0.01022
16	153	0.004639	0.01772	0.005710	0.00971
17	171	0.004628	0.01762	0.005754	0.01059
18	190	0.004638	0.01784	0.005778	0.01079
19	210	0.004648	0.01794	0.005776	0.01058
20	231	0.004650	0.01792	0.005773	0.01046
21	253	0.004649	0.01788	0.005769	0.01047
22	276	0.004648	0.01788	0.005773	0.01051
23	300	0.004648	0.01789	0.005774	0.01048
24	325	0.004649	0.01789	0.005774	0.01047
25	351	0.004649	0.01789	0.005774	0.01049
26	378	0.004649	0.01789	0.005774	0.01050
27	406	0.004649	0.01790	0.005774	0.01049
28	435	0.004649	0.01790	0.005774	0.01049
29	465	0.004649	0.01790	0.005774	0.01049
30	496	0.004649	0.01790	0.005774	0.01049
Analytic[15]		0.004648	0.01791	0.005771	0.01050

1% can be achieved when the global approximation order m is increased to 18. It is seen clearly that the errors are reduced with the increase of the global approximation order. It is worth mentioning that the localized TDQ ($m' = 9$) is implemented in all computations for the sake of consistency and coding simplicity. As reported by Zhong et al., (2003), the loss of efficiency or accuracy is insignificant as long as the local approximation order is higher than 5. On the other hand, the localized implementation leads to remarkable numerical stability.

The radial variation of deflection and the radial variation of bending moment are shown in Figs. 3.7 and 3.8. The shear modulus of foundation are taken as the following different values $G_f = 0.3 \times 10^4 \text{kN/m}, 2.82843 \times 10^5 \text{kN/m}$ and 3×10^6 kN/m in order to investigate the effects of the shear modulus on the behavior of plates. Excellent agreement with the available analytical solution (Yu, 1957) is achieved for both simply supported and clamped plates resting on Pasternak foundation as demonstrated in Figs. 3.7 and 3.8.

(b) Results for clamped and simply supported elliptic plates

To investigate the effect of thickness-to-width ratio, the central deflection and moments for clamped and simply supported elliptic plates are shown in Tables 3.3 and 3.4. Also listed is the least order of global approximation m from which results with four stabilized digits are obtained. It is clearly seen that more grid points are needed to attain convergent results for thin

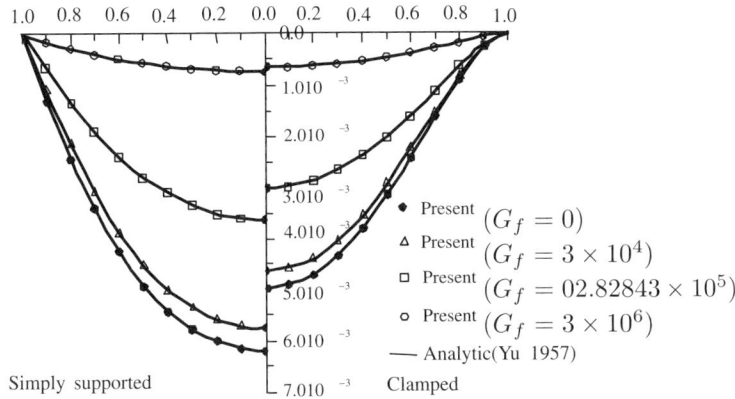

FIGURE 3.7: Deflections of the circular plate.

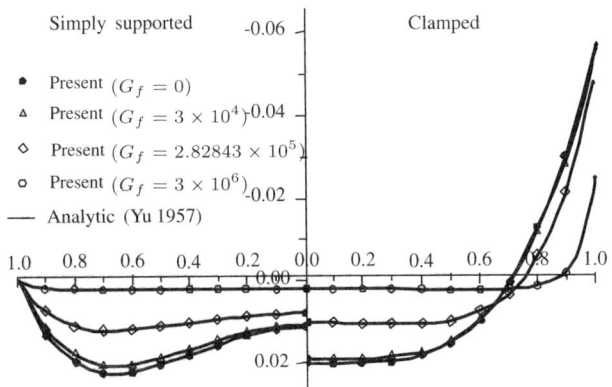

FIGURE 3.8: Radial bending moments of the circular plate.

plates. It is worth mentioning that $m + 1$ merely denotes the number of grid points on an edge of the triangular domain to furnish convergent results with four stabilized digits and a localized TDQ approximation of order $m' = 9$ is adopted in all cases. For clamped elliptic plates, the normalized deflection and the two moments decrease with the increase of thickness-to-width ratio except for circular plates. For simply supported plates, the normalized central deflection decreases with the increase of thickness-to-width ratio while the two moments at the center of elliptic plates show the reverse trend. It should be pointed out that the relative extremes of moments M_x and M_y may not occur at the center of the elliptic plates even for circular plates in the presence of elastic foundation, as shown in Fig. 3.8.

The effect of ellipticity on the deflection and moments is also investigated. Moderately thick elliptic plates ($h/b = 0.1$) with clamped or simply supported edges resting on Pasternak foundations are studied for plates with ellipticity $a/b = 1.5$, 2.0, 2.5, 3.0 and 5.0. The distributions of normalized deflection and moments along the major and the minor axes are plotted for simply supported and clamped boundary conditions (see Figs. 3.9–3.14). It is noted that the convergence rate drops with the increase of ellipticity a/b of plates. In addition, the convergence rate for simply supported elliptic plates is somewhat slower than that for fully clamped elliptic plates and the convergence rate of bending moments is slower than that of the deflection of plates. In the case of elliptic plates with $a/b = 5.0$, four significant digits are stabilized at $m = 20$ and $m = 27$ for the deflection and bending moments at the center of fully clamped elliptic plates, respectively; whereas $m = 22$ and $m = 28$ are required to keep the same level of convergence of the deflection and bending moments at the center of the simply supported elliptic plates.

3.5 Conclusions

The TDQM introduced in this chapter overcomes the difficulties associated with the direct DQ method in dealing with problems on a triangular domain. Transformation of the physical domain is avoided, resulting in simple algebraic equations. Singularity does not appear in the process of the TDQ analysis. In the two examples provided, excellent agreement with the theoretical values is achieved with rather fewer grid points, indicating that the present method is an attractive new numerical tool. In view of the great geometric flexibility of triangles, various geometric domains may be decomposed into a number of triangles. Therefore, the potential applications of the present method lie in the incorporation of domain decomposition technique into the TDQM.

Herein an effort has been made to extend the TDQM to problems with arbitrary geometric domains. A triangular serendipity transformation with

TABLE 3.3:   Computed results at the center of clamped elliptic plates resting on a Pasternak foundation

	m	w_0	M_{x0}	M_{y0}
h/b		$a/b = 1.0$		
0.01	27	0.004649	0.01790	0.01789
0.05	18	0.004653	0.01769	0.01769
0.10	21	0.004668	0.01706	0.01706
0.15	14	0.004689	0.01611	0.01611
		$a/b = 2.0$		
0.01	35	0.005153	0.004976	0.01390
0.05	28	0.005149	0.004926	0.01375
0.10	23	0.005141	0.004782	0.01328
0.15	22	0.005129	0.004561	0.01256
		$a/b = 3.0$		
0.01	40	0.005111	0.004170	0.01314
0.05	28	0.005108	0.004126	0.01300
0.10	25	0.005102	0.003993	0.01255
0.15	23	0.005093	0.003790	0.01187

TABLE 3.4:   Computed results at the center of simply supported elliptic plates resting on a Pasternak foundation

	m	w_0	M_{x0}	M_{y0}
h/b		$a/b = 1.0$		
0.01	26	0.005774	0.01049	0.01048
0.05	16	0.005765	0.01051	0.01051
0.10	16	0.005745	0.01053	0.01053
0.15	14	0.005714	0.01056	0.01056
		$a/b = 2.0$		
0.01	37	0.005571	0.001425	0.005602
0.05	29	0.005564	0.001449	0.005666
0.10	29	0.005551	0.001524	0.005775
0.15	27	0.005531	0.001641	0.005940
		$a/b = 3.0$		
0.01	39	0.005494	0.001520	0.005233
0.05	30	0.005487	0.001534	0.005299
0.10	29	0.005476	0.001573	0.005398
0.15	28	0.005458	0.001634	0.005549

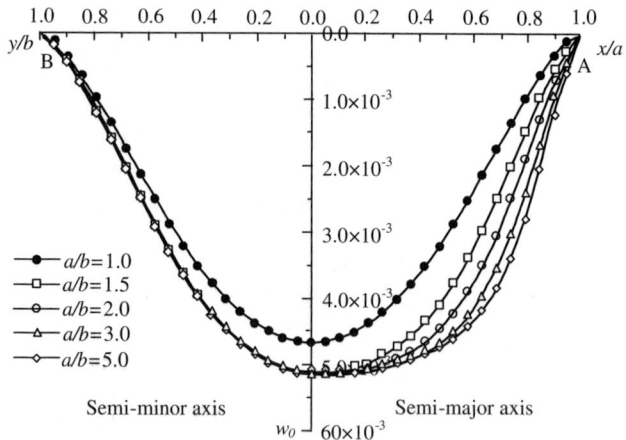

FIGURE 3.9:   Deflections of clamped elliptic plates.

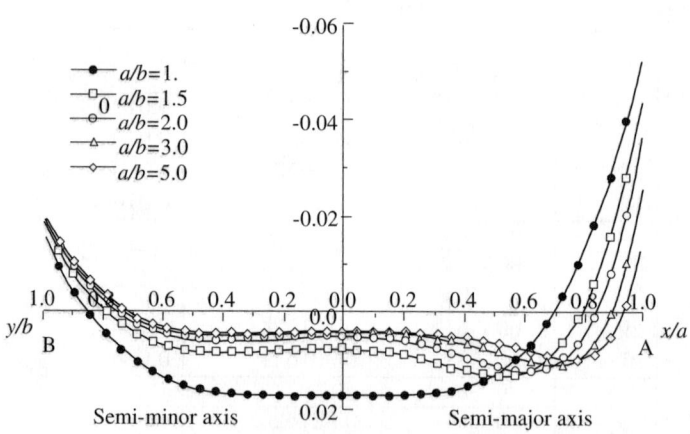

FIGURE 3.10:   Bending moment M_{x0} of clamped elliptic plates.

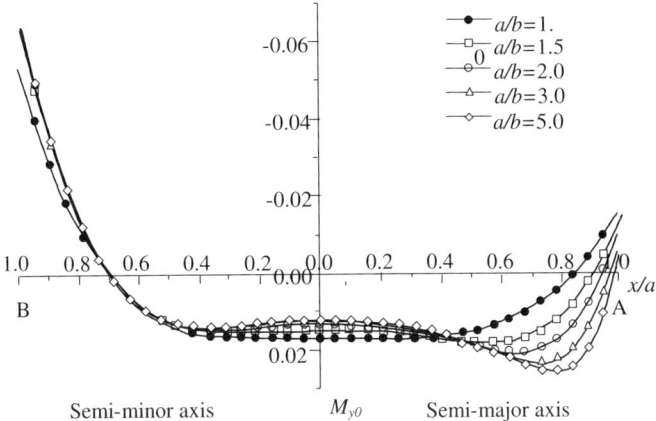

FIGURE 3.11: Bending moment M_{y0} of clamped elliptic platese.

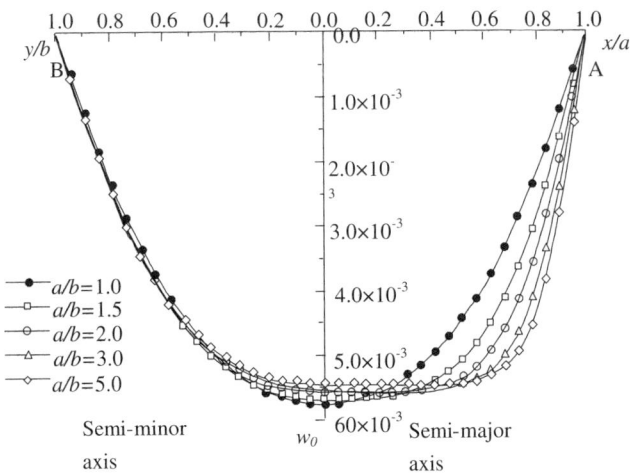

FIGURE 3.12: Deflections of simply supported elliptic plates.

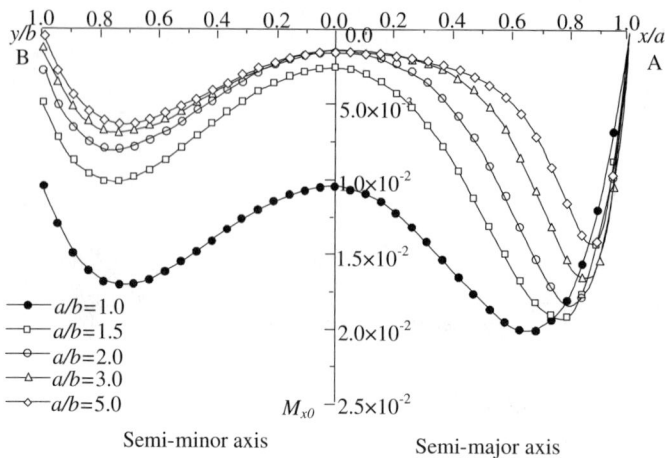

FIGURE 3.13: Bending moment M_{x0} of simply supported elliptic plates.

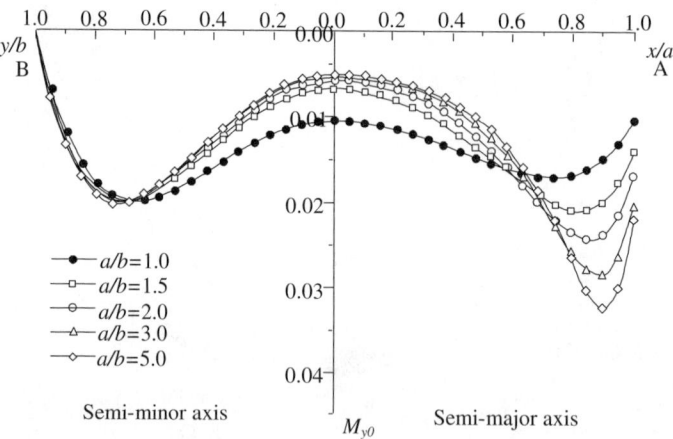

FIGURE 3.14: Bending moment M_{y0} of simply supported elliptic plates.

no inner points is introduced. Consequently, the sensitivity associated with the usual geometric transformations with inner points is removed. Although the triangular serendipity transformation of any order can be developed in the same manner, the quintic order transformation is accurate enough in all cases. Static flexural analysis of elliptic Reissner-Mindlin plates resting on a Pasternak foundation is investigated using the TDQM. Parametric study of elliptic plates with various aspect ratios is performed. The excellent agreement with the available analytical solution for circular plates verifies the effectiveness of the present technique.

Of course, the present strategy may only be applied successfully to the problems with simple and relatively regular geometry. However, it should be pointed out that the geometric transformation is not the only way to make TDQM applicable to problems with curvilinear boundaries. By using the localized TDQM (Zhong et al., 2003), the TDQM can be applied directly to the problems with curvilinear boundaries without the use of geometric transformation. In view of this, the TDQM is more attractive in the analysis of problems with irregular geometric domain.

Appendix A

Although the selection of transformation points on the boundary of a curvilinear domain is a simple matter, uniform distribution is desirable to avoid severe distortion. The functions for the transformation (see Fig. 3.5) are given as

$$f_4 = \frac{3125}{24} L_1 L_2 (L_1 - \frac{1}{5})(L_1 - \frac{2}{5})(L_1 - \frac{3}{5}), \tag{A1}$$

$$f_5 = -\frac{3125}{12} L_1 L_2 (L_1 - \frac{1}{5})(L_1 - \frac{2}{5})(L_1 - \frac{4}{5}) \tag{A2}$$

$$f_6 = \frac{3125}{12} L_1 L_2 (L_1 - \frac{1}{5})(L_1 - \frac{3}{5})(L_1 - \frac{4}{5}) \tag{A3}$$

$$f_7 = -\frac{3125}{24} L_1 L_2 (L_1 - \frac{2}{5})(L_1 - \frac{3}{5})(L_1 - \frac{4}{5}) \tag{A4}$$

$$f_8 = \frac{3125}{24} L_2 L_3 (L_2 - \frac{1}{5})(L_2 - \frac{2}{5})(L_2 - \frac{3}{5}) \tag{A5}$$

$$f_9 = -\frac{3125}{12} L_2 L_3 (L_2 - \frac{1}{5})(L_2 - \frac{2}{5})(L_2 - \frac{4}{5}) \tag{A6}$$

$$f_{10} = \frac{3125}{12} L_2 L_3 (L_2 - \frac{1}{5})(L_2 - \frac{3}{5})(L_2 - \frac{4}{5}) \tag{A7}$$

$$f_{11} = -\frac{3125}{24} L_2 L_3 (L_2 - \frac{2}{5})(L_2 - \frac{3}{5})(L_2 - \frac{4}{5}) \tag{A8}$$

$$f_{12} = \frac{3125}{24} L_3 L_1 (L_3 - \frac{1}{5})(L_3 - \frac{2}{5})(L_3 - \frac{3}{5}) \qquad (A9)$$

$$f_{13} = -\frac{3125}{12} L_3 L_1 (L_3 - \frac{1}{5})(L_3 - \frac{2}{5})(L_3 - \frac{4}{5}) \qquad (A10)$$

$$f_{14} = \frac{3125}{12} L_3 L_1 (L_3 - \frac{1}{5})(L_3 - \frac{3}{5})(L_3 - \frac{4}{5}) \qquad (A11)$$

$$f_{15} = -\frac{3125}{24} L_3 L_1 (L_3 - \frac{2}{5})(L_3 - \frac{3}{5})(L_3 - \frac{4}{5}) \qquad (A12)$$

$$f_1 = L_1 - \frac{4}{5}(f_4 + f_{15}) - \frac{3}{5}(f_5 + f_{14}) - \frac{2}{5}(f_6 + f_{13}) - \frac{1}{5}(f_7 + f_{12}) \qquad (A13)$$

$$f_2 = L_2 - \frac{4}{5}(f_7 + f_8) - \frac{3}{5}(f_6 + f_9) - \frac{2}{5}(f_5 + f_{10}) - \frac{1}{5}(f_4 + f_{11}) \qquad (A14)$$

$$f_3 = L_3 - \frac{4}{5}(f_{11} + f_{12}) - \frac{3}{5}(f_{10} + f_{13}) - \frac{2}{5}(f_9 + f_{14}) - \frac{1}{5}(f_8 + f_{15}) \qquad (A15)$$

where $L_1 = 1 - \xi - \eta$, $L_2 = \xi$ and $L_3 = \eta$ are the three area coordinates of an arbitrary point in the standard triangle.

Appendix B

The entries in matrix $[K]$ of Eq. (3.50) are given as

$$K_{1,1} = \xi_x^2 + c_1\xi_y^2, K_{1,2} = 2\xi_x\eta_x + 2c_1\xi_y\eta_y, K_{1,3} = \eta_x^2 + c_1\eta_y^2,$$
$$K_{1,4} = \xi_{xx} + c_1\xi_{yy}, K_{1,5} = \eta_{xx} + c_1\eta_{yy}, K_{1,6} = -c_3,$$
$$K_{1,7} = c_2\xi_x\xi_y, K_{1,8} = c_2\left(\xi_x\eta_y + \xi_y\eta_x\right), K_{1,9} = c_2\eta_x\eta_y, \qquad (B1)$$
$$K_{1,10} = c_2\xi_{xy}, K_{1,11} = c_2\eta_{xy}, K_{1,12} = K_{1,13} = K_{1,14} = K_{1,15} = 0,$$
$$K_{1,16} = c_3\xi_x, K_{1,17} = c_3\eta_x, K_{1,18} = 0,$$

$$K_{2,1} = c_2\xi_x\xi_y, K_{2,2} = c_2\left(\xi_x\eta_y + \xi_y\eta_x\right), K_{2,3} = c_2\eta_x\eta_y, K_{2,4} = c_2\xi_{xy},$$
$$K_{2,5} = c_2\eta_{xy}, K_{2,6} = 0, K_{2,7} = c_1\xi_x^2 + \xi_y^2, K_{2,8} = 2(c_1\xi_x\eta_x + \xi_y\eta_y),$$
$$K_{2,9} = c_1\eta_x^2 + \eta_y^2, K_{2,10} = c_1\xi_{xx} + \xi_{yy}, K_{2,11} = c_1\eta_{xx} + \eta_{yy}, K_{2,12} = -c_3,$$
$$K_{2,13} = K_{2,14} = K_{2,15} = 0, K_{2,16} = c_3\xi_y, K_{2,17} = c_3\eta_y, K_{2,18} = 0,$$

$$(B2)$$

$$K_{3,1} = K_{3,2} = K_{3,3} = K_{3,6} = 0, K_{3,4} = c_3\xi_x, K_{3,5} = c_3\eta_x,$$
$$K_{3,7} = K_{3,8} = K_{3,9} = K_{3,12} = 0, K_{3,10} = c_3\xi_y, K_{3,11} = c_3\eta_y,$$
$$K_{3,13} = -\left(c_3 + \frac{G_f}{D}\right)\left(\xi_x^2 + \xi_y^2\right), K_{3,14} = -2\left(c_3 + \frac{G_f}{D}\right)\left(\xi_x\eta_x + \xi_y\eta_y\right), \qquad (B3)$$
$$K_{3,15} = -\left(c_3 + \frac{G_f}{D}\right)\left(\eta_x^2 + \eta_y^2\right), K_{3,16} = -\left(c_3 + \frac{G_f}{D}\right)\left(\xi_{xx} + \xi_{yy}\right),$$
$$K_{3,17} = -\left(c_3 + \frac{G_f}{D}\right)\left(\eta_{xx} + \eta_{yy}\right), K_{3,18} = -\left(c_3 + \frac{G_f}{D}\right)\frac{k_f}{D}$$

with

$$c_1 = \frac{1-\nu}{2}; \quad c_2 = \frac{1+\nu}{2}; \quad c_3 = \frac{6\kappa(1-\nu)}{h^2} \qquad (B4)$$

Chapter 4

Multiple Scale Differential Quadrature Method

Direct DQ method is a highly efficient method, being able to yield very accurate results by using a small number of nodes. But most of its applications are limited to one- and two-dimensional problems. Recent years saw considerable efforts to formulate three-dimensional DQ methods. Two approaches are commonly used in developing a three dimensional DQ method. The direct way is to discretize the physical quantity in all three dimensions in the same manner as that for two-dimensional problems. However, paramount difficulties arise in the direct discretization since DQ methods are global schemes with nearly full resultant matrices. It can be expected that the mesh size must be small if direct solution methods such as Gauss elimination are used. This is evidenced from the mesh size reported in the literature. For example, the maximum meshes used in the three-dimensional DQ calculations by Zhou et al., (2002) and Liew et al., (2001) were $8 \times 8 \times 6$ and $11 \times 11 \times 11$, respectively. The three digits indicate the corresponding mesh nodes adopted in the three dimensions. So $8 \times 8 \times 6$ means that two dimensions are discretized by 8 nodes, respectively, and the other by 6 nodes. The mesh sizes in these applications are small. It is also found that increasing the mesh size to $18 \times 18 \times 18$ will invalidate the computation on a computer of 128 MB memory. The restriction of the mesh size due to the employment of direct solution methods will be demonstrated later in the present chapter.

Another difficulty associated with three-dimensional DQ methods is that mesh size exerts great influence on computational speed. A slight increase in mesh size may result in significant increase in CPU time. For example, the CPU time for the mesh $15 \times 15 \times 15$ is 800 times than that for the mesh $8 \times 8 \times 8$.

In order to overcome the above mentioned difficulties, four different solution methods (including direct Gauss elimination, one-dimensional band storage, direct iterative method and a multi-scale iterative method) are compared in the first part of this chapter. Both Gauss elimination and one-dimensional band storage suffer from limitations in the mesh size. The mesh size must be smaller than $18 \times 18 \times 18$ (in approximate meaning) for the problems studied here using a Pentium 4 computer. Iterative methods such as SOR (successive over-relaxation) greatly increase the mesh size, but the convergence rate is very slow. Based on these studies, a multi-scale iterative method proposed by

Zong (2004) is introduced in this chapter. Numerical examples have shown that it has exhibited a tenfold increase in convergence speed over the direct iterative method.

The second approach to handle a three-dimensional problem is to transform a three-dimensional problem into a nearly two-dimensional problem. For example, separate a three-dimensional quantity $f(x, y, z)$ into the product of two different functions, that is, $g(x, y)h(z)$ (Malik and Bert, 1998; Shu et al., 1996a,b), and then apply direct DQ method to approximate the two-dimensional quantity $g(x, y)$. This is in fact a two-dimensional problem, and thus it does not increase computational efforts significantly when compared with those for a two-dimensional problem. Zong (2003b) and Wu and Tsai (2004) smartly employed multi-scale analysis which dramatically reduces the computational demand. Detailed introduction on the multi-scale DQ methods will be presented in the latter part of this chapter.

Last in this chapter, the multi-scale approach is applied to find the equations governing quasi-static behaviors of a poroelastic medium at the macroscopic scale. Then direct DQ-based solution strategy is applied to solve these equations.

4.1 Multi-scale DQ method for potential problems

We take three-dimensional heterogeneous Laplace equation as an example to elucidate the formulation. Quite a few physical phenomena can be described by a heterogeneous Laplace equation. Take blood flow in heterogeneous tissues for instance. The blood flow is assumed to obey Darcy's law (Lang et al., 1999)

$$\mathbf{v} = -c(x, y, z)\nabla p \qquad (4.1)$$

where $\mathbf{v}$ is fluid flow velocity, $c(x, y, z)$ is the permeability coefficient dependent on location (x, y, z) and p is the pressure. The equation postulates that the permeable velocity results from pressure gradient.

It is clinically observed that the permeability coefficient of normal tissue is higher than that of tumor (Lang et al., 1999). Hence the blood circulation around a tumor is poor. Based on mass conservation, the divergence of the flow velocity should be zero, that is,

$$\nabla \cdot \mathbf{v} = 0 \qquad (4.2)$$

By substituting Eq. (4.1) into Eq. (4.2), we obtain the heterogeneous Laplace equation

$$c(x, y, z)\left(\frac{\partial^2 p}{\partial x^2} + \frac{\partial^2 p}{\partial y^2} + \frac{\partial^2 p}{\partial z^2}\right) + \frac{\partial c}{\partial x}\frac{\partial p}{\partial x} + \frac{\partial c}{\partial y}\frac{\partial p}{\partial y} + \frac{\partial c}{\partial z}\frac{\partial p}{\partial z} = 0 \qquad (4.3)$$

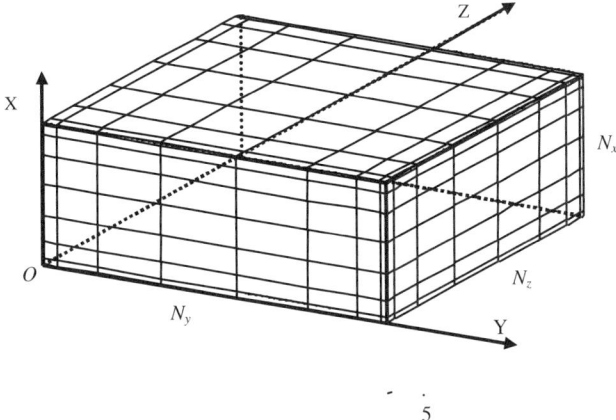

FIGURE 4.1: Discretization of a cube.

subject to Dirichlet or Neumann boundary conditions.

Consider a cube of dimensions $L \times W \times B$. It is discretized by $N_x \times N_y \times N_z$ Chebyshev nodes as defined by Eq. (1.48) in Chapter 1 as shown in Fig. 4.1. On each node, pressure and permeability coefficient are denoted by p_{ijk} and c_{ijk} $(i = 1,\ldots,N_x,\ j = 1,\ldots,N_y,\ k = 1,\ldots,N_z)$. Then the velocity in each direction can be expressed in the following form by means of DQ

$$q_x^{ijk} = \sum_{l=1}^{N_x} a_x^{il} c_{ljk} \tag{4.4a}$$

$$q_y^{ijk} = \sum_{m=1}^{N_y} a_y^{jm} c_{imk} \tag{4.4b}$$

$$q_z^{ijk} = \sum_{n=1}^{N_z} a_z^{kn} c_{ijn} \tag{4.4c}$$

where the symbols a_x^{ij}, a_y^{jm} and a_z^{kn} are DQ weighting coefficients in the three directions, respectively. By substituting Eqs. (4.4) into Eq. (4.3) and using the expression of coefficients of the first-and second-order derivatives, the discrete Laplace equation turns out to be

$$c_{ijk} \left[\sum_{l=1}^{N_x} b_x^{il} p_{ljk} + \sum_{m=1}^{N_y} b_y^{jm} p_{imk} + \sum_{n=1}^{N_z} b_z^{kn} p_{ijn} \right]$$

$$+ q_x^{ijk} \sum_{l=1}^{N_x} a_x^{il} p_{ljk} + q_y^{ijk} \sum_{m=1}^{N_y} a_y^{jm} p_{imk} + q_z^{ijk} \sum_{n=1}^{N_z} b_z^{kn} p_{ijn} = 0 \tag{4.5}$$

On the Dirichlet boundaries, we have

$$p_{ijk} = \hat{p}_{ijk} \tag{4.6a}$$

for Dirichlet boundary condition where the "hat" symbol denotes given values. On Neumann boundaries, we have

$$-c_{ijk} \sum_{l=1}^{N_x} a_x^{il} p_{ljk} = \hat{u}_{ijk},$$

$$-c_{ijk} \sum_{m=1}^{N_y} a_y^{mj} p_{imn} = \hat{v}_{ijk}, \tag{4.6b}$$

$$-c_{ijk} \sum_{n=1}^{N_z} a_z^{kn} p_{ijn} = \hat{w}_{ijk}$$

In the following, we will apply four methods to solve Eq. (4.5) with the two different boundary conditions given in Eq. (4.6).

4.2   Solutions of potential problems

In this section, three methods, including Gauss elimination, one-dimensional band storage and successive over-relaxation (SOR) method, are employed to solve the potential problems formulated in the previous section. The fourth method, SOR-based multi-scale DQ method, will be discussed in detail in Section 4.3.

4.2.1   Gauss elimination

To apply Gauss elimination, we introduce the following indices

$$\alpha = (i-1)N_y N_z + (j-1)N_z + k, i = 1, \ldots N_x, \ j = 1, \ldots, N_y, \ k = 1, \ldots N_z$$
$$\beta = (l-1)N_y N_z + (m-1)N_z + n, l = 1, \ldots N_x, \ m = 1, \ldots, N_y, \ n = 1, \ldots N_z \tag{4.7}$$

and the matrices

$$d_{\alpha\beta} = c_\alpha \left(b_x^{il} \delta^{mj} \delta^{kn} + \delta^{il} b_y^{jm} \delta^{kn} + \delta^{il} \delta^{jm} b_z^{kn} \right)$$
$$+ q_x^{ijk} a_x^{il} \delta^{jm} \delta^{kn} + q_y^{ijk} \delta^{il} \delta^{kn} a_y^{jm} (y_j) + q_z^{ijk} \delta^{il} \delta^{jm} a_z^{kn} \tag{4.8}$$

where the delta function possesses the property defined by $\delta^{ij} = 1$ if $i = j$ and $\delta^{ij} = 0$ if $i \neq j$. We can form a similar matrix **b** to denote boundary conditions. Then we obtain the following system of linear algebraic equations

$$\mathbf{DP} = \mathbf{b}, \quad \mathbf{P} = \{p_\beta\}_N, \quad \mathbf{D} = \{d_{\alpha\beta}\}_{N \times N} \qquad (4.9)$$

where $N = N_x \times N_y \times N_z$ is the total number of nodes, $\mathbf{D}$ is an $N \times N$ matrix, which may be very large. For example, for a mesh of $10 \times 10 \times 10$, $\mathbf{D}$ would be a matrix having 10^6 elements. This would require very large system memory. A code based on Gauss elimination was written by Zong (2004) and run on IBM SP2 with 8 nodes. It has 10GB RAM, but each user is only assigned 128 MB Memory.

Consider a homogeneous Laplace equation with the permeability coefficient $c(x, y, z) = 1$ everywhere in the problem domain. The domain is assumed as a simple cube of size $0.2 \times 0.3 \times 0.2\text{m}^3$. The boundary conditions are given by

$$p(x, y = 0, z) = 1, \quad p(x, y = 0.3, z) = 0 \qquad (4.10a)$$

$$\frac{\partial p(0, y, z)}{\partial n} = 0 \quad \text{on the rest of the boundaries} \qquad (4.10b)$$

The exact solution to the above equation can be readily found to be $p = 1 - 10y/3$.

Five cases using different node numbers (meshes) have been computed. The results are shown in Table 4.1. From Table 4.1, it is readily found that the CPU time for mesh $8 \times 8 \times 8$ is 3 seconds, whereas the CPU time for mesh $15 \times 15 \times 15$ is 42 minutes 37 seconds! As node number increases from $8 \times 8 \times 8$ to $15 \times 15 \times 15$, CPU time increases 800 fold! This is a little surprising. We may try to make a simple theoretical estimate. The total number of operations (addition, subtraction, multiplication and division) by Gauss elimination is proportional to the cube of unknown number, N^3 (Atkinson, 1989). The increase in node number is about 8 fold from $8 \times 8 \times 8$ to $15 \times 15 \times 15$, and thus the increase in the number of operations is about $8^3 = 512$, close to 800 in terms of order. Thus, an 800 fold increase in CPU time is theoretically reasonable. Upon further increase to $17 \times 17 \times 17$, the CPU time is about five and half hours, approximately 8-fold longer than that for $15 \times 15 \times 15$, in agreement of the abovementioned theoretical estimates.

Because CPU time is proportional to the cube of node number, a slight increase in node number would result in significant increase in CPU time. When the node number is $18 \times 18 \times 18$, the assigned memory limit (128 MB) of the computer is reached and thus the computation is terminated.

From Table 4.1, it is concluded that the accuracy of Gauss elimination is high, up to 10^{-8}. But the method is computationally expensive and limited to small mesh size. It might be questioned that more powerful computers can be used to speed up the convergence rate. However, the objective of the study is to compare the efficiency of the different methods. To make the solutions comparable, it is crucial to compute the solutions using different methods on the identical computational platform. Therefore, it does not matter whether a powerful computer is used or not.

TABLE 4.1:   CPU Time (H:M:S)

Node number	CPU Time (H:M:S)	Maximum error
$8 \times 8 \times 8$	00:00:03	4×10^{-8}
$10 \times 10 \times 10$	00:00:27	4×10^{-8}
$12 \times 12 \times 12$	00:03:10	4×10^{-8}
$15 \times 15 \times 15$	00:42:37	4×10^{-8}
$17 \times 17 \times 17$	05:36:28	4×10^{-8}
$18 \times 18 \times 18$	Reached the memory limit assigned	

4.2.2   One-dimensional band storage

The assembly matrix formed in a finite element method (FEM) is sparse, thus enabling one-dimensional band storage to work well. Together with Gauss elimination or other solution procedures, one-dimensional band storage can save CPU time greatly when a FEM is used. This was particularly true over two or three decades ago, when workstations were not as powerful as today's.

Matrix $\mathbf{D}$ is not full. From Eq. (4.8), the band width is estimated to be $N_x + N_y + N_z$. A code based on one-dimensional storage and Gauss elimination was written to do the test. The CPU time and maximum errors by using different node numbers are given in Table 4.2.

The maximum mesh size the method can handle is still $17 \times 17 \times 17$. It does not enhance the capability for larger problems when compared with direct Gauss elimination. This is due to the fact that matrix $\mathbf{D}$ is not sparse. But with regard to computational time, this method is much better. The accuracy is, however, lower as shown in Table 4.2. This may result from the truncation errors. In the test code, all elements which are smaller than a prescribed number are treated as zeros and neglected in the computations.

Based on the foregoing discussion, it is clearly seen that direct solution procedures suffer great limitations due to their computational intensity and computer memory demand.

TABLE 4.2:   CPU time and maximum error for band storage method

Node number	CPU Time(H:M:S)	Maximum error
$8 \times 8 \times 8$	00:00:01	5×10^{-5}
$10 \times 10 \times 10$	00:00:03	2.5×10^{-4}
$12 \times 12 \times 12$	00:00:19	6×10^{-4}
$15 \times 15 \times 15$	00:03:02	6×10^{-3}
$17 \times 17 \times 17$	00:13:00	6×10^{-3}
$18 \times 18 \times 18$	Reached the memory limit assigned	

4.2.3 Successive over-relaxation (SOR) method

The demand of large memory by direct solution procedures such as Gauss elimination suggests that iterative methods may be a better option for solving Eq. (4.9) due to its large band width. In this section, a SOR method-based code was developed to solve Eq. (4.9). In the form of Eq. (4.9), SOR formulation can be expressed as (Atkinson, 1989)

$$
P_\alpha^{(t+1)} = P_\alpha^{(t)} + \omega \left(b_\alpha - \sum_{\beta=1}^{\alpha-1} d_{\alpha\beta} P_\beta^{(t+1)} - \sum_{\beta=\alpha}^{N} d_{\alpha\beta} P_\beta^{(t)} \right) \Big/ d_{\alpha\alpha} \qquad (4.10c)
$$

where ω is the relaxation factor and the superscript "t" denotes iteration step ($t = 1, 2 \ldots, n_t$). Note that the form of the above equation is slightly different from the general SOR formula because $\omega P_\alpha^{(t)}$ is put inside the bracket here.

The results obtained from the SOR method for the previous example are given in Table 4.3. The relaxation factor used in the simulations is $\omega=1.8$. When comparng the results in Table 4.3 with those in Tables 4.1 and 4.2, it is readily seen that the size of the mesh SOR method that can be used is dramatically increased. This is encouraging in the sense that a three-dimensional DQ method can be effectively solved by using the SOR method. More details on the SOR method can be found in Chapter 8.

The iteration is stopped if the maximum difference between any two consecutive steps is smaller than a prescribed small tolerance ε, or the iteration number is greater than 10000. Throughout this chapter, $\varepsilon = 10^{-5}$. Note that the accuracy might be higher than this magnitude because the accuracy (defined by maximum error in Table 4.3) is the difference between the final results and the exact solution. As shown in Table 4.3, the maximum error for SOR method is identical to that achieved by Gauss elimination.

TABLE 4.3: CPU time and maximum error for SOR Method with $\omega = 1.8$

Node number	CPU Time (H:M:S)	Maximum error
$8 \times 8 \times 8$	00:00:05	4×10^{-8}
$10 \times 10 \times 10$	00:00:24	4×10^{-8}
$12 \times 12 \times 12$	00:02:05	4×10^{-8}
$15 \times 15 \times 15$	00:14:36	4×10^{-8}
$20 \times 20 \times 20$	02:50:16	4×10^{-8}
$30 \times 30 \times 30$	97:38:09	4×10^{-8}

It is well known that the convergence speed of an SOR method is heavily dependent on relaxation factor ω, which must vary from 0 to 2 to guarantee convergence. For a particular problem, there is generally no theoretical

TABLE 4.4:   Dependence of convergence rate
on the relaxation factor (mesh $10 \times 10 \times 10$)

Relaxation factor ω	CPU Time (H:M:S)	Iteration steps	Reached accuracy
0.1	00:28:13	10000	1×10^{-2}
0.5	00:19:22	5584	1×10^{-5}
1.0	00:05:41	1907	1×10^{-5}
1.5	00:01:22	655	1×10^{-5}
1.7	00:00:43	342	1×10^{-5}
1.8	00:00:24	190	1×10^{-5}
1.9	00:01:18	621	1×10^{-5}
1.99		Divergent	

guideline to find the optimal relaxation factor for the fastest convergence. For the present example, numerical tests have been performed for the mesh $10 \times 10 \times 10$ and the dependence of convergence speed on the relaxation factor is obtained as shown in Table 4.4. The convergence speed increases slowly with increasing ω until the peak point is reached $\omega \approx 1.8$. After that the convergence speed deceases rapidly and finally diverges at $\omega \approx 1.99$.

SOR method greatly increases the mesh size. But the computational speed remains an obstacle. Table 4.3 shows that the CPU time for mesh $20 \times 20 \times 20$ is approximately three hours, about 10-fold that of mesh $10 \times 10 \times 10$. And the mesh $30 \times 30 \times 30$ takes 30-fold CPU time of the mesh $20 \times 20 \times 20$. In the following section, a multi-scale SOR method is employed to enhance the computational efficiency.

4.3   SOR-based multi-scale DQ method

The convergence speed of SOR method is still slow. Because the number of operations is proportional to the cube of the node number N^3 (Atkinson, 1989), a double increase of node number in each of the three directions would result in an $8^3 = 512$-fold increase of computational time. Even though SOR method improves the solution capability than direct Gauss elimination and one-dimensional band storage methods, intensive computation remains a challenge to implementing a three-dimensional DQ method.

Iteration number is evidently dependent on the choice of initial values. If the initial values are well chosen, the convergence speed would undoubtedly be greatly increased. If we use the results obtained from a coarse mesh as the initial values for a fine mesh, a faster convergence speed can be expected. And if the results obtained from the fine mesh are used for the initial values for an even finer mesh, a faster convergence speed and higher accuracy iteration would be expected. This is the main idea of the multi-scale SOR method proposed by Zong (2004). It will be introduced in this section.

Putting it in the mathematical way, we assign $\mathbf{0}$ as the initial value for a coarse discretization mesh $N_x^{(1)} \times N_y^{(1)} \times N_z^{(1)}$, where the superscript "(1)" denotes the first-scale. After some iterations until the step at which the maximum error is smaller than the prescribed tolerance, we obtain the first-scale solution $\mathbf{P}^{(1)}$. Based on a finer mesh $N_x^{(2)} \times N_y^{(2)} \times N_z^{(2)}$, where the superscript "(2)" denotes the second scale and $N_x^{(2)} > N_x^{(1)}$, $N_y^{(2)} > N_y^{(1)}$, $N_z^{(2)} > N_z^{(1)}$, we use the following Lagrange interpolating polynomial to obtain the initial values for the second-scale:

$$P(x_i, y_j, z_k) = \sum_{l=1}^{N_x^{(1)}} \sum_{m=1}^{N_y^{(1)}} \sum_{n=1}^{N_z^{(1)}} \lambda_x^l(x_i) \lambda_y^m(y_j) \lambda_z^n(z_k) P_{lmn}^{(1)} \tag{4.11a}$$

$$i = 1, ..., N_x^{(2)}, \quad j = 1, ..., N_y^{(2)}, \quad k = 1, ..., N_z^{(2)}$$

where

$$\lambda_x^k(x) = \frac{L_k(x)}{L_k(x_k)}, \quad L_k(x) = \prod_{\substack{j=1 \\ j \neq k}}^{N_x^{(1)}} (x - x_j) \tag{4.11b}$$

Similarly, the formulas for y- and z-axes can be obtained with the corresponding superscripts and subscripts replaced by y and z. With the results given from Eq. (4.11) as initial values, the new estimate $\mathbf{P}^{(2)}$ for the second scale can be found through use of the SOR method. With $\mathbf{P}^{(2)}$ as the initial values for an even finer mesh, a better estimate $\mathbf{P}^{(3)}$ is obtained. The procedure can be repeated until n_s-th scale.

Table 4.5 shows the solution of the previous example by use of the multi-scale method for the mesh $20 \times 20 \times 20$. If one-scale procedure is used, the CPU time is 2 hours 50 minutes and 15 seconds. In two-scale procedure, the initial value for the first scale is $\mathbf{0}$ with a mesh $10 \times 10 \times 10$. 24 seconds of CPU time are needed to yield the solution for the first-scale. The mesh for the second scale is $20 \times 20 \times 20$. The total CPU time for the two scales is considerably reduced to 36 minutes and 20 seconds.

It is clearly demonstrated in Table 4.5 that the multi-scale approach with scale number larger than one can reduce CPU time significantly, but the amount of reduction is not linearly proportional to the scale numbers. In other words, over-large scale number may not lead to a faster computational speed.

Table 4.6 shows the results for another mesh $20 \times 30 \times 20$. One-scale procedure took 27 hours to produce the results of desired accuracy. A two-scale procedure, however, produces the results of same accuracy in less than three hours. And a three-scale procedure produces the results in less than two hours. The reduction in CPU time is remarkable, indicating the effectiveness of the multi-scale method. The prescribed tolerance is again 10^{-5}.

TABLE 4.5: CPU time for multi-scale DQ (Mesh $20 \times 20 \times 20$, $\omega = 1.8$)

Scale number	CPU Time (H:M:S)	Prescribed tolerance
One scale	02:50:16	1×10^{-5}
Two scales	00:36:20	1×10^{-5}
Three scale	01:44:33	1×10^{-5}
Four scales	01:03:14	1×10^{-5}
Five scales	02:01:47	1×10^{-5}

TABLE 4.6: CPU Time for multi-scale DQ (Mesh $20 \times 30 \times 20$, $\omega = 1.8$)

Scale number	CPU Time (H:M:S)	Meshes used
One scale	27:05:33	$20 \times 30 \times 20$
Two scales	02:52:27	$10 \times 15 \times 10, 20 \times 30 \times 20$
Three scale	01:56:53	$10 \times 20 \times 10, \quad 15 \times 25 \times 15, \quad 20 \times 30 \times 20$

TABLE 4.7: CPU Time for multi-scale DQ (Mesh $30 \times 30 \times 30$, ωg 1.8)

Method	CPU Time (H:M:S)	Meshes used
One-scale SOR	97:38:09	$30 \times 30 \times 30$
Three-scale SOR	08:25:29	$10 \times 10 \times 10, \quad 20 \times 20 \times 20, \quad 30 \times 30 \times 30$

In Table 4.7 another example is presented for mesh $30 \times 30 \times 30$. The CPU time for one-scale is nearly 100 hours but that of the multi-scale procedure is less than 10 hours, a 10 times reduction in CPU time. Again, the efficiency of the multi-scale method is demonstrated.

Finally, a more complicated example is considered. Suppose a tissue cube of size $0.2 \times 0.3 \times 0.4 \text{m}^3$, in which there is a tumor located at $(0.1, 0.15, 0.2)$. The permeability coefficient $c(x, y, z)$ follows the following spatial pattern

$$c(x, y, z) = \begin{cases} 10^{-1} & r = 3(y - 0.15)^2 + z^2 \leq 0.01 \\ 10^{-1} \left[1 + \tanh(200r)\right] & r = 3(y - 0.15)^2 + z^2 > 0.01 \end{cases} \quad (4.12)$$

Because of symmetry, only half domain symmetric about z axis is considered. The boundary conditions are

$$p(x, 0, z) = 1, \quad p(x, 0.3, z) = 0 \quad (4.13a)$$

$$\frac{\partial p(0, y, z)}{\partial n} = 0 \text{ on the rest of boundaries} \quad (4.13b)$$

A mesh $10 \times 10 \times 10$ is used at first-scale approximation, from which optimal relaxation factor is found to be $\omega = 1.75$. However, $\omega = 1.8$ is adopted in the calculations. At the second-scale, a mesh $20 \times 20 \times 20$ is employed while $30 \times 30 \times 30$ is used for third-scale.

Figure 4.2 shows the pressure contours obtained from these three scales at the section $x = L/2 = 0.1\text{m}$. The results obtained from the second (dotted line) and third scales (thick solid line) are almost identical, showing negligible differences. Slight differences are observed between the first scale (thin solid line) and the other two scales, particularly around the tumour.

In Fig. 4.3(a), two contour lines ($w = 0.043$ and $w = -0.043$) are plotted. Qualitatively speaking, the fluid has a positive vertical velocity on the left side of the tumor ($w > 0$) and a negative one on the right side of the tumor. The flow is scattered by the tumor.

Figure 4.3(b) shows the contours of flow velocity in the Y-axis. Three sets of contour for $v = 0.5, 0.6, 0.7$ m/s obtained from the three-scales are plotted. In general, the results obtained from the second scale are in acceptable agreement with those obtained from the third scale, while those from the first scale show significant differences from the other two scales. Both in front of and behind the tumor, waves are observable because of the scattering effects of the tumor.

Further improvements on the method may be anticipated. The method in the present form has several scales of meshes covering the same computational domains. This is not computationally effective. In the finite element method we have fine mesh where the quantities under consideration change dramatically while coarse mesh where the quantities under consideration change mildly. The same is true of the multi-scale DQ method to be improved. The first scale acts as coarse mesh covering the whole computational domain. The second scale acts as fine mesh covering only those regions

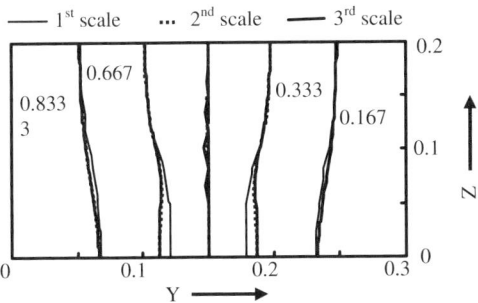

FIGURE 4.2: Pressure contour plots at the section $x = L/2$.

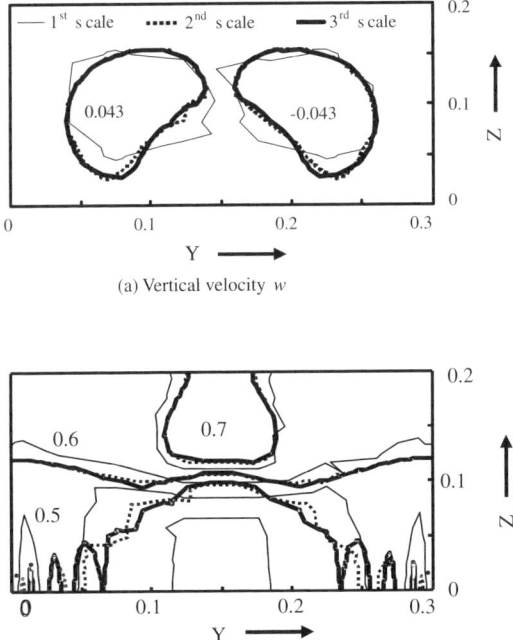

(a) Vertical velocity w

(b) Lateral velocity v

FIGURE 4.3: Velocity contour plots for vertical and lateral velocities.

in the whole computational domain where the quantities under consideration experience significant changes. The third scale acts as even finer mesh
covering part of the regions already under the second scale covering. Using
this technique we may anticipate high accuracy with reduced computational
efforts.

To implement the methodology, however, we need to assess and find the
regions where quantities under consideration may experience fast changes. A
measure is needed to do the job. It tells which regions require finer mesh and
which do not. In the simple cases, the measure is a derivative-like quantity.
We have some mature techniques in finite element method and the interested
reader is referred to Bath (1996) for more details.

4.4   Asymptotic multi-scale DQ method

Multi-scale DQ method introduced in the preceding section is characterized
by applying direct DQ method to each scale. In applications, such cases are
available in which one dimension is much smaller than the other two dimensions for a three-dimensional problem. This allows us to employ perturbation
method to simplify a three-dimensional analysis. We take functionally graded
(FG) annular spherical shells for instance to show the applicability of multi-
scale DQ method. Before that, we brief the recent progress in the field of FG
material.

In recent years, considerable studies have been carried out to analyze the
thermoelastic, dynamic and buckling behaviors of functionally graded (FG)
isotropic plates and shells (Shen, 2002; Yang and Shen, 2001; Ma and Wang,
2003). This is attributed to the wide applications of functionally graded materials (FGMs) as the components of structures in the advanced engineering.
When compared with the traditional and fiber-reinforced laminated composite
materials, FGMs can sustain harsh environments with high temperature gra-
dients while maintaining their structural integrity. It is well known that the
fiber-reinforced laminated composite materials have a mismatch of mechanical properties across an interface due to two different layers bonded together.
By contrast, FGMs gradually and continuously vary the mechanics properties
along the thickness direction by changing the volume fractions of two materials as desired. The weakness of the fiber-reinforced laminated composite
materials such as debonding, huge residual stress, locally large plastic deformations, etc., can be avoided or reduced in FGMs (Noda, 1991; Tanigawa,
1995). Hence, FGMs hold great potentials for engineering applications, especially in the high-temperature environments such as aerospace structures,
fusion reactors and nuclear industry.

Several approximate 2D solutions for the thermoelastic, dynamic and buckling analyses of FG plates and shells have been presented in the literature (Reddy et al., 1999; Yang and Shen, 2001; Pradhan et al., 2000). Wu and his colleagues have established a number of 3D solutions of thermoelastic, dynamic and buckling problems of laminated composite doubly curved and conical shells (Wu et al., 1996a, 1996b; Wu and Chiu, 2001) by using the method of asymptotic expansion. Other 3D linear analyses of laminated composite plates and shells can also be found in the literature (Fan and Zhang, 1992; Huang and Tauchert, 1992, Bhimaraddi, 1993). Various computational models of laminates have been developed (Kapania, 1989; Noor and Burton, 1990).

In the present chapter we study the static problem of FG annular spherical shells by combining the numerical method of DQ with the analytical asymptotic expansion approach. By introducing a suitable perturbation parameter and a set of dimensionless field variables in the 3D elasticity formulation, we recast the 3D problems as various orders of classical shear theory (CST) problems. The materials consisting of a mixture of metal and ceramic are considered to be isotropic and inhomogeneous. The material properties vary in the thickness direction according to a power law and are dependent on the volume fractions of the constituents. Direct DQ method is used to obtain the asymptotic solutions. In the computations, the relevant integrations are estimated by the method of Gaussian quadrature. The shell considered is artificially divided by NL layers. The field variables at each order level are interpolated using the nominal polynomials in each layer in order to avoid the calculation of convolution integration. According to the recurrence property, the solution procedure for the leading-order problem can be repeatedly used for solving the higher-order problems in a consistent and hierarchic manner (Wu and Tsai, 2003).

4.4.1 Basic three-dimensional equations

Consider a FG isotropic annular spherical shell shown in Fig. 4.4. A system of spherical polar coordinates (ϕ, θ, ρ) located on the middle surface is adopted. The radial coordinate, ρ, is also represented as $\rho = R + \zeta$ where R is the radius of the annular spherical shell at the mid-surface and ζ is the normal coordinate measured from the mid-surface of the shell along the thickness direction. $2h$ denotes the thickness of the shell. The stress-strain relations for isotropic materials are given by

$$
\left\{
\begin{array}{c}
\sigma_\phi \\
\sigma_\theta \\
\sigma_\rho \\
\tau_{\theta\rho} \\
\tau_{\phi\rho} \\
\tau_{\phi\theta}
\end{array}
\right\}
=
\begin{bmatrix}
c_{11} & c_{12} & c_{12} & 0 & 0 & 0 \\
c_{12} & c_{11} & c_{12} & 0 & 0 & 0 \\
c_{12} & c_{12} & c_{11} & 0 & 0 & 0 \\
0 & 0 & 0 & \frac{(c_{11}-c_{12})}{2} & 0 & 0 \\
0 & 0 & 0 & 0 & \frac{(c_{11}-c_{12})}{2} & 0 \\
0 & 0 & 0 & 0 & 0 & \frac{(c_{11}-c_{12})}{2}
\end{bmatrix}
\left\{
\begin{array}{c}
\varepsilon_\phi \\
\varepsilon_\theta \\
\varepsilon_\rho \\
\gamma_{\theta\rho} \\
\gamma_{\phi\rho} \\
\gamma_{\phi\theta}
\end{array}
\right\},
\quad (4.14)
$$

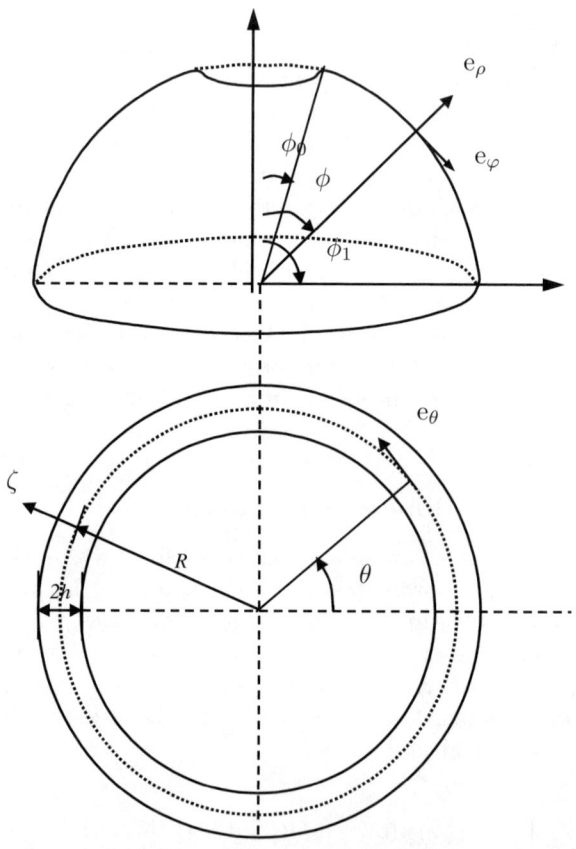

FIGURE 4.4: The geometry and coordinate system for an annular spherical shell.

where σ_ϕ, σ_θ, σ_ρ, $\tau_{\phi\rho}$, $\tau_{\theta\rho}$, $\tau_{\phi\theta}$ and ε_ϕ, ε_θ, ε_ρ, $\gamma_{\phi\rho}$, $\gamma_{\theta\rho}$, $\gamma_{\phi\theta}$ are the stress and strain components, respectively; c_{ij} denotes the stiffness coefficients and can be rewritten in terms of engineering constants as $c_{11} = (1 - v) E/(1 + v)(1 - 2v)$, $c_{12} = v E/(1 + v)(1 - 2v)$ where E is Young's modulus, v is Poisson's ratio. In FGMs, the material properties are assumed to be nonhomogeneous through the thickness direction only (i.e., $E = E(\rho)$ and $v = v(\rho)$).

The kinematics relations in terms of the spherical polar coordinates ϕ, θ and ρ can be expressed as

$$\begin{Bmatrix} \varepsilon_\phi \\ \varepsilon_\theta \\ \varepsilon_\rho \\ \gamma_{\theta\rho} \\ \gamma_{\phi\rho} \\ \gamma_{\phi\theta} \end{Bmatrix} = \begin{bmatrix} \partial_\phi/\rho & 0 & 1/\rho \\ c_\phi/\rho\,s_\phi & \partial_\theta/\rho\,s_\phi & 1/\rho \\ 0 & 0 & \partial_\rho \\ 0 & \partial_\rho - (1/\rho) & \partial_\theta/\rho\,s_\phi \\ \partial_\rho - (1/\rho) & 0 & \partial_\phi/\rho \\ \partial_\theta/\rho\,s_\phi & (\partial_\phi/\rho) - (c_\phi/\rho\,s_\phi) & 0 \end{bmatrix} \begin{Bmatrix} u_\phi \\ u_\theta \\ u_\rho \end{Bmatrix}, \qquad (4.15)$$

in which $\partial_\phi = \partial/\partial\phi$, $\partial_\theta = \partial/\partial\theta$, $\partial_\rho = \partial/\partial\rho$, $c_\phi = \cos\phi$, $s_\phi = \sin\phi$; u_ϕ, u_θ and u_ρ are the displacement components.

The stress equilibrium equations in the absence of body forces are given by

$$\sigma_{\phi,\phi}/\rho + \tau_{\phi\theta,\theta}/\rho\,s_\phi + \tau_{\phi\rho,\rho} + (\sigma_\phi - \sigma_\theta)\,c_\phi/\rho\,s_\phi + 3\tau_{\phi\rho}/\rho = 0, \qquad (4.16)$$

$$\tau_{\phi\theta,\phi}/\rho + \sigma_{\theta,\theta}/\rho\,s_\phi + \tau_{\theta\rho,\rho} + 2\,\tau_{\phi\theta}c_\phi/\rho\,s_\phi + 3\tau_{\theta\rho}/\rho = 0, \qquad (4.17)$$

$$\tau_{\phi\rho,\phi}/\rho + \tau_{\theta\rho,\theta}/\rho\,s_\phi + \sigma_{\rho,\rho} + \tau_{\phi\rho}c_\phi/\rho\,s_\phi - (\sigma_\phi + \sigma_\theta - 2\sigma_\rho)/\rho = 0. \qquad (4.18)$$

The traction boundary conditions on the inner and outer surfaces are specified as follows:

$$[\tau_{\phi\rho}, \tau_{\theta\rho}, \sigma_\rho] = [0, 0, q_\rho^+] \quad on \quad \zeta = h \qquad (4.19a)$$

$$[\tau_{\phi\rho}, \tau_{\theta\rho}, \sigma_\rho] = [0, 0, q_\rho^-] \quad on \quad \zeta = -h \qquad (4.19b)$$

where $q_\rho^\pm$ are the external loads at the inner (-) and outer (+) surfaces in the thickness directions, respectively.

The edge boundary conditions require that one member of each pair of the following quantities be satisfied,

$$n_1\,\sigma_\phi + n_2\,\tau_{\phi\theta} = \bar{p}_\phi, \quad or \quad u_\phi = \bar{u}_\phi \qquad (4.20a)$$

$$n_1\,\tau_{\phi\theta} + n_2\,\sigma_\theta = \bar{p}_\theta, \quad or \quad u_\theta = \bar{u}_\theta \qquad (4.20b)$$

$$n_1\,\tau_{\phi\rho} + n_2\,\tau_{\theta\rho} = \bar{p}_\rho, \quad or \quad u_\rho = \bar{u}_\rho \qquad (4.20c)$$

where $\bar{p}_\phi$, $\bar{p}_\theta$ and $\bar{p}_\rho$ are applied edge loads; $\bar{u}_\phi$, $\bar{u}_\theta$ and $\bar{u}_\rho$ are the prescribed edge displacements; n_1 and n_2 denote the outward normal at a point along the edge.

4.4.2   Nondimensionalization

A set of dimensionless field variables is defined in the present formulation as follows:

$$x = \phi/\varepsilon, \quad y = \theta/\varepsilon, \quad z = \zeta/R\varepsilon^2; \tag{4.21a}$$

$$u = u_\phi/R\varepsilon, \quad v = u_\theta/R\varepsilon, \quad w = u_\rho/R; \tag{4.21b}$$

$$\sigma_x = \sigma_\phi/Q, \quad \sigma_y = \sigma_\theta/Q, \quad \tau_{xy} = \tau_{\phi\theta}/Q; \tag{4.21c}$$

$$\tau_{xz} = \tau_{\phi\rho}/Q\varepsilon, \quad \tau_{yz} = \tau_{\theta\rho}/Q\varepsilon, \quad \sigma_z = \sigma_\rho/Q\varepsilon^2. \tag{4.21d}$$

in which $\varepsilon^2 = h/R$ and Q stands for a reference elastic modulus.

Herein, the displacements u_ϕ, u_θ, u_ρ and transverse stresses $\tau_{\phi\rho}$, $\tau_{\theta\rho}$, σ_ρ are regarded as the primary variables; the in-surface stresses σ_ϕ, σ_θ, $\tau_{\phi\theta}$ as the dependent variables. After eliminating the in-surface stresses from Eqs. (4.14) to (4.18) and then substituting Eq. (4.21) into the resulting equations, we can reformulate the basic 3D equations in the dimensionless form of

$$w_{,z} = -\varepsilon^2 \mathbf{L}_1\,\mathbf{u}\; -\varepsilon^2\,l_3\,w + \varepsilon^4(Q/c_{11})\,\sigma_z, \tag{4.22}$$

$$\mathbf{u}_{,z} = -\mathbf{D}\,w + \varepsilon^2\mathbf{L}_2\,\mathbf{u} + \varepsilon^2\,\mathbf{S}\,\sigma_s + \varepsilon^4 z\,\mathbf{S}\,\sigma_s, \tag{4.23}$$

$$\sigma_{s,z} = -\mathbf{L}_3\,\mathbf{u} - \mathbf{L}_4\,w - \varepsilon^2\mathbf{L}_5\,\sigma_s - \varepsilon^2\mathbf{L}_6\sigma_z - \varepsilon^4\mathbf{L}_7\,\sigma_s - \varepsilon^4\mathbf{L}_8\sigma_z, \tag{4.24}$$

$$\begin{aligned}\sigma_{z,z} = {} & \mathbf{L}_9\,\mathbf{u} + l_{19}\,w - \mathbf{D}^{\mathrm{T}}\,\sigma_s - l_{20}\tau_{xz} - \varepsilon^2 z\mathbf{D}^{\mathrm{T}}\,\sigma_s \\ & - \varepsilon^2 z l_{20}\tau_{xz} - \varepsilon^2 l_{21}\sigma_z - \varepsilon^4 l_{22}\sigma_z,\end{aligned} \tag{4.25}$$

$$\sigma_m = \mathbf{L}_{10}\,\mathbf{u} + \mathbf{L}_{11}\,w + \varepsilon^2\mathbf{L}_{12}\sigma_z, \tag{4.26}$$

where $\mathbf{u} = \{u\;\; v\}^{\mathrm{T}}, \sigma_s = \{\tau_{xz}\;\; \tau_{yz}\}^{\mathrm{T}}, \sigma_m = \{\sigma_x\;\; \sigma_y\;\; \tau_{xy}\}^{\mathrm{T}}, \mathbf{D} = \{\partial_x\;\; \partial_y/s_\phi\}^{\mathrm{T}}, \mathbf{S} = Q\begin{bmatrix}(c_{11}-c_{12})/2 & 0 \\ 0 & (c_{11}-c_{12})/2\end{bmatrix}^{-1}, \mathbf{L}_i(i=1\sim12);$ $\mathbf{L}_i(i=1\sim12)$ are given in Appendix A.

The dimensionless forms of lateral boundary conditions on the inner and outer surfaces are specified as follows:

$$[\tau_{xz},\; \tau_{yz},\; \sigma_z] = [0,\; 0,\; q_z^+] \quad \text{on } z = 1, \tag{4.27a}$$

$$[\tau_{xz},\; \tau_{yz},\; \sigma_z] = [0,\; 0,\; q_z^-] \quad \text{on } z = -1. \tag{4.27b}$$

where $q_z^\pm = q_\rho^\pm/Q\varepsilon^2$.

The edge boundary conditions require that one member of each pair of the following quantities be satisfied,

$$n_1 \sigma_x + n_2 \tau_{xy} = \bar{p}_x, \quad \text{or} \quad u = \bar{u}, \tag{4.28a}$$

$$n_1 \tau_{xy} + n_2 \sigma_y = \bar{p}_y, \quad \text{or} \quad v = \bar{v}, \tag{4.28b}$$

$$n_1 \tau_{xz} + n_2 \tau_{yz} = \bar{p}_z, \quad \text{or} \quad w = \bar{w}, \tag{4.28c}$$

where $\bar{p}_x = \bar{p}_\phi/Q, \bar{p}_y = \bar{p}_\theta/Q, \bar{p}_z = \bar{p}_\rho/Q\varepsilon; \bar{u} = \bar{u}_\phi/R\varepsilon, \bar{v} = \bar{u}_\phi/R\varepsilon, \bar{w} = \bar{u}_\rho/R.$

4.4.3 Asymptotic expansion

Equations (4.22) to (4.26) contain only even power terms of ε. In view of this, we can we expand the displacements and stresses in the form of

$$f(x, y, z, \varepsilon) = f_{(0)}(x, y, z) + \varepsilon^2 f_{(1)}(x, y, z) + \varepsilon^4 f_{(2)}(x, y, z) + \dots \tag{4.29}$$

By substituting Eq. (4.29) into Eqs. (4.22) to (4.26) and collecting coefficients of equal powers of ε, we then obtain the sets of equations for various orders as follows.

Order ε^0:

$$w_{(0),z} = 0, \tag{4.30}$$

$$\mathbf{u}_{(0),z} = -\mathbf{D}\, w_{(0)}, \tag{4.31}$$

$$\sigma_{s(0),z} = -\mathbf{L}_3\, \mathbf{u}_{(0)} - \mathbf{L}_4\, w_{(0)}, \tag{4.32}$$

$$\sigma_{z(0),z} = \mathbf{L}_9\, \mathbf{u}_{(0)} + l_{19}\, w_{(0)} - \mathbf{D}^\mathrm{T} \sigma_{s(0)} - l_{20}\tau_{xz(0)}, \tag{4.33}$$

$$\sigma_{m(0)} = \mathbf{L}_{10}\, \mathbf{u}_{(0)} + \mathbf{L}_{11}\, w_{(0)}, \tag{4.34}$$

Order ε^{2k} ($k=1, 2, \dots$):

$$w_{(k),z} = -\mathbf{L}_1\, \mathbf{u}_{(k-1)} - l_3\, w_{(k-1)} + (Q/c_{11})\, \sigma_{z(k-2)}, \tag{4.35}$$

$$\mathbf{u}_{(k),z} = -\mathbf{D}\, w_{(k)} + \mathbf{L}_2\, \mathbf{u}_{(k-1)} + \mathbf{S}\, \sigma_{s(k-1)} + z\mathbf{S}\, \sigma_{s(k-2)}, \tag{4.36}$$

$$\sigma_{s(k),z} = -\mathbf{L}_3\, \mathbf{u}_{(k)} - \mathbf{L}_4\, w_{(k)} - \mathbf{L}_5\, \sigma_{s(k-1)} - \mathbf{L}_6\sigma_{z(k-1)} \\ - \mathbf{L}_7\, \sigma_{s(k-2)} - \mathbf{L}_8\sigma_{z(k-2)}, \tag{4.37}$$

$$\sigma_{z(k),z} = \mathbf{L}_9\, \mathbf{u}_{(k)} + l_{19}\, w_{(k)} - \mathbf{D}^\mathrm{T} \sigma_{s(k)} - l_{20}\tau_{xz(k)} - z\mathbf{D}^\mathrm{T} \sigma_{s(k-1)} \\ - zl_{20}\tau_{xz(k-1)} - l_{21}\sigma_{z(k-1)} - l_{22}\sigma_{z(k-2)}, \tag{4.38}$$

$$\sigma_{m_{(k)}} = \mathbf{L}_{10}\,\mathbf{u}_{(k)} + \mathbf{L}_{11}\,w_{(k)} + \mathbf{L}_{12}\sigma_{z_{(k-1)}}\,, \tag{4.39}$$

where $f_{(i)} = 0$ when i is a negative number and f stands for the field variables.

By substituting Eq. (4.29) into Eq. (4.27), we arrive at the lateral boundary conditions on the inner and outer surfaces for various orders as follows.

Order $O(\varepsilon^0)$:

$$\left[\tau_{xz(0)}, \tau_{yz(0)}, \sigma_{z(0)}\right] = \left[0, 0, q_z^+\right] \quad on \quad z = 1, \tag{4.40a}$$

$$\left[\tau_{xz(0)}, \tau_{yz(0)}, \sigma_{z(0)}\right] = \left[0, 0, q_z^-\right] \quad on \quad z = -1. \tag{4.40b}$$

Order $O(\varepsilon^{2k})$ $(k=1, 2, \ldots)$:

$$\left[\tau_{xz(k)}, \tau_{yz(k)}, \sigma_{z(k)}\right] = [0, 0, 0] \quad on \quad z = 1, \tag{4.41a}$$

$$\left[\tau_{xz(k)}, \tau_{yz(k)}, \sigma_{z(k)}\right] = [0, 0, 0] \quad on \quad z = -1. \tag{4.41b}$$

4.4.4 Successive integration and CST

The asymptotic equations in Eqs. (4.30) to (4.33) can be integrated with respect to z in succession. As a result, we obtain at the leading order

$$w_{(0)} = w_0(x, \ y)\,, \tag{4.42}$$

$$\mathbf{u}_{(0)} = \mathbf{u}_0\,(x, \ y) \ - \ z\,\mathbf{D}\,w_0, \tag{4.43}$$

$$\sigma_{s_{(0)}} = -\int_{-1}^{z} \left[\mathbf{L}_3\,(\mathbf{u}_0 \ - \ \eta\,\mathbf{D}\,w_0) + \mathbf{L}_4\,w_0\right]\,d\eta, \tag{4.44}$$

$$\begin{aligned} \sigma_{z_{(0)}} = \int_{-1}^{z} \left[\mathbf{L}_9\,(\mathbf{u}_0 - \eta\,\mathbf{D}\,w_0) + l_{19}\,w_0\right]d\eta \\ + \int_{-1}^{z} \left\{(z - \eta)\,\mathbf{D}^{\mathrm{T}}\,\left[\mathbf{L}_3\,(\mathbf{u}_0 - \eta\,\mathbf{D}\,w_0) + \mathbf{L}_4\,w_0\right]\right\}d\eta \\ + l_{20}\int_{-1}^{z} \left\{(z - \eta)\,\left[\mathbf{L}_{13}\,(\mathbf{u}_0 - \eta\,\mathbf{D}\,w_0) + l_9 w_0\right]\right\}\,d\eta \ + \ q_z^-\,, \end{aligned} \tag{4.45}$$

where $w_0(x,y)$, $\mathbf{u}_0 = \{u_0(x,y) \quad v_0(x,y)\}^{\mathrm{T}}$ represent the displacements on the middle surface and $\mathbf{L}_{13} = [l_5 \quad l_6]$.

By substituting the lateral boundary conditions at the upper surface in Eq. (4.40a) into Eqs. (4.44) to (4.45) and making a simple manipulation, we obtain

$$K_{11}\,u_0 + K_{12}\,v_0 + K_{13}\,w_0 = 0\,, \tag{4.46}$$

$$K_{21}\,u_0 + K_{22}\,v_0 + K_{23}\,w_0 = 0\,, \tag{4.47}$$

$$K_{31}\,u_0 + K_{32}\,v_0 + K_{33}\,w_0 = q_z^+ - q_z^-\,, \tag{4.48}$$

where

$$K_{11} = -\left[A_{11}\partial_{xx} + (A_{11} - A_{12})\partial_{yy}/2s_\phi^2 + A_{11}\bar{c}_\phi\partial_x/s_\phi - A_{12}h/R - A_{11}\bar{c}_\phi^2/s_\phi^2\right]$$

$$K_{12} = -\left[(A_{11} + A_{11})\partial_{yy}/2s_\phi - (3A_{11} - A_{12})\bar{c}_\phi\partial_y/2s_\phi^2\right],$$

$$K_{13} = B_{11}\partial_{xxx} + B_{11}\partial_{xyy}/s_\phi^2 + B_{11}\bar{c}_\phi\partial_{xx}/s_\phi - 2B_{11}\bar{c}_\phi\partial_{yy}/s_\phi^3$$
$$- (A_{11} + A_{12} + B_{12}h/R + B_{11}\bar{c}_\phi^2/s_\phi^2)\partial_x$$

$$K_{21} = -\left[(A_{11} + A_{12})\partial_{xy}/2s_\phi + (3A_{11} - A_{12})\bar{c}_\phi\partial_y/2s_\phi^2\right]$$

$$K_{22} = -\left[(A_{11} - A_{12})\partial_{xx}/2 + A_{11}\partial_{yy}/s_\phi^2 + (A_{11} - A_{12})\bar{c}_\phi\partial_x/2s_\phi\right.$$
$$\left. + (A_{11} - A_{12})(h/R - 2\bar{c}_\phi^2)/2s_\phi^2\right]$$

$$K_{23} = B_{11}\partial_{xxy}/s_\phi + B_{11}\partial_{yyy}/s_\phi^3 + B_{11}\bar{c}_\phi\partial_{xy}/s_\phi^2$$
$$- \left[(A_{11} + A_{12} - B_{11}h/R + B_{12}h/R)/s_\phi\right]\partial_y$$

$$K_{31} = -B_{11}\partial_{xxx} - B_{11}\partial_{xyy}/s_\phi^2 - 2B_{11}\bar{c}_\phi\partial_{xx}/s_\phi - B_{11}\bar{c}_\phi\partial_{yy}/s_\phi^3$$
$$+ (A_{11} + A_{12} + B_{11}h/R + B_{12}h/R + B_{11}\bar{c}_\phi^2/s_\phi^2)\partial_x$$
$$+ (A_{11} + A_{12})\bar{c}_\phi/s_\phi - B_{11}(2\bar{c}_\phi h/R - \bar{c}_\phi^3)/s_\phi^3 + B_{12}\bar{c}_\phi h/Rs_\phi$$

$$K_{32} = -B_{11}\partial_{xxy}/s_\phi - B_{11}\partial_{yyy}/s_\phi^3 + B_{11}\bar{c}_\phi\partial_{xy}/s_\phi^2$$
$$+ \left[(A_{11} + A_{12})/s_\phi - B_{11}h/Rs_\phi^3 - B_{11}h/Rs_\phi + B_{12}h/Rs_\phi\right]\partial_y$$

$$K_{33} = D_{11}\partial_{xxxx} + 2D_{11}\partial_{xxyy}/s_\phi^2 + D_{11}\partial_{yyyy}/s_\phi^4 + 2D_{11}\bar{c}_\phi\partial_{xxx}/s_\phi$$
$$- 2D_{11}c_\phi\partial_{xyy}/s_\phi^3 - (2B_{11} + 2B_{12} + D_{11}h/R + D_{12}h/R + D_{11}\bar{c}_\phi^2/s_\phi^2)\partial_{xx}$$
$$- \left[2(B_{11} + B_{12})s_\phi^2 + D_{12}(h/R - \bar{c}_\phi^2) - D_{11}(3h/R + \bar{c}_\phi^2)\right]\partial_{yy}/s_\phi^4$$
$$+ [2A_{11} + 2A_{12}] - \left[(2B_{11} + 2B_{12} + D_{12}h/R)\bar{c}_\phi/s_\phi\right.$$
$$\left. - D_{11}(2\bar{c}_\phi h/R - \bar{c}_\phi^3)/s_\phi^3\right]\partial_x$$

$$A_{ij} = \int_{-1}^{1} \tilde{Q}_{ij}\, dz\,, \quad B_{ij} = \int_{-1}^{1} z\tilde{Q}_{ij}\, dz\,, \quad D_{ij} = \int_{-1}^{1} z^2\, \tilde{Q}_{ij}\, dz\,.$$

From the governing equations for displacements in Donnell's CST (Donnell, 1976), it is found that the Donnell's CST theory can be used as a first-order approximation to the 3D theory.

Carrying on the analysis to order ε^{2k} by integrating Eqs. (4.35) to (4.38) in succession, we obtain

$$w_{(k)} = w_k(x,\, y) + \phi_{3\kappa}(x,\, y,\, z), \qquad (4.49)$$

$$\mathbf{u}_{(k)} = \mathbf{u}_k(x,\, y) - z\, \mathbf{D}\, w_k + \phi_k(x,\, y,\, z), \qquad (4.50)$$

$$\sigma_{s(k)} = -\int_{-1}^{z} [\mathbf{L}_3(\mathbf{u}_k - \eta\,\mathbf{D}\;w_k) + \mathbf{L}_4 w_k]\,d\eta - \mathbf{f}_k\,(x,\,y,\,z)\,, \qquad (4.51)$$

$$\sigma_{z(k)} = \int_{-1}^{z} [\mathbf{L}_9\,(\mathbf{u}_k - \eta\;\mathbf{D}\;w_k) + l_{19}\,w_k]\,d\eta$$

$$+\int_{-1}^{z}\left\{(z-\eta)\,\mathbf{D}^{\mathrm{T}}\,[\mathbf{L}_3\,(\mathbf{u}_k - \eta\,\mathbf{D}\,w_k) + \mathbf{L}_5\,w_k]\right\} d\eta \qquad (4.52)$$

$$+\,l_{20}\int_{-1}^{z}\left\{(z-\eta)\,[\mathbf{L}_{13}\,(\mathbf{u}_k - \eta\,\mathbf{D}\,w_k) + l_{43}w_k]\right\} d\eta - f_{3k}(x,\,y,\,z),$$

where w_k and $\mathbf{u}_k$ represent the higher-order modifications to the displacements on the middle surface, and

$$\mathbf{u}_k = \{u_k(x,y)\quad v_k(x,y)\}^{T}$$

$$f_{3k}(x,\,y,\,z) = -\int_{-1}^{z} [\mathbf{D}^{\mathrm{T}}\mathbf{f}_k + l_{20}f_{1k} + \mathbf{L}_9\phi_k + l_{19}\phi_{3k} - \eta\mathbf{D}^{\mathrm{T}}\sigma_s^{(k-1)}$$

$$-\,\eta\,l_{20}\,\tau_{xz}^{(k-1)} - l_{21}\,\sigma_z^{(k-1)} - l_{22}\,\sigma_z^{(k-2)}]d\eta$$

$$\mathbf{f}_k = \left\{\begin{array}{c} f_{1k}(x,y,z) \\ f_{2k}(x,y,z) \end{array}\right\}$$
$$= \int_{-1}^{z} \left[\mathbf{L}_3\phi_k + \mathbf{L}_4\phi_{3k} + \mathbf{L}_5\sigma_s^{(k-1)} + \mathbf{L}_6\sigma_z^{(k-1)} + \mathbf{L}_7\sigma_s^{(k-2)} + \mathbf{L}_8\sigma_z^{(k-2)}\right]d\eta,$$

$$\phi_{3k}(x,\,y,\,z) = -\int_{0}^{z} \left[\mathbf{L}_1\,\mathbf{u}_{(k-1)} + l_3\,w_{(k-1)} + (Q/c_{33})\,\sigma_{z(k-2)}\right] d\eta\,,$$

$$\phi_k = \left\{\begin{array}{c} \phi_{1k}(x,\,y,\,z) \\ \phi_{2k}(x,\,y,\,z) \end{array}\right\} = \int_{0}^{z} \left[\mathbf{L}_2\,\mathbf{u}_{(k-1)} + \mathbf{S}\sigma_{s(k-1)} + \eta\mathbf{S}\sigma_{s(k-2)} - \mathbf{D}\,\phi_{3k}\right] d\eta$$

Upon imposing the associated lateral boundary conditions in Eq. (4.41) on Eq. (4.51) and Eq. (4.52), we obtain the CST-type equations with nonhomogeneous terms carried over from the lower-order solution.

$$K_{11}\,u_k + K_{12}\,v_k + K_{13}\,w_k = f_{1k}\,(x,\,y,\,1)\,, \qquad (4.53)$$

$$K_{21}\,u_k + K_{22}\,v_k + K_{23}\,w_k = f_{2k}\,(x,\,y,1)\,, \qquad (4.54)$$

$$K_{31}\,u_k + K_{32}\,v_k + K_{33}\,w_k = f_{3k}\,(x,\,y,1) + \mathbf{D}^{\mathrm{T}}\mathbf{f}_k(x,\,y,1) + l_{20}\,f_{1k}(x,\,y,1).$$
$$\qquad (4.55)$$

4.4.5 Edge conditions

The governing equations at each order level associated with the appropriate edge boundary conditions will compose a well-posed boundary-value problem. However, it is noted that the solution procedure for various order problems can hardly be accomplished by satisfying the edge conditions in Eq. (4.28) point by point on the edge surfaces. A common treatment in the literature is to satisfy the edge conditions in their resultant forms. According to the variational principles for finite deformations (Washizu, 1982), we derive the resultant forms of the edge conditions at each order level instead.

The edge boundary conditions are turned out from

$$\oint_\Gamma \int_{-h}^h [(n_1 \sigma_\phi + n_2 \tau_{\phi\theta} - \bar{p}_\phi)\, \delta u_\phi + (n_1 \tau_{\phi\theta} + n_2 \sigma_\theta - \bar{p}_\theta)\, \delta u_\theta,$$
$$+ (n_1 \tau_{\phi\rho} + n_2 \tau_{\theta\rho} - \bar{p}_\rho)\, \delta u_\rho]\, d\zeta\, d\Gamma = 0 \tag{4.56}$$

where Γ denotes the edge contour of the shell.

The dimensionless form of Eq. (4.56) can be written as

$$\oint_{\bar{\Gamma}} \int_{-1}^1 [(n_1 \gamma_\theta \sigma_x + n_2 \gamma_\phi \tau_{xy} - \gamma_t \bar{p}_x)\delta u + (n_1 \gamma_\theta \tau_{xy} + n_2 \gamma_\phi \sigma_y - \gamma_t \bar{p}_y)\delta v$$
$$+ (n_1 \gamma_\theta \tau_{xz} + n_2 \gamma_\phi \tau_{yz} - \gamma_t \bar{p}_z)\delta w] dz d\bar{\Gamma} = 0 \tag{4.57}$$

where $\bar{\Gamma}$ denotes the edge contour of the mid-surface of the shell, $n_1\, d\Gamma = \gamma_\theta n_1\, d\bar{\Gamma}$, $n_2\, d\Gamma = n_2 \gamma_\phi\, d\bar{\Gamma}$ and $\gamma_\phi = \gamma_\theta = \gamma_t = 1 + h z/R$ in the set of spherical polar coordinates.

By asymptotically expanding the field variables in the form of Eq. (4.29), we can decompose Eq. (4.57) into

$$\oint_{\bar{\Gamma}} \int_{-1}^1 \Big[\big(n_1 \gamma_\theta \sigma_{x(0)} + n_2 \gamma_\phi \tau_{xy(0)} - \gamma_t \bar{p}_x \big)\, \delta u$$
$$+ \big(n_1 \gamma_\theta \tau_{xy(0)} + n_2 \gamma_\phi \sigma_{y(0)} - \gamma_t \bar{p}_y \big)\, \delta v \tag{4.58}$$
$$+ \big(n_1 \gamma_\theta \tau_{xz(0)} + n_2 \gamma_\phi \tau_{yz(0)} - \gamma_t \bar{p}_z \big)\, \delta w \Big]\, dz\, d\bar{\Gamma} = 0$$

$$\oint_{\bar{\Gamma}} \int_{-1}^1 \Big[\big(n_1 \gamma_\theta \sigma_{x(k)} + n_2 \gamma_\phi \tau_{xy(k)} \big)\, \delta u$$
$$+ \big(n_1 \gamma_\theta \tau_{xy(k)} + n_2 \gamma_\phi \sigma_{y(k)} \big)\, \delta v \tag{4.59}$$
$$+ \big(n_1 \gamma_\theta \tau_{xz(k)} + n_2 \gamma_\phi \tau_{yz(k)} \big)\, \delta w \Big]\, dz\, d\bar{\Gamma} = 0$$

At the ε^0-order, based on Eqs. (4.29) and (4.30), the variations of the displacements can be given by

$$\delta u = \delta u_0 - z\delta w_{0,x} \tag{4.60a}$$

$$\delta v = \delta v_0 - z\delta w_{0,y} \tag{4.60b}$$

$$\delta w = \delta w_0 \tag{4.60c}$$

By substituting Eq. (4.60) into Eq. (4.58), we obtain

$$
\oint_{\bar{\Gamma}} \left(n_1 N_{x(0)} + n_2 N_{yx(0)} - \bar{N}_{nx(0)}\right) \delta u_0 \, d\bar{\Gamma}
$$

$$
+ \oint_{\bar{\Gamma}} \left(n_1 N_{xy(0)} + n_2 N_{y(0)} - \bar{N}_{ny(0)}\right) \delta v_0 \, d\bar{\Gamma}
$$

$$
- \oint_{\bar{\Gamma}} \left(n_1 M_{x(0)} + n_2 M_{yx(0)} - \bar{M}_{nx(0)}\right) \delta w_{0,x} \, d\bar{\Gamma} \tag{4.61}
$$

$$
+ \oint_{\bar{\Gamma}} \left(n_1 Q_{x(0)} + n_2 Q_{y(0)} - \bar{Q}_{nz(0)}\right) \delta w_0 d\bar{\Gamma}
$$

$$
- \oint_{\bar{\Gamma}} \left(n_1 M_{xy(0)} + n_2 M_{y(0)} - \bar{M}_{ny(0)}\right) \delta w_{0,y} \, d\bar{\Gamma} = 0
$$

where

$$
[N_{x(0)} \; N_{xy(0)} \; N_{yx(0)} \; N_{y(0)}] = \int_{-1}^{1} [\gamma_\theta \sigma_{x(0)} \; \gamma_\theta \tau_{xy(0)} \; \gamma_\phi \tau_{xy(0)} \; \gamma_\phi \sigma_{y(0)}] dz
$$

$$
[M_{x(0)} \; M_{xy(0)} \; M_{yx(0)} \; M_{y(0)}] = \int_{-1}^{1} z[\gamma_\theta \sigma_{x(0)} \; \gamma_\theta \tau_{xy(0)} \; \gamma_\phi \tau_{xy(0)} \; \gamma_\phi \sigma_{y(0)}] dz
$$

$$
[Q_{x(0)} \; Q_{y(0)}] = \int_{-1}^{1} [\gamma_\theta \tau_{xz(0)} \; \gamma_\phi \tau_{yz(0)}] \, dz
$$

$$
[\bar{N}_{nx(0)} \; \bar{N}_{ny(0)} \; \bar{Q}_{nz(0)}] = \int_{-1}^{1} \gamma_t [\bar{p}_x \; \bar{p}_y \; \bar{p}_z] \, dz
$$

$$
[\bar{M}_{nx(0)} \; \bar{M}_{ny(0)}] = \int_{-1}^{1} z\,\gamma_t [\bar{p}_x \; \bar{p}_y] \, dz
$$

By using the transformation matrix relating the field variables in x and y directions to those in the normal and tangential directions, we can rewrite Eq. (4.61) as

$$
\oint_{\bar{\Gamma}} \left(N_{n(0)} - \bar{N}_{n(0)}\right) \delta u_{n0} \, d\bar{\Gamma} + \oint_{\bar{\Gamma}} \left(N_{s(0)} - \bar{N}_{s(0)}\right) \delta u_{s_0} \, d\bar{\Gamma}
$$

$$
- \oint_{\bar{\Gamma}} \left(M_{n(0)} - \bar{M}_{n(0)}\right) \delta w_{0,n} \, d\bar{\Gamma} - \oint_{\bar{\Gamma}} \left(M_{s(0)} - \bar{M}_{s(0)}\right) \delta w_{0,s} \, d\bar{\Gamma} \tag{4.62}
$$

$$
+ \oint_{\bar{\Gamma}} \left(Q_{n(0)} - \bar{Q}_{n(0)}\right) \delta w_0 \, d\bar{\Gamma} = 0
$$

According to Eq. (4.62), it is observed that each term in Eq. (4.62) represents admissible boundary conditions at each point along the edge contour.

Based on physical considerations, we usually combine the last two boundary conditions of Eq. (4.62) into a single boundary condition. As a result, once the moment is continuous along a smooth contour, Eq. (4.62) can be converted to

$$\oint_{\bar{\Gamma}} \left(N_{n(0)} - \bar{N}_{n(0)} \right) \delta u_{n0} \, d\bar{\Gamma} + \oint_{\bar{\Gamma}} \left(N_{s(0)} - \bar{N}_{s(0)} \right) \delta u_{s_0} \, d\bar{\Gamma} \\ - \oint_{\bar{\Gamma}} \left(M_{n(0)} - \bar{M}_{n(0)} \right) \delta w_{0,n} \, d\bar{\Gamma} + \oint_{\bar{\Gamma}} \left(V_{n(0)} - \bar{V}_{n(0)} \right) \delta w_0 \, d\bar{\Gamma} = 0 \quad (4.63)$$

where $V_{n(0)} = Q_{n(0)} + M_{s(0),s}$, $\bar{V}_{n(0)} = \bar{Q}_{n(0)} + \bar{M}_{s(0),s}$ and the boundary condition $V_n = \bar{V}_n$ is known as the Kirchhoff free-edge condition.

For an annular spherical shell, Eq. (4.63) turns out that the edge conditions at $x=$ constants for the ε^0-order are of the form

$$\text{Specify} \quad u_0 \quad \text{or} \quad N_{x(0)} = \bar{N}_{x(0)}, \quad (4.64a)$$

$$\text{Specify} \quad v_0 \quad \text{or} \quad N_{xy(0)} = \bar{N}_{xy(0)}, \quad (4.64b)$$

$$\text{Specify} \quad w_{0,x} \quad \text{or} \quad M_{x(0)} = \bar{M}_{x(0)}, \quad (4.64c)$$

$$\text{Specify} \quad w_0 \quad \text{or} \quad V_{x(0)} = \bar{V}_{x(0)}. \quad (4.64d)$$

At the higher-order level, the variations of the modifications to the displacements are expressed as

$$\delta u = \delta u_k - z \delta w_{k,x} \quad (4.65a)$$

$$\delta v = \delta v_k - z \delta w_{k,y} \quad (4.65b)$$

$$\delta w = \delta w_k \quad (4.65c)$$

By substituting Eq. (4.65) into Eq. (4.59) and following the similar derivation at the leading order level, we obtain

$$\oint_{\bar{\Gamma}} N_{n(k)} \delta u_{nk} \, d\bar{\Gamma} + \oint_{\bar{\Gamma}} N_{s(k)} \delta u_{s_k} \, d\bar{\Gamma} \\ - \oint_{\bar{\Gamma}} M_{n(k)} \delta w_{k,n} \, d\bar{\Gamma} + \oint_{\bar{\Gamma}} V_{n(k)} \delta w_k \, d\bar{\Gamma} = 0 \quad (4.66)$$

In view of Eq. (4.66), the edge conditions at $x=$ constants for the ε^k-order are of the form

$$\text{Specify} \quad u_k \quad \text{or} \quad N_{x(k)} = 0, \quad (4.67a)$$

$$\text{Specify} \quad v_k \quad \text{or} \quad N_{xy(k)} = 0, \quad (4.67b)$$

$$\text{Specify} \quad w_{k,x} \quad \text{or} \quad M_1^{(k)} = 0, \quad (4.67c)$$

$$\text{Specify} \quad w_k \quad \text{or} \quad Q_1^{(k)} = 0. \quad (4.67d)$$

In the present formulation, it is shown that the 3D theory is reformulated as the recursive sets of the CST governing equations for various orders (Eqs. (4.46) to (4.48) and Eqs. (4.53) to (4.55)) where the differential operators K_{ij} $(i, j= 1, 2, 3)$ remain unchanged at various levels and the nonhomogeneous terms for higher-order level can be calculated from the lower-order solutions. The edge boundary conditions for various orders are derived in the forms of resultants. With the appropriate boundary conditions at edges in Eqs. (4.64) and (4.67), the 3D analysis finally is asymptotically expanded as a series of CST problems to be determined.

Example 4.1 *Isotropic Annular Spherical Shell(I)*
Here we consider the FG isotropic annular spherical shells with simply supported edges under lateral loads for the sake of illustration. The lateral boundary conditions at the inner and outer surfaces of the shell are specified by

$$\{\tau_{xz} \quad \tau_{yz} \quad \sigma_z\} = \{0 \quad 0 \quad \tilde{q}_z^+(x)\cos \tilde{m}y\} on z = 1, \tag{4.68a}$$

$$\{\tau_{xz} \quad \tau_{yz} \quad \sigma_z\} = \{0 \quad 0 \quad 0\} on z = -1, \tag{4.68b}$$

where $\tilde{m} = m\sqrt{h/R}$ and m is a positive integer.

The boundary conditions on two edges, the small edge at $\phi = \phi_0$ and the large edge at $\phi = \phi_1$, are specified as ε^{2k} − order (k= 0, 1, 2, ...):

$$v_k = w_k = N_{x\,(k)} = M_{x\,(k)} = 0. \tag{4.69}$$

When the spherical shell is not truncated, the symmetric conditions at the apex ($\phi = 0$) should be taken into account and given by ε^{2k} − order (k=0, 1, 2, ...):

$$u_k = v_k = w_{k,x} = V_{x\,(k)} = 0. \tag{4.70}$$

We let the displacements of the leading order be of the form

$$u_0 = \tilde{u}_0(x)\cos \tilde{m}y \tag{4.71}$$

$$v_0 = \tilde{v}_0(x)\sin \tilde{m}y \tag{4.72}$$

$$w_0 = w_0(x)\sin \tilde{m}z \tag{4.73}$$

By substituting Eqs.(4.71) to (4.73) into Eqs.(4.46) to (4.48), we have the leading-order equations:

$$\begin{bmatrix} k_{11} & k_{12} & k_{13} \\ k_{21} & k_{22} & k_{23} \\ k_{31} & k_{32} & k_{33} \end{bmatrix} \left\{ \begin{array}{c} \tilde{u}_0(x) \\ \tilde{v}_0(x) \\ \tilde{w}_0(x) \end{array} \right\} = \left\{ \begin{array}{c} 0 \\ 0 \\ \tilde{q}_z^+(x) \end{array} \right\}, \tag{4.74}$$

where

$$k_{11} = -(A_{11}\partial_{xx} + A_{11}\bar{c}_\phi\partial_x/s_\phi - \tilde{m}^2\,(A_{11} - A_{12})\,/2s_\phi^2 - A_{12}h/R - A_{11}\bar{c}_\phi^2/s_\phi^2),$$

$$k_{12} = -\left[\,\tilde{m}\,(A_{11} + A_{12})\,\partial_x/2s_\phi - \tilde{m}\,(3A_{11} - A_{12})\bar{c}_\phi/2s_\phi^2\,\right],$$

$$k_{13} = B_{11}\partial_{xxx} + B_{11}\bar{c}_\phi\,\partial_{xx}/s_\phi - \tilde{m}^2 B_{11}\,\partial_x/s_\phi^2$$
$$-(A_{11} + A_{12} + B_{12}h/R + B_{11}\bar{c}_\phi^2/s_\phi^2)\partial_x + 2\tilde{m}^2 B_{11}\bar{c}_\phi/s_\phi^3,$$

$$k_{21} = \tilde{m}\,\left[\,(A_{11} + A_{12})\,\partial_x/2s_\phi + (3A_{11} - A_{12})\bar{c}_\phi/2s_\phi^2\,\right],$$

$$k_{22} = -[(A_{11} - A_{12}) \, \partial_{xx}/2 + (A_{11} - A_{12}) \, \bar{c}_\phi \partial_x/2s_\phi - \tilde{m}^2 A_{11}/s_\phi^2$$
$$+ (A_{11} - A_{12}) \, (h/R - 2\bar{c}_\phi^2)/2s_\phi^2],$$

$$k_{23} = -\tilde{m} \, B_{11} \, \partial_{xx}/s_\phi - \tilde{m} \, B_{11}\bar{c}_\phi \partial_x/s_\phi^2 + \tilde{m}^3 B_{11}/s_\phi^3$$
$$+ \tilde{m} \, [(A_{11} + A_{12} - B_{11}h/R + B_{12}h/R)/s_\phi],$$

$$k_{31} = -B_{11} \, \partial_{xxx} - 2B_{11}\bar{c}_\phi \, \partial_{xx}/s_\phi + (A_{11} + A_{12} + B_{11}h/R + B_{12}h/R$$
$$+ B_{11}\bar{c}_\phi^2/s_\phi^2)\partial_x + \tilde{m}^2 B_{11} \, \partial_x/s_\phi^2 + (A_{11} + A_{12}) \, \bar{c}_\phi/s_\phi + \tilde{m}^2 B_{11}\bar{c}_\phi/s_\phi^3$$
$$- B_{11}(2\bar{c}_\phi h/R - \bar{c}_\phi^3)/s_\phi^3 + B_{12}\bar{c}_\phi/Rs_\phi,$$

$$k_{32} = -\tilde{m} \, B_{11}\partial_{xx}/s_\phi + \tilde{m} \, B_{11}\bar{c}_\phi\partial_x/s_\phi^2 + \tilde{m}^3 B_{11}/s_\phi^3$$
$$+ \tilde{m} \, \left[(A_{12} + A_{22})/s_\phi - B_{11}h/Rs_\phi^3 - B_{11}h/Rs_\phi + B_{12}h/Rs_\phi \right],$$

$$k_{33} = D_{11} \, \partial_{xxxx} + 2D_{11}\bar{c}_\phi\partial_{xxx}/s_\phi - (2 \, B_{11} + 2B_{12} + D_{11}h/R + D_{12}h/R$$
$$+ D_{11}\bar{c}_\phi^2/s_\phi^2)\partial_{xx} - 2\tilde{m}^2 D_{11} \, \partial_{xx}/s_\phi^2 - [2\tilde{m}^2 D_{11}c_\phi/s_\phi^3 + (2B_{11} + 2B_{12}$$
$$+ D_{12}h/R)\bar{c}_\phi/s_\phi - D_{11}(2\bar{c}_\phi h/R - \bar{c}_\phi^3)/s_\phi^3]\partial_x + \tilde{m}^4 D_{11}/s_\phi^4$$
$$+ \tilde{m}^2 [2(B_{11} + B_{12})s_\phi^2 + D_{12}(h/R - \bar{c}_\phi^2) - D_{11}(3h/R + \bar{c}_\phi^2)]/s_\phi^4$$
$$+ [2A_{11} + 2A_{12}].$$

It is noted that Eq. (4.74) consists of a set of differential equations with variable coefficients. The direct DQ method is used for solving these equations.

The first two governing equations in Eq. (4.74) are applied at the interior points ($i= 2, 3, \ldots, N$-1) and the third governing equation at the interior points ($i= 3, 4, \ldots, N$-2) where N is the total number of the sampling points. Boundary conditions in Eq. (4.69) at the leading order ($k= 0$) are applied at the edges. Finally, a system of $3N$ algebraic equations in terms of $3N$ unknowns (i.e., the mid-surfaces u_0, v_0 and w_0 at all the sampling points) is set up. The resulting equations can be solved by using the Gaussian elimination method. Once u_0, v_0 and w_0 are determined, the displacements of ε^0 order can be furnished by Eqs. (4.42) to (4.43), the transverse shear and normal stresses by Eqs. (4.44) to (4.45), and the in-surface stresses by Eq. (4.34).

Carrying on the solution to order ε^{2k} ($k= 1, 2, 3, \ldots$), we find that the non-homogeneous terms in the ε^{2k}-order equations are

$$f_{1k}(x, \ y, \ 1) = \tilde{f}_{1k}(x, \ z = 1) \ \cos \tilde{m}y, \tag{4.75}$$

$$f_{2k}(x, \ y, \ 1) = \tilde{f}_{2k}(x, \ z = 1) \ \sin \tilde{m}y, \tag{4.76}$$

$$f_{3k}(x, \ y, \ 1) = \tilde{f}_{3k}(x, \ z = 1) \ \cos \tilde{m}y. \tag{4.77}$$

In view of the recurrence of the equations, we let the displacements of ε^{2k}-order be of the following form

$$u_k = \tilde{u}_k(x) \cos \tilde{m}y \tag{4.78}$$

$$v_k = \tilde{v}_k(x) \sin \tilde{m}y \tag{4.79}$$

$$w_k = \tilde{w}_k(x) \cos \tilde{m}y \tag{4.80}$$

By substituting Eqs. (4.71) to (4.73) into Eqs. (4.53) to (4.55), we have

$$
\begin{bmatrix} k_{11} & k_{12} & k_{13} \\ k_{21} & k_{22} & k_{23} \\ k_{31} & k_{32} & k_{33} \end{bmatrix} \begin{Bmatrix} \tilde{u}_k \\ \tilde{v}_k \\ \tilde{w}_k \end{Bmatrix} = \begin{Bmatrix} \tilde{f}_{1k} \\ \tilde{f}_{2k} \\ \tilde{f}_{3k} + \tilde{f}_{3k,x} + \tilde{m}\,\tilde{f}_{2k}/s_\phi + \bar{c}_\phi \tilde{f}_{1k}/s_\phi \end{Bmatrix} \tag{4.81}
$$

By comparing Eq. (4.81) with (4.74), it is found that they have the same coefficient matrix but different nonhomogeneous terms. Since the nonhomogeneous terms in Eq. (4.81) can be calculated from the lower-order solutions, Eqs.(4.81) and (4.74) belong to the same type of mathematics problems. Hence, solution procedure of the DQ method for solving the leading-order problem can be continued to the higher-order problems.

Example 4.2 *Isotropic Annular Spherical Shells (II)*
The static problems of FG isotropic annular spherical shells with simply supported edges are considered again. The annular spherical shells are subjected to the lateral loads (i.e., $q_\zeta^+ = q_\phi^+(\phi)\cos m\theta$) at the outer surface where $q_\phi^+(\phi)$ are uniformly and sinusoidally distributed in the meridian direction. The FGMs consist of stainless steel and silicon nitride. Touloukian's formula (Touloukian, 1967) relating the material properties to the environment temperature is adopted and given by

$$
P = P_0(P_{-1}\,T^{-1} + 1 + P_1\,T + P_2\,T^2 + P_3\,T^3) \tag{4.82}
$$

where $P = E$ or v; P_{-1}, P_0, P_1, P_2 and P_3 are temperature coefficients and those values of stainless steel and silicon nitride are listed in Table 4.8. T denotes the temperature in the working environment. As it is known, the temperature change between the fabrication and working environments will approach the steady state for a long time. Hence, the heat conduction and the temperature-dependent material properties are neglected in the present analysis for simplicity. The stresses and displacements of functionally graded annular spherical shells under the mechanical loads in a particular working temperature are considered.

TABLE 4.8: Properties of materials*

Coefficients	Stainless steel		Silicon nitride	
	$E(N/m^2)$	v	$E(N/m^2)$	v
P_0	201.04×10^9	0.3262	348.43×10^9	0.24
P_{-1}	0	0	0	0
P_1	4.079×10^{-4}	-2.002×10^{-4}	-4.070×10^{-4}	0
P_2	-6.534×10^{-7}	4.790×10^{-7}	2.160×10^{-7}	0
P_3	0	0	-8.946×10^{-11}	0
	207.82×10^9	0.3177	322.4×10^9	0.24

*The properties were evaluated at $T = 300$ K.

The material properties of the FG shell are expressed as

$$E_{fgm} = (E_2 - E_1)\left(\frac{\zeta + h}{2h}\right)^n + E_1 \tag{4.83}$$

$$\upsilon_{fgm} = (\upsilon_2 - \upsilon_1)\left(\frac{\zeta + h}{2h}\right)^n + \upsilon_1 \tag{4.84}$$

where n is the power law exponent.

It is clearly indicated in Eqs. (4.83)-(4.84) that at the inner surface ($\zeta = -h$) the material properties are the same as those of material 1 (i.e., silicon nitride) while at the outer surface ($\zeta = h$) the material properties are the same as those of material 2 (i.e., stainless steel). In addition, the material properties of the FG shell are graded in the thickness direction by following a power law distribution of the volume fractions of the constituents.

The normalized parameters of deflection, stress, force resultant and moment resultant are defined as

$$\bar{w} = \left[10^2 E_1 (2h)^3 / q_0 R^4 (\phi_1 - \phi_0)^4\right] u_\rho \tag{4.85a}$$

$$\bar{N}_\phi = 10^4 N_\phi (2h)^3 / q_0 R^4 (\phi_1 - \phi_0)^4 \tag{4.85b}$$

$$\bar{M}_\phi = 10^4 M_\phi (2h)^2 / q_0 R^4 (\phi_1 - \phi_0)^4 \tag{4.85c}$$

$$(\bar{\sigma}_\phi, \bar{\sigma}_\theta, \bar{\sigma}_\rho) = (\sigma_\phi, \sigma_\theta, \sigma_\rho) / q_0 \tag{4.85d}$$

$$(\bar{\tau}_{\phi\theta}, \bar{\tau}_{\phi\rho}, \bar{\tau}_{\theta\rho}) = (\tau_{\phi\theta}, \tau_{\phi\rho}, \tau_{\theta\rho}) / q_0 \tag{4.85e}$$

The power law exponents are taken as $n = 0, 0.1, 0.5, 1, 5, \infty$, the geometry parameters are $\phi_0 = 10^0$, $\phi_1 = 90^0$ and the ratios of radius-thickness are taken as $R/2h = 5, 10, 20$.

To enhance the efficiency of the asymptotic approach, we develop a multi-layer computational scheme where the FG shell is artificially divided into NL layers. Once the asymptotic solutions of the field variables for various orders are obtained, they are interpolated as NL piecewisely continuous polynomials so that the complicated computation for evaluating the convolution integrals in the present formulation can be avoided and the computation time is saved. The accuracy and convergence of the NL-layer computational scheme are examined below.

(a) Axisymmetric FG Annular Spherical Shells

We consider the axisymmetric static problems of FG annular spherical shells under lateral loads. The external lateral loads are assumed to be independent of the circumferential coordinate and the wave number m is taken as zero. The asymptotic multi-scale DQ solutions are presented in Tables 4.9 to 4.12. Table 4.9 shows asymptotic multi-scale DQ results for middle-surface deflection at the center along the meridian direction with $R/2h = 10$, $n = 0.1$, $q_\phi^+ = -q_0$, and q_0 is constant. It is observed from Table 4.9 that the convergence of the results can be achieved when the number of sampling points $N = 35$ and Gauss points $NG = 26$ in a two-layer computational scheme whereas $N = 35$ and $NG = 20$ in a four-scale computational scheme. The present ε^0-order results are identical to those of Donnell's CST. The relative error between the present convergent solution and the CST one is up to 18%. Tables 4.10 to 4.12 show the CST and the present asymptotic solutions of displacement $(\bar{W})$, moment resultant $(\bar{M}_\phi)$ and force resultant $(\bar{N}_\phi)$ at the center of FG annular spherical shells for different radius-thickness ratio. It is observed that the convergent solutions are furnished at the ε^2-order level for thin shells (i.e., $R/2h = 20$), at the ε^4-order level for moderately thick shells (i.e., $R/2h = 10$) and at the ε^6-order level for thick shells (i.e., $R/2h = 5$). The magnitudes of central deflection of FG annular spherical shells are shown not to lie in between those of isotropic annular spherical shells. The central deflection of FG annular spherical shells increase as the power law exponent n increases from 0 to 0.5, and then decrease as n is larger than 0.5. This is mainly due to the fact that extensional and bending stiffnesses decrease as n increases and coupling stiffness initially increases and later decreases as n increases. This is also reported in the literature (Pradhan et al., 2000).

(b) FG annular spherical shells

The simply supported FG annular spherical shells subjected to the sinusoidally distributed lateral load $q_\rho^+ = q_0 \sin[(\phi - \phi_0)\pi/(\phi_1 - \phi_0)]\cos\theta$, are considered. The geometry parameters of the shell are $\phi_0 = 10^0$, $\phi_1 = 90^0$, $R/2h = 10$. The material properties are the same as those adopted in the above section and $n = 0.2$. The distributions of field variables through the thickness direction are plotted in Figs. 4.5 and 4.6. It is shown that the present leading-order solutions are significantly improved as calculation is performed to the ε^2-order level. It can be observed from Figs. 4.5 and 4.6 that the asymptotic solutions converge rapidly. Also as shown in Figs. 4.5 and 4.6, the field variables are continuous through the thickness direction due to the fact that FGMs gradually and continuously vary the mechanical properties along the thickness direction. The traction conditions at the outer and inner surfaces are satisfied exactly. The magnitude of in-surface stresses is much lager than that of transverse shear stress.

TABLE 4.9: Convergence of displacements of FG annular spherical shells under uniform lateral loads ($n = 0.1R/2h = 10$)

NL	Theory	$NG=16$	$NG=18$	$NG=20$	$NG=22$	$NG=24$	$NG=26$
2	Donnell's CST	2.73863	2.75370	2.77359	2.77873	2.78932	2.79892
	Present ($N= 31$)						
	ε^0	2.73863	2.75370	2.77359	2.77873	2.78932	2.79892
	ε^2	4.34235	4.36074	4.38591	4.39127	4.40420	4.41592
	ε^4	4.36098	4.37951	4.38258	4.41028	4.42329	4.43515
	ε^6	4.35024	4.36874	4.36285	4.39943	4.41241	4.42418
	Present ($N= 33$)						
	ε^0	2.73863	2.75370	2.77359	2.77873	2.78932	2.79892
	ε^2	4.34235	4.36074	4.38591	4.39127	4.40420	4.41592
	ε^4	4.36097	4.37950	4.38257	4.41027	4.42328	4.43510
	ε^6	4.35020	4.36870	4.36215	4.39939	4.41237	4.42414
	Present ($N= 35$)						
	ε^0	2.73863	2.75370	2.77359	2.77873	2.78932	2.78992
	ε^2	4.34235	4.36074	4.38591	4.39127	4.40420	4.41592
	ε^4	4.36096	4.37950	4.38257	4.41026	4.42328	4.43508
	ε^6	4.35017	4.36867	4.36180	4.39937	4.41235	4.42412
4	Donnell's CST	2.78334	2.79742	2.81125			
	Present ($N= 31$)						
	ε^0	2.78334	2.79742	2.81125			
	ε^2	4.40000	4.41720	4.43424			
	ε^4	4.42003	4.43735	4.44029			
	ε^6	4.41022	4.42751	4.42475			
	Present ($N= 33$)						
	ε^0	2.78334	2.79742	2.81125			
	ε^2	4.40000	4.41720	4.43424			
	ε^4	4.42002	4.43734	4.44029			
	ε^6	4.41018	4.42747	4.42422			
	Present ($N= 35$)						
	ε^0	2.78334	2.79742	2.81125			
	ε^2	4.40000	4.41720	4.43424			
	ε^4	4.42001	4.43734	4.44028			
	ε^6	4.41016	4.42745	4.42396			

Note: $\overline{W} = 10^2 E_1 (2h)^3 u_\rho / q_0 R^4 (\phi_1 - \phi_0)^4$

TABLE 4.10: Displacements, moment resultants and force resultants of FG annular spherical shells under uniform lateral loads $(R/2h=20)$

Variables	Theory	Material 1 ($n=\infty$)	Material 2 ($n=0$)	$n=0.1$	$n=0.5$	$n=1$	$n=5$
$\overline{W}(\frac{5\pi}{18}, \theta, 0)$	Donnell's CST	0.94030	0.94820	1.14441	1.22172	1.15355	1.01906
	Present ε^0	0.94030	0.94820	1.14441	1.22172	1.15355	1.01906
	ε^2	1.03630	1.04981	1.26618	1.34902	1.27202	1.12171
	ε^4	1.03455	1.04780	1.26371	1.34638	1.26957	1.11970
	ε^6	1.03378	1.04687	1.26261	1.34528	1.26858	1.11889
$\overline{M}_\phi(\frac{5\pi}{18})$	Donnell's CST	0.01333	0.01627	0.0187	0.02250	0.02290	0.01835
	Present ε^0	0.01333	0.01627	0.0187	0.02250	0.02290	0.01835
	ε^2	-0.01143	-0.01139	-0.00858	-0.00349	-0.00222	-0.00578
	ε^4	-0.01411	-0.01458	-0.01174	-0.00649	-0.0051	-0.00845
	ε^6	-0.01408	-0.01454	-0.01170	-0.00646	-0.00507	-0.00842
$\overline{N}_\phi(\frac{5\pi}{18})$	Donnell's CST	-0.11687	-0.11702	-0.11704	-0.11690	-0.11672	-0.11650
	Present ε^0	-0.11687	-0.11702	-0.11704	-0.11690	-0.11672	-0.11650
	ε^2	-0.12476	-0.12524	-0.12534	-0.12524	-0.12500	-0.12451
	ε^4	-0.12310	-0.12336	-0.12348	-0.12347	-0.12329	-0.12288
	ε^6	-0.12291	-0.12314	-0.12326	-0.12326	-0.12309	-0.12269

Note: $\overline{W} = 10^2 E_1(2h)^3 u_\rho / q_0 R^4 (\phi_1 - \phi_0)^4$; $\overline{M}_\phi = 10^4 M_\phi (2h)^2 / q_0 R^4 (\phi_1 - \phi_0)^4$; $\overline{N}_\phi = 10^4 N_\phi (2h)^3 / q_0 R^4 (\phi_1 - \phi_0)^4$

TABLE 4.11: Displacements, moment resultants and force resultants of FG annular spherical shells under uniform lateral loads ($R/2h = 10$)

Variables	Theory	Material 1 ($n = \infty$)	Material 2 ($n = 0$)	$n = 0.1$	$n = 0.5$	$n = 1$	$n = 5$
$\overline{W}(\frac{5\pi}{18}, \theta, 0)$	Donnell's CST	2.30531	2.32233	2.79742	2.97884	2.81329	2.49486
	Present ε^0	2.30531	2.32233	2.79742	2.97884	2.81329	2.49486
	ε^2	2.79379	2.84081	4.41720	4.62367	4.41244	4.01497
	ε^4	2.80904	2.85814	4.43734	4.64333	4.43008	4.03012
	ε^6	2.80209	2.84979	4.42745	4.63343	4.42117	4.02278
$\overline{M}_\phi(\frac{5\pi}{18})$	Donnell's CST	0.47214	0.51790	0.53792	0.55690	0.54739	0.49260
	Present ε^0	0.47214	0.51790	0.53792	0.55690	0.54739	0.49260
	ε^2	0.30149	0.33023	0.35627	0.38843	0.38457	0.33056
	ε^4	0.24788	0.26694	0.29393	0.32968	0.32842	0.27800
	ε^6	0.24487	0.26298	0.28997	0.32594	0.32492	0.27496
$\overline{N}_\phi(\frac{5\pi}{18})$	Donnell's CST	-1.01350	-1.01343	-1.01130	-1.00688	-1.00544	-1.00777
	Present ε^0	-1.01350	-1.01343	-1.01130	-1.00688	-1.00544	-1.00777
	ε^2	-1.20687	-1.21842	-1.21664	-1.20955	-1.20458	-1.19973
	ε^4	-1.18627	-1.19550	-1.19434	-1.18862	-1.18428	-1.17983
	ε^6	-1.17857	-1.18631	-1.18532	-1.18018	-1.17625	-1.17231

Note: $\overline{W} = 10^2 E_1 (2h)^3 u_\rho / q_0 R^4 (\phi_1 - \phi_0)^4$; $\overline{M}_\phi = 10^4 M_\phi (2h)^2 / q_0 R^4 (\phi_1 - \phi_0)^4$;
$\overline{N}_\phi = 10^4 N_\phi (2h)^3 / q_0 R^4 (\phi_1 - \phi_0)^4$

TABLE 4.12: CPU Time and maximum error for Gauss elimination

Variables	Theory	Material 1 ($n = \infty$)	Material 2 ($n = 0$)	$n = 0.1$	$n = 0.5$	$n = 1$	$n = 5$
$\overline{W}(\frac{5\pi}{18},\theta,0)$	Donnell's CST	4.79901	4.97702	5.98201	6.36833	6.03315	5.39775
	Present ε^0	4.79901	4.97702	5.98201	6.36833	6.03315	5.39775
	ε^2	7.24606	7.36911	8.83240	9.32345	7.78034	7.79820
	ε^4	7.58387	7.75030	9.28073	9.77185	9.18767	8.14658
	ε^6	7.55732	7.71987	9.24461	9.73523	9.15429	8.11874
$\overline{M}_\phi(\frac{5\pi}{18})$	Donnell's CST	6.86724	7.22234	7.32013	7.36058	7.25941	6.90312
	Present ε^0	6.86724	7.22234	7.32013	7.36058	7.25941	6.90312
	ε^2	7.02803	7.43021	7.58904	7.71356	7.61678	7.16067
	ε^4	6.26603	6.56591	6.74684	6.92853	6.86418	6.43658
	ε^6	6.08639	6.34432	6.52882	6.72527	6.67222	6.26104
$\overline{N}_\phi(\frac{5\pi}{18})$	Donnell's CST	-6.84559	-6.75335	-6.72398	-6.72411	-6.76832	-6.88397
	Present ε^0	-6.84559	-6.75335	-6.72398	-6.72411	-6.76832	-6.88397
	ε^2	-10.58589	-10.66312	-10.61243	-10.53931	-10.53188	-10.57946
	ε^4	-10.82339	-10.93870	-10.89200	-10.81415	-10.79667	-10.82165
	ε^6	-10.62367	-10.70958	-10.66947	-10.60727	-10.59822	-10.63017

Note: $\overline{W} = 10^2 E_1(2h)^3 u_\rho / q_0 R^4 (\phi_1 - \phi_0)^4$; $\overline{M}_\phi = 10^4 M_\phi(2h)^2 / q_0 R^4 (\phi_1 - \phi_0)^4$;
$\overline{N}_\phi = 10^4 N_\phi(2h)^3 / q_0 R^4 (\phi_1 - \phi_0)^4$;

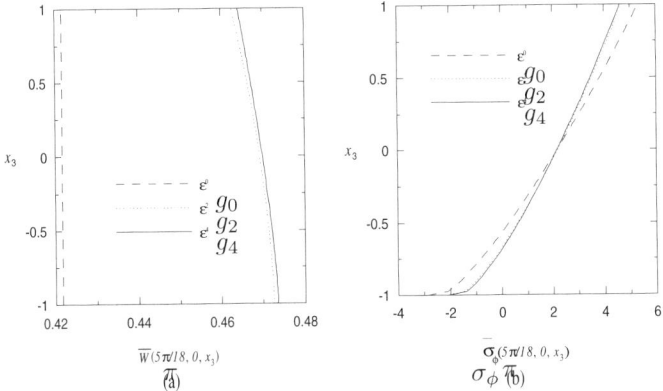

FIGURE 4.5: Distribution of the results through the thickness of a FG annular spherical shell under a uniform lateral load: (a) the central deflection (b) in-surface normal stress.

Herein the asymptotic multi-scale DQ solutions for the static analysis of FG isotropic annular spherical shells are presented to verify the applicability of multi-scale DQ method. By using the asymptotic expansion method, we obtain the recursive sets of Donnell's CST governing equations for various orders. Then the DQ method is adopted to determine the asymptotic solutions. In the illustrative examples, it is shown that the present asymptotic multi-scale DQ solutions converge rapidly. Similar to the case of axisymmetric FG annular spherical shells examined above, the convergent solutions for the general FG annular spherical shells are obtained at the ε^2-order level for thin shells (i.e., $R/2h= 20$), at the ε^4-order level for moderately thick shells (i.e., $R/2h= 10$) and at the ε^6-order level for thick shells (i.e., $R/2h= 5$). Since FGMs change the mechanical properties along the thickness direction in a gradual and continuous manner, the present asymptotic theory yields the continuous distributions of all the field variables along the thickness direction. The traction conditions at the inner and outer surfaces are also exactly satisfied.

It is noted that the FG annular spherical shells would possibly undergo nonlinear buckling and the deformation should be beyond a linear range once the critical buckling loads are reached. More works are required to extend the present asymptotic approach to the analysis of nonlinear problems.

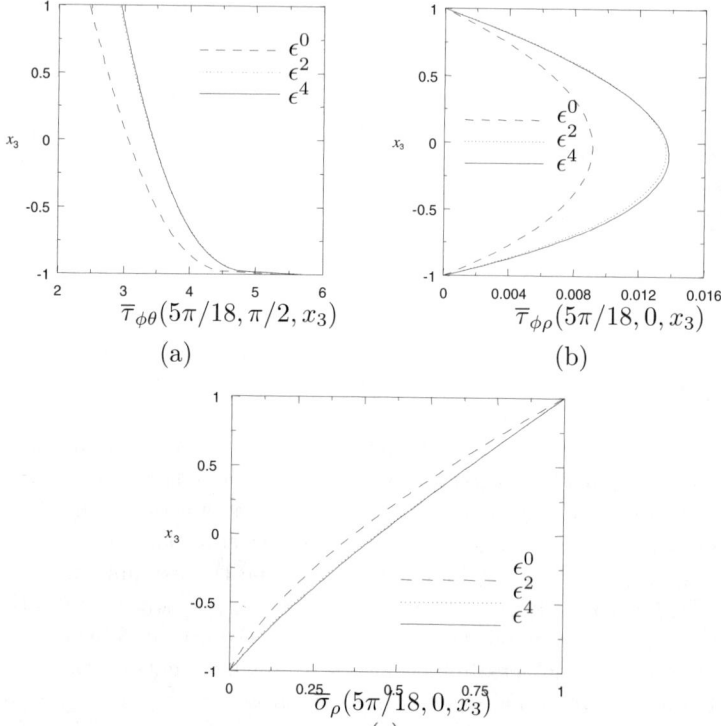

FIGURE 4.6: Distribution of the stresses through the thickness direction of a FG annular spherical shell under a uniform lateral load: (a) the in-surface shear stress (b) the transverse shear stress (c) the transverse normal stress.

4.5   DQ solution to multi-scale poroelastic problems

A poroelastic material consists of an elastic matrix containing interconnected fluid-saturated pores. Poroelasticity describes the deformable behaviors of poroelastic material under loading. It postulates that when a porous material is subjected to loading, the resulting matrix deformation leads to volumetric changes in the pores. Since the pores are fluid-filled, the presence of the fluid not only acts as a stiffener of the material, but also results in the flow of the pore fluid (diffusion) from regions of higher to those of lower pore pressure. If the fluid is viscous the behavior of the material becomes time dependent. On studying soil consolidation, Biot (1941, 1956) proposed a phenomenological model for quasi-static poroelasticity. Since then the theory has been applied to broader fields covering industrial filtering, deformation of arteries and articular cartilage (Cowin, 1999). The literature dealing with poroelasticity based on Biot's classical model is voluminous. The interested reader is referred to Cowin (1999). In this section, we apply multiple scale approach to re-find the governing equations for consolidation process (quasi-static poroelasticity).

Besides fundamental assumptions of fluid mechanics and elasticity, Biot used two more key assumptions in his constructing poroelastic theory. The first assumption is the so-called effective stress principle and the second the famed Darcy's law. The effective stress principle introduced by Terzaghi postulates that the stress at any point in a poroelastic material is the sum of solid stress and fluid pressure. The Darcy's law states that the permeable flow velocity is proportional to the gradient of pore pressure. Although strongly supported by experiments, these two assumptions make some authors feel unsatisfactory. Theoretically speaking, these two assumptions should not have come into poroelastic theory as basic assumptions. The fundamental assumptions of fluid mechanics and elasticity, which are the elements of poroelasticity, should be sufficient to develop the theory. It would be good if we could derive these conclusions rather than accept them as the basic assumptions of the theory. Some attempts are thus directed to deriving the poroelastic equations based on more basic assumptions. Among these attempts are the theory of mixture and the homogenization. In this section we describe a multiple scale approach to deal with poroelasticity.

4.5.1   The multi-scale approach

Multi-scale approach is a very effective method to deal with physical problems which are characterized by the presence of multiple temporal or spatial scales. Taking a poroelastic medium for instance, we are able to roughly define two scales. At the small or microscopic scale, fluid and solid are well separated. There exists clear boundary between them, and both fluid and solid

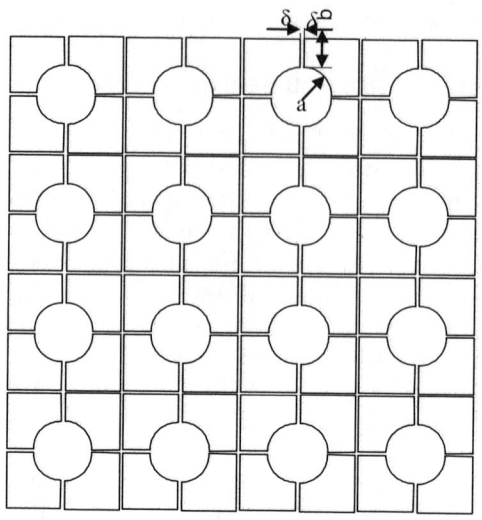

FIGURE 4.7:   Poroelastic medium viewed at two scales.

obey their own physical laws as shown in Fig. 4.7. The typical dimension at this scale is the mean fluid pore size d.

At the large or macroscopic scale, fluid and solid cannot be separated. At a spatial point, both fluid and solid co-exist. Most physical quantities of interest, such as solid deformation and consolidation, happen to be on this scale. At each point at this scale, there are in fact a lot of pores. An analogy to poroelastic medium is a city. Each house in the city is just like a pore. We can go in and go out. But when viewed from the sky, each house becomes a point. We use L to denote the typical dimension at macroscopic scale. The fundamental assumption in a multi-scale analysis is that the ratio of microscopic scale d over the macroscopic scale L is a small quantity, that is,

$$\varepsilon = \frac{d}{L} << 1 \tag{4.86}$$

This small parameter ε helps us to define two sets of coordinate systems

$$X_n = \varepsilon^n x \quad Y_n = \varepsilon^n y \quad n = 0, 1, \cdots \tag{4.87}$$

so a function can be expanded in the form of

$$f(x, y) = f^0(X_0, Y_0) + \varepsilon f^1(X_1, Y_1) + \cdots \tag{4.88}$$

Here we restrict ourselves to two-dimensional theory. A three-dimensional theory can be obtained in the similar way. Hence the changes of a function at two different scales are reflected by their respective components in the two coordinate systems. Further down the road we may obtain the derivatives

$$\frac{\partial}{\partial x} = \frac{\partial}{\partial X_0} + \varepsilon \frac{\partial}{\partial X_1} + \cdots \tag{4.89a}$$

$$\frac{\partial^2}{\partial x^2} = \frac{\partial^2}{\partial X_0^2} + 2\varepsilon \frac{\partial^2}{\partial X_0 \partial X_1} + \cdots \tag{4.89b}$$

Similar equations are obtainable for the derivatives with respect to y. Simply by combining the above equations, we have alternative forms of differentiations

$$\nabla = \left(\frac{\partial}{\partial x}, \frac{\partial}{\partial y} \right), \nabla^0 = \left(\frac{\partial}{\partial X_0}, \frac{\partial}{\partial Y_0} \right) \tag{4.90a}$$

$$\nabla = \nabla^0 + \varepsilon \nabla^1 + \cdots \tag{4.90b}$$

The same argument applies to integration, too. Suppose the integration is performed at the microscopic scale. So fluid and solid are well separated. We use A to denote the integration domain, A_F and A_S denote fluid and solid domain, respectively. It is clear that $A = A_F + A_S$ and their intersection $S = A_F \cap A_S$ is a line set. Note that this is only true at the microscopic scale. At the large or macroscopic scale, these two relations are not valid any more. Therefore the following relation holds

$$O(A_F) = O(A_S) = O(A) = O(\varepsilon^2) \tag{4.91}$$

The porosity is then given by the integration over the fluid domain over the integration domain as shown in Fig. 4.7

$$\phi = \frac{1}{A} \int_{A_F} dA_F \tag{4.92}$$

The integration of a function f over the fluid domain divided by the integration domain area A is

$$\frac{1}{A} \int_{A_F} f dA_F = \frac{1}{A} \int_{A_F} \left[f^0 + \varepsilon f^1 + \cdots \right] dA_F \tag{4.93}$$

The first term on the right hand side can be taken outside of the integral sign since it is a quantity on the large scale independent of the small scale A_F.

$$\frac{1}{A} \int_{A_F} f dA_F = \frac{f^0(X_0, Y_0)}{A} \int_{A_F} dA_F = \phi f^0(X_0, Y_0) + \cdots \tag{4.94}$$

This equation is in fact equivalent to homogenization process, a frequently used approach in biomechanics for poroelastic analysis. But such process is performed automatically and formally through the above formulas. This process is an up-scaling one. The same procedure leads to the following upscaling of integration of derivatives:

$$\frac{1}{A}\int_{A_F}\frac{\partial f}{\partial x}dA_F = \frac{\partial\phi f^0(X_0,Y_0)}{\partial X_0}+\cdots, \quad \frac{1}{A}\int_{A_F}\frac{\partial^2 f}{\partial x^2}dA_F = \frac{\partial^2\phi f^0(X_0,Y_0)}{\partial X_0^2}+\cdots$$
(4.95)

Similar formulas are obtainable for y. These are important equations. Because what we are interested in is the up-scaling process, we will deliberately neglect the superscript "0" in the following sub-sections, but the quantities should be understood as macroscopic ones.

4.5.2 Governing equations of poroelasticity

At the microscopy, fluid and solid are assumed homogeneous and isotropic. Take a microscopic element $A = A_f + A_s$ from the medium. Inside the fluid domain A_F encircled by its boundary S the following mass conservation holds

$$\frac{d}{dt}\int_{A_F}\rho_f dA_F = \int_S \rho_f \dot{u}_n dS$$
(4.96)

where ρ_f is fluid density and $\mathbf{u}$ is solid displacement vector. The over-dot over displacement vector $\mathbf{u}$ denotes derivative with respect to time. Then $\dot{u}_n$ denotes the normal velocity of the solid. During the deformation course, both fluid and solid domains change with time, that is, the integration domain in the above equation changes with time. With this in mind, the transport theorem should be applied when total differentiation operator is removed inside the integral sign, that is,

$$\frac{d}{dt}\int_{A_F}\rho_f dA_F = \int_{A_F}\left[\frac{\partial\rho_f}{\partial t}+\mathbf{v}\bullet\nabla\rho_f\mathbf{v}\right]dA_F$$
(4.97)

where $\mathbf{v}$ is fluid velocity vector. The term on the right hand side of Eq. (4.96) is a surface integral, which can be converted to volume integral through use of the Gauss theorem on surface-volume integral relation,

$$\int_S \rho_f \dot{u}_n dS = -\int_{A_F}\rho_f\nabla\bullet\mathbf{u}dA_F$$
(4.98)

Note that a negative sign is inserted before the integral sign in the above equation due to the fact that S is defined positive for fluid and negative for solid if it points from fluid to solid. Substituting Eqs. (4.97) and (4.98) into Eq. (4.96) yields

$$\int_{A_F}\left[\frac{\partial\rho_f}{\partial t}+\mathbf{v}\bullet\nabla\rho_f\mathbf{v}\right]dA_F = -\int_{A_F}\nabla\bullet\rho_f\dot{\mathbf{u}}dA_F$$
(4.99)

Consolidation process is a very slow one. It is thus reasonable to assume that fluid is incompressible and its motion is linear. Up-scaling of Eq. (4.99), we obtain

$$\frac{\partial \phi}{\partial t} + \nabla \phi \mathbf{v} = -\nabla \phi \dot{\mathbf{u}} \tag{4.100}$$

If steady flow is further assumed, we have

$$\nabla \phi \mathbf{v} = -\nabla \phi \dot{\mathbf{u}} \tag{4.101}$$

which is Biot's mass conservation equation (Biot, 1956).

Besides mass conservation, momentum also conserves during the whole process. The momentum balance equation reads

$$\frac{d}{dt} \int_{A_F} \rho_f \mathbf{v} dA_F = \int_S (-p + \mu \rho_f \nabla^2 \mathbf{v}) dS + \int_S \rho_f \dot{u}_n \dot{\mathbf{u}} dS \tag{4.102}$$

where p is fluid pressure. It states the fact that the acceleration of a fluid element equals to the pressures acted upon the sides of the element and the momentum transferred through the surface S. Applying the transport theorem and Gauss theorem to the above equation, we obtain

$$\int_{A_F} \left[\frac{\partial \mathbf{v}}{\partial t} + v \bullet \nabla \mathbf{v} \right] dA_F = \int_{A_F} (-\frac{\nabla p}{\rho_f} + \mu \nabla^2 \mathbf{v}) dS - \int_{A_F} \nabla \bullet \dot{u} \dot{u} dA_F \tag{4.103}$$

We may go directly to up-scale the above equation as we have done for mass conservation. But our purpose here is to derive Darcy's law. So instead of up-scaling the equation, we proceed still at the microscopic scale. From fluid mechanics, the integrands under the integral signs on both sides of the above equation are identical. Neglecting non-linearity and time-dependence, we obtain the differential form of viscous flow inside the fluid domain

$$\rho_f \mu \nabla^2 \mathbf{v} = \nabla p \quad in \quad A_F \tag{4.104}$$

This is a Poisson equation for velocity $\mathbf{v}$. Its particular solution is

$$\mathbf{v} = \frac{1}{2\pi \rho_f \mu} \int_{A_F} \nabla p \log \sqrt{(x-\xi)^2 + (y-\eta)^2} dA_F(\xi, \eta) \tag{4.105a}$$

Applying the up-scaling formula yields

$$\mathbf{v} = \frac{\nabla p}{2\pi \rho_f \mu} \int_{A_F} \log \sqrt{(X_0-\xi)^2 + (Y_0-\eta)^2} dA_F(\xi, \eta) = -k\nabla p \tag{4.105b}$$

where k is permeability coefficient. This is Darcy's law. Note that Darcy's law is no longer an experimentally determined one. It is obtainable from the above expressions once the microscopic geometries of a poroelastic medium is given. The advantage to do so is that we can mathematically rather than

experimentally determine k. For the poroelastic medium of microscopic structures as shown in Fig. 4.7, it is either numerically or analytically estimated
that

$$k = -\frac{1}{\rho_f \mu} \left[\frac{\pi a^2}{4} (4 \log a - 1) + b\delta(\log b - 1) \right] \tag{4.106}$$

If $a=1$ mm, $b=1$ mm, $\delta=0.01$ mm, $\mu=10^{-6}$, $\rho_f= 1000$ kg/m^3, the permeability coefficient $k \approx 0.3$ mm/s and $\phi \approx 0.2$. It should be pointed out, however,
that Eq. (4.106) holds only for oversimplified cases. A more reasonable model
accounting for random distributions of pore sizes should be used instead for
real cases. It is more helpful to view Eq. (4.106) as an enlightening approximation rather than an exact expression of the physics behind the complicated
permeable flow.

Let's turn to solid skeleton. The solid is assumed isotropic and homogeneous
obeying the following Hook's law

$$\sigma_{ij} = \frac{2G\nu}{1 - 2\nu}\varepsilon\delta_{ij} + G \left(\frac{\partial u_i}{\partial x_j} + \frac{\partial u_j}{\partial x_i} \right) \tag{4.107}$$

The static equilibrium states that the all forces acting on the solid surface
must be zero, in other words,

$$\int_S \sigma_{ij} n_j dS = \int_S p n_i dS \tag{4.108}$$

Applying the transport theorem again we obtain

$$\int_{A_S} \sigma_{ij,j} dA_S = - \int_{A_S} \nabla p dA_S \tag{4.109}$$

Upscaling by use of Eq.(4.95) yields

$$\frac{\partial \sigma_{ij}(1 - \phi)}{\partial x_j} = -\frac{\partial(1 - \phi)p}{\partial x_j} \tag{4.110}$$

Substituting Eq. (4.107) into the above equation and assuming that the porosity is a constant, we obtain

$$\frac{G}{1 - 2\nu}\nabla\varepsilon + G\nabla^2\mathbf{u} + \nabla p = 0 \tag{4.111}$$

where $\varepsilon = \nabla \cdot \mathbf{u}$ is the divergence of displacement vector. This is one of the
governing equations in Biot's theory. To obtain the other one, we need to
employ the following equations given in most of textbooks on elasticity, say
Timoshenko and Goodier (1970). If the pressure gradient is taken as a body
force, it is obtained from Eq. (4.111)

$$\nabla^2\sigma = \nabla^2(\sigma_{11} + \sigma_{22}) = -\frac{1}{1 - \nu}\nabla^2 p \tag{4.112}$$

where σ is the total stress associated with total strain through the following equation

$$\sigma = \kappa \varepsilon \quad \kappa = \begin{cases} \frac{E}{1-\nu} & \text{for plane stress} \\ \frac{E}{(1-2\nu)(1+\nu)} & \text{for plane strain} \end{cases} \quad (4.113)$$

Equation (4.112) permits σ and p being identical up to a harmonic function, that is,

$$\sigma = -\frac{1}{1-\nu}p + \theta \quad (4.114)$$

where θ satisfies the Laplace equation. This resembles Terzaghi's effective stress principle except the coefficient before pressure p. Because the boundary conditions of both total stress σ and pressure p are fixed with time for the current problem, the boundary conditions determined by Eq. (4.114) for θ are also fixed with time. Therefore, θ is a constant independent. Substituting Eqs. (4.113) and (4.114) into the mass conservation equation (4.101), we obtain the following equation governing pressure releasing process

$$\frac{\partial p}{\partial t} = \kappa(1-\nu)\nabla \cdot k\nabla p \quad (4.115)$$

This is a diffusion equation with diffusion (consolidation) constant $C_v = \kappa(1-\nu)k/\mu$. Up to now, we re-obtained all the equations in Biot's theory.

We obtain some interesting conclusions different from previous studies on poroelasticity. First of these is that we have not used Darcy's law and Terzaghi's effective stress principle. On the contrary, Darcy's law is derived from the basic equations. The whole theory is based on the fundamental laws of fluid mechanics and elasticity. All the derivations are formally obtained from the multi-scale approach.

Another pleasant result is that a mathematical expression is obtained for the permeability coefficient k provided that the microstructure of a poroelastic medium is given. Even if the microscopic structure is complicated, a numerical integration may be employed to find the coefficient, at least this theoretically being possible.

Equations (4.111) and (4.115) are the equations governing quasi-static consolidation process based on poroelasticity.

4.5.3 DQ solution of poroelastic governing equations

Suppose the porosity is a constant throughout the field. Then the equations to be solved are those obtained from Eqs (4.111) and (4.115). Here we follow the solution strategy proposed by Terzaghi. Permeation is a slow process. At the beginning when a load is applied to a poroelastic medium, the excess pore pressure sustains all the normal loading force. After that, this excess pore pressure diffuses with time, transferring the loading force to the solid skeleton. Based on these arguments, a three-step solution procedure is proposed.

(a)Pressure initiation

Consider a poroelastic medium domain Ω with boundary Γ composing of three parts. On the first part Γ_s of it, normal loading S is applied. On the second part Γ_q (like free surface), pressure is prescribed. On the third part Γ_n, permeation velocity or pressure gradient along the normal direction is given. Mathematically, we have

$$\nabla \bullet \left(\frac{k}{\mu}\nabla p\right) = 0 \tag{4.116}$$

$$\frac{\partial p}{\partial n} = v_n \text{ on } \Gamma_n \quad p = s \text{ on } \Gamma_s \quad \text{and } p = q \quad \text{on } \Gamma_q \tag{4.117}$$

Solving this Laplace Equation gives the initial pressure distribution on

$$\mathbf{P}^0 = p^0(x, y, 0). \tag{4.118}$$

Discretize

$$p_{ij} = p(x_i, y_j) = \sum_{k=1}^{M}\sum_{l=1}^{N} \lambda_{ik}\lambda_{jl}p_{kl} \tag{4.119}$$

Apply differentiation formulas and we obtain

$$\nabla^2 p_{ij} = p(x_i, y_j) = \sum_{k=1}^{M}\sum_{l=1}^{N} [b_{ik}\lambda_{jl} + \lambda_{ik}b_{jl}]p_{kl} \tag{4.120a}$$

$$\frac{\partial p_{ij}}{\partial n} = \sum_{k=1}^{M}\sum_{l=1}^{N} [a_{ik}\lambda_{jl}\cos(n, x) + \lambda_{ik}b_{jl}\cos(n, y)]p_{kl} = v_{n,ij} \tag{4.120b}$$

$$p_{ij} = \sum_{k=1}^{M}\sum_{l=1}^{N} \lambda_{ik}\lambda_{jl}p_{kl} = s_{ij} \tag{4.120c}$$

Define new index to change the two-dimensional problems into one-dimensional

$$\alpha = (i-1)N + j \quad \beta = (k-1)N + l \quad a, \beta = 1, 2, \cdots MN \tag{4.121a}$$

$$C_{\alpha\beta} = b_{ik}\lambda_{jl} + \lambda_{ik}b_{jl} \quad H_\alpha = 0 \tag{4.121b}$$

$$C_{\alpha\beta} = a_{ik}\lambda_{kl}\cos(n, x) + \lambda_{ik}a_{jl}\cos(n, y) \quad H_\alpha = v_{n,ij} \tag{4.121c}$$

$$C_{\alpha\beta} = \lambda_{ik}\lambda_{jl} \quad H_\alpha = s_{ij} \tag{4.121d}$$

In matrix form, Eq. (4.120) becomes

$$\mathbf{CP} = \mathbf{H} \tag{4.122}$$

where

$$\mathbf{P} = \{P_\alpha\} = \{p_{ij}\} \tag{4.123}$$

Solving Eq (4.122) yields the initial excess pressure distribution $\mathbf{P}^0$.

(b) Pressure diffusion

Both pressure p and displacement u are approximated by their nodal values at the following N nodes. Then the discrete forms of Eq. (4.115) are

$$\frac{\partial p_{ij}}{\partial t} = \sum_{k=1}^{M} \sum_{l=1}^{N} \left[b_{ik}^x \lambda_{jl}^y + \lambda_{ik}^x b_{jl}^y \right] p_{kl} \quad i,k = 1,\dots,M. \quad j,l = 1,\dots,N \tag{4.124a}$$

$$\frac{\partial p_{ij}}{\partial n} = \sum_{k=1}^{M} \sum_{l=1}^{N} [a_{ik}\lambda_{jl}\cos(n,x) + \lambda_{ik}b_{jl}\cos(n,y)]p_{kl} = v_{n,ij} \tag{4.124b}$$

$$p_{ij} = \sum_{k=1}^{M} \sum_{l=1}^{N} \lambda_{ik}\lambda_{jl}p_{kl} = s_{ij} \tag{4.124c}$$

where the weighting coefficients and the Kronecker Delta functions are defined in the same as before. For a time-dependent equation, we also need to discretize time. There have been attempts to apply DQ approach for temporal discretization, but in this section we employ simpler temporal discretization method, fourth-order Runge–Kutta (RK) scheme (see Chapter 8 for more details). In the above equations, DQ approximation was applied to the second order derivative with respect to x. Equation (4.124a) enables us to use RK easily.

(c) Solid skeleton deformation

Once pressure p is determined, the following equilibrium determines the evolution of solid deformation

$$\frac{G}{1-2\nu}\nabla\varepsilon + G\nabla^2\mathbf{u} = -\nabla p^t \tag{4.125}$$

Applying DQ approximation leads to the following discrete form of governing equations

$$\nabla\varepsilon = \left(\sum_{k=1}^{M} \sum_{l=1}^{N} [b_{ik}\lambda_{jl}u_{kl} + a_{ik}a_{jl}v_{kl}], \sum_{k=1}^{N} \sum_{l=1}^{N} [a_{ik}a_{jl}u_{kl} + \lambda_{ik}b_{jl}v_{kl}] \right) \tag{4.126}$$

$$\nabla^2 \mathbf{u} = \sum_{k=1}^{M} \sum_{l=1}^{N} [b_{ik}\lambda_{jl} + \lambda_{ik}b_{jl}]\mathbf{u} \tag{4.127}$$

Denoting

$$D_{\alpha\beta}^{11} = \frac{2G(1-\nu)}{1-2\nu}b_{ik}\lambda_{jl} + G\lambda_{ik}b_{jl} \quad D_{\alpha\beta}^{12} = \frac{G}{1-2\nu}a_{ik}a_{jl} \tag{4.128a}$$

$$D_{\alpha\beta}^{21} = \frac{G}{1-2\nu}a_{ik}a_{jl} \quad D_{\alpha\beta}^{22} = \frac{2G(1-\nu)}{1-2\nu}\lambda_{ik}b_{jl} + Gb_{ik}\lambda_{jl} \tag{4.128b}$$

$$Q_{\alpha}^{1} = \sum_{k=1}^{N} \sum_{l=1}^{N} a_{ik}\lambda_{jl}p_{kl}^{t} \quad Q_{\alpha}^{2} = \sum_{k=1}^{N} \sum_{l=1}^{N} \lambda_{ik}a_{jl}p_{kl}^{t} \tag{4.128c}$$

we have the governing equations for displacements

$$\begin{bmatrix} \mathbf{D}^{11} & \mathbf{D}^{12} \\ \mathbf{D}^{21} & \mathbf{D}^{22} \end{bmatrix} \begin{pmatrix} \mathbf{u}^t \\ \mathbf{v}^t \end{pmatrix} = \begin{pmatrix} \mathbf{Q}^1 \\ \mathbf{Q}^2 \end{pmatrix} \tag{4.129}$$

Solving this equation yields the displacement value at time t. Stress values may be obtained from the following interpolating equations

$$\sigma_{ij}^{x} = \sum_{k=1}^{M} \sum_{l=1}^{N} \left[a_{ik}\lambda_{jl}u_{kl}^{t} - \nu\lambda_{ik}a_{jl}v_{kl}^{t} \right] \tag{4.130a}$$

$$\sigma_{ij}^{y} = \sum_{k=1}^{M} \sum_{l=1}^{N} \left[-\nu a_{ik}\lambda_{jl}u_{kl}^{t} + \lambda_{ik}a_{jl}v_{kl}^{t} \right] \tag{4.130b}$$

$$\tau_{ij} = \frac{G}{2} \sum \sum \left[\lambda_{ik}a_{jl}u_{ij}^{t} + a_{ik}\lambda_{jl}v_{ij}^{t} \right] \tag{4.130c}$$

Example 4.3 *One-dimensional poroelastic material subjected to uniform pressure loading*

Consider a one-dimensional poroelastic material of finite thickness H subjected to uniform pressure p_0 at the top; see Fig. 4.8. As the pressure p_0 is applied, consolidation occurs with pressure inside the poroelastic medium redistributed. The problem is formulated in the following mathematical form

$$p = p_0, \; for \; t = 0 \; and \; 0 \le z \le H \tag{4.131a}$$

$$p = 0, \; for \; 0 < t < \infty \; and \; z = 0 \tag{4.131b}$$

$$\frac{\partial p}{\partial z} = 0, \; for \; 0 \le t < \infty \; and \; z = H \tag{4.131c}$$

$$p = 0, \; for \; t \to \infty \; and \; 0 \le z \le H \tag{4.131d}$$

Applying Fourier transform, we may obtain the exact solution to the above equations

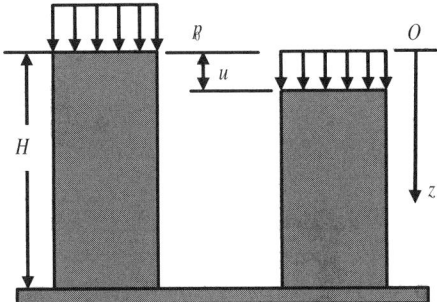

FIGURE 4.8: 1-D consolidation mode.

$$p(z,t) = \frac{4p_0}{\pi} \sum_{m=1,3,5,\ldots}^{\frac{1}{m}} \sin\left(\frac{m\pi z}{2H}\right) \exp\left(-m^2 \frac{\pi^2}{4} \frac{C_V}{H^2} t\right) \qquad (4.132)$$

Consider DQ solution to the problem. Suppose the modulus of elasticity $E = 10^7$ (Pa), Poisson ratio $\nu = 0.3$. Following the procedures introduced in the above, this 1-D poroelastic problem can be solved with numerical results (solid lines) given in Fig. 4.9(a). Figure 4.9(a) shows the slow change of fluid pressure with time. Three cases for pressure at the locations $H=0.109$, 0.05 and 0.0125 are considered. $N =15$ nodes are used. Analytical results (dot) obtained from Eq. (4.132) for these three cases are also plotted in the figure for comparison. The time histories of displacement for the three cases are plotted in Fig. 4.9(b).

Example 4.4 *Two-dimensional consolidation problem*

Consider the consolidation of a two-dimensional poroelastic material occupying the domain $-a \le x \le a; 0 \le y \le H$ as shown in Fig. 4.10. Suppose the material is subjected to a Dirac delta function loading $p_0\delta(x)$, that is, the loading is applied to the point $x = 0$.

The equation governing the pore fluid flow is given in Eq. (4.115) subject to the following boundary conditions:

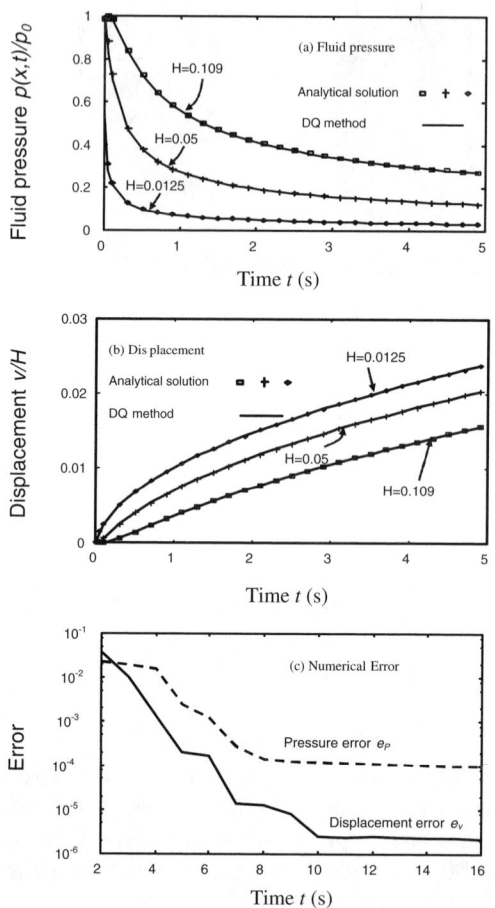

FIGURE 4.9: 1-D poroelastic material subjected to uniform pressure.

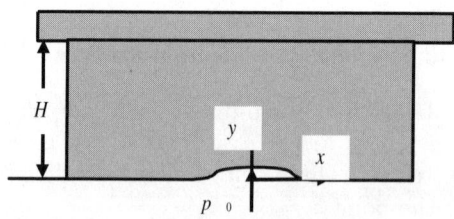

FIGURE 4.10: 2-D consolidation model with delta-function loading.

$$p = p_0 \delta(x), for\, t = 0\, and\, y = 0 \tag{4.133a}$$

$$p = 0, \quad for\, 0 < t < \infty\, and\, y = 0 \tag{4.133b}$$

$$\frac{\partial p}{\partial y} = 0, \quad for\, 0 \leq t < \infty\, and\, y = H \tag{4.133c}$$

$$\frac{\partial p}{\partial x} = 0, \quad for\, 0 \leq t < \infty\, and\, x = \pm a \tag{4.133d}$$

$$p = 0, \quad for\, t \to \infty\, and\, 0 \leq z \leq H \tag{4.133e}$$

The equation governing the horizontal and vertical displacements is given by Equation (4.111) subject to the following boundary condition

$$\frac{\partial u}{\partial y} = \frac{\partial v}{\partial y} = 0, \quad for\, 0 \leq t < \infty y = 0 \tag{4.134a}$$

$$u = v = 0, \quad for\, 0 \leq t < \infty\, and\, y = H \tag{4.134b}$$

$$u = v = 0, \quad for\, 0 \leq t < \infty\, and\, x = \pm a \tag{4.134c}$$

In the above, the numerical values for the known quantities are $p_0 = 4 \times 10^7$Pa, $a = 0.5$ m and $H = 1$ m, $E = 10^7$Pa and $\nu = 0.3$.

Apply the numerical procedures to the above problem and we obtain the corresponding numerical results shown in Fig. 4.11. The three figures in Fig. 4.11 are fluid pressure, horizontal and vertical displacements. Results for pressure and displacements at three points inside the domain are plotted in the figure.

The time histories of the two points on the central vertical axis are similar, but those of the point (0.283,0.095) exhibit significant delay in response to the loading, exhibiting the important role of fluid in transmitting the fluid force from one point to the other.

4.6   Conclusions

Differential quadrature (DQ) method is a highly efficient method that has been efficiently applied to solve many one- or two-dimensional engineering problems. But for the three-dimensional problems, direct extension of DQ method from two-dimension to three-dimension gives rise to great difficulties. From the analysis of potential problem governed by the three-dimensional heterogeneous Laplace equations, it is found that the applicability of the three-dimensional DQ methods is strongly dependent on the choice of methods in the solution procedure (Gauss elimination, one-dimensional band storage and successive over-relaxation (SOR) method and SOR-based multi-scale DQ

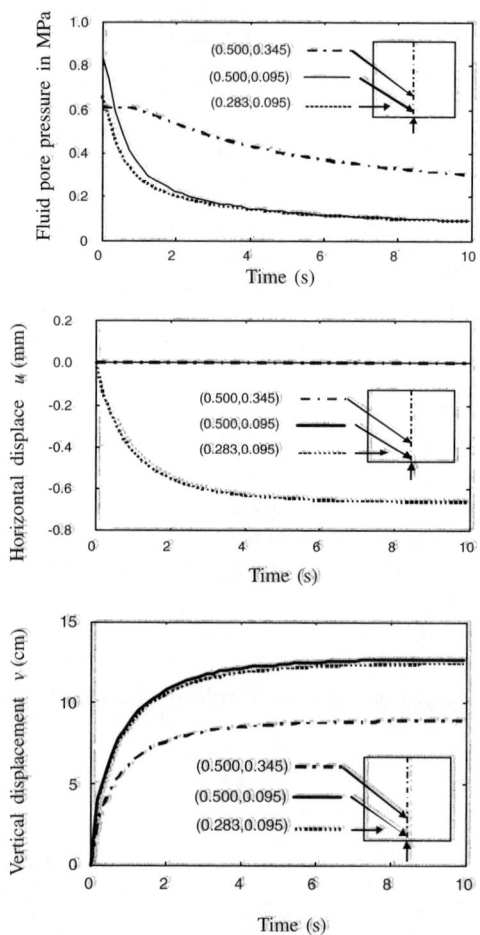

FIGURE 4.11: A two-dimensional poroelastic example.

method). The numerical test has illustrated that direct solution methods such as Gauss elimination are restricted to problems of small mesh size while iterative methods are more capable of solving a problem which needs a large mesh to discretize. In terms of this, the multi-scale DQ method introduced in this chapter shed light on overcoming the computational difficulties. It is evidenced from the numerical solutions that the multi-scale DQ method can speed up iterative convergence rate and two- or three-scales can produce results with high accuracy in much less CPU time.

An alternative approach to solving three-dimensional problems by use of DQ method is to reduce a three-dimensional problem into a two-dimensional one through asymptotic analysis. This applies only to cases where the variation along one direction is much smaller than the other two. Thus the asymptotic multi-scale DQ method is useful for FG analysis.

Poroelasticity itself is a multiple scale problem, whose governing equations can be obtained theoretically if a multiple scale approach is employed. Although we have not solved a three-dimensional proroelastic problem herein, we develop poroelasticity theory in this chapter using multiple scale method.

The poroelastic theory dealt with in this chapter is quasi-static. A dynamic poroelastic theory will be mentioned in Chapter 7.

Appendix A: Expressions of the differential operators

The expressions of the differential operators $\mathbf{L}_i$ ($i= 1\sim12$) are given by

$$\mathbf{L}_1 = [l_1 \quad l_2], \quad \mathbf{L}_2 = \begin{bmatrix} l_4 & 0 \\ 0 & l_4 \end{bmatrix}, \quad \mathbf{L}_3 = \begin{bmatrix} l_5 & l_6 \\ l_7 & l_8 \end{bmatrix}, \quad \mathbf{L}_4 = \begin{bmatrix} l_9 \\ l_{10} \end{bmatrix} \quad (A1)$$

$$\mathbf{L}_5 = \begin{bmatrix} l_{11} & 0 \\ 0 & l_{11} \end{bmatrix}, \quad \mathbf{L}_6 = \begin{bmatrix} l_{12} \\ l_{13} \end{bmatrix}, \quad \mathbf{L}_7 = \begin{bmatrix} l_{14} & 0 \\ 0 & l_{14} \end{bmatrix}, \quad \mathbf{L}_8 = \begin{bmatrix} l_{15} \\ l_{16} \end{bmatrix}, \quad (A2)$$

$$\mathbf{L}_9 = [l_{17} \quad l_{18}], \quad \mathbf{L}_{10} = \begin{bmatrix} l_{23} & l_{24} \\ l_{25} & l_{26} \\ l_{27} & l_{28} \end{bmatrix}, \quad \mathbf{L}_{11} = \begin{bmatrix} l_{29} \\ l_{29} \\ 0 \end{bmatrix}, \quad \mathbf{L}_{12} = \begin{bmatrix} c_{12}/c_{11} \\ c_{12}/c_{11} \\ 0 \end{bmatrix},$$
$$(A3)$$

where

$$l_1 = (c_{12}/c_{11}\,\gamma)\,\partial_x + (c_{12}\bar{c}_\phi/c_{11}\,\gamma\,s_\phi) \qquad (A4a)$$

$$l_2 = (c_{12}/c_{11}\,\gamma\,s_\phi)\partial_y, \quad l_3 = 2c_{12}/(c_{11}\,\gamma), \quad l_4 = 1 - z\partial_z, \qquad (A4b)$$

$$l_6 = (\tilde{Q}_{11} + \tilde{Q}_{12})\partial_{xy}/2s_\phi - (3\tilde{Q}_{11} - \tilde{Q}_{12})\bar{c}_\phi\partial_y/2s_\phi^2, \qquad (A4c)$$

$$l_7 = (\tilde{Q}_{11} + \tilde{Q}_{12})\partial_{xy}/2s_\phi + (3\tilde{Q}_{11} - \tilde{Q}_{12})\bar{c}_\phi\partial_y/2s_\phi^2, \qquad (A4d)$$

$$l_8 = \left(\tilde{Q}_{11} - \tilde{Q}_{12}\right)\partial_{xx}/2 + \tilde{Q}_{11}\partial_{yy}/s_\phi^2 + \left(\tilde{Q}_{11} - \tilde{Q}_{12}\right)\bar{c}_\phi\partial_x/2r \\ - \left(\tilde{Q}_{11} - \tilde{Q}_{12}\right)(h/R - 2\bar{c}_\phi^2)/2s_\phi^2, \qquad (A4e)$$

$$l_9 = (\tilde{Q}_{11} + \tilde{Q}_{12})\,\partial_x, \quad l_{10} = (\tilde{Q}_{11} + \tilde{Q}_{12})\,\partial_y/s_\phi, \quad l_{11} = 3 - 2\,z\,\partial_z \quad (A4f)$$

$$l_{12} = (c_{12}/c_{11})\partial_x, \quad l_{13} = (c_{12}/c_{11}s_\phi)\partial_y, \quad l_{14} = 3z - z^2\partial_z,$$

$$l_{15} = z\,l_{12}, \quad l_{16} = zl_{13}, \quad l_{17} = (\tilde{Q}_{11} + \tilde{Q}_{12})\partial_x + (\tilde{Q}_{11} + \tilde{Q}_{12})\bar{c}_\phi/s_\phi,$$

$$l_{18} = (\tilde{Q}_{11} + \tilde{Q}_{12})\,\partial_y/s_\phi, \quad l_{19} = 2\tilde{Q}_{11} + 2\tilde{Q}_{12}, \quad l_{20} = \bar{c}_\phi/s_\phi,$$

$$l_{21} = 2 - (1/c_{11}) + 2\,z\,\partial_z, \quad l_{22} = 2z - (z/c_{11}) + z^2\partial_z,$$

$$l_{23} = \left[\tilde{Q}_{11}\partial_x + \tilde{Q}_{12}\bar{c}_\phi/s_\phi\right]/\gamma, \quad l_{24} = (\tilde{Q}_{12}/s_\phi)\partial_y/\gamma,$$

$$l_{25} = \left[\tilde{Q}_{12}\partial_x + \tilde{Q}_{11}\bar{c}_\phi/s_\phi\right]/\gamma, \quad l_{26} = (\tilde{Q}_{11}/s_\phi)\partial_y/\gamma,$$

$$l_{27} = \left(\tilde{Q}_{11} - \tilde{Q}_{12}\right)\partial_y/2\,s_\phi\,\gamma, \quad l_{28} = \left[\left(\tilde{Q}_{11} - \tilde{Q}_{12}\right)\partial_x/2 - \left(\tilde{Q}_{11} - \tilde{Q}_{12}\right)\bar{c}_\phi/2s_\phi\right]/\gamma,$$

$$l_{29} = (\tilde{Q}_{11} + \tilde{Q}_{12})/\gamma, \quad \tilde{Q}_{ij} = Q_{ij}/Q, \quad \bar{c}_\phi = \sqrt{h/R}\,\cos\phi, \quad \gamma = 1 + hz/R.$$

Chapter 5

Variable Order DQ Method

Ever since their introduction, DQ methods have attracted attention from researchers worldwide and have triggered a variety of applications to engineering problems (Bellman et al. 1972; Shu, 2000a; Bert et al., 1996a,b,c). Most of the applications are relevant to static or free vibrations. Recent years saw increasing applications of DQ methods to dynamic problems (Fung, 2001; Shu and Kha, 2002). In the dynamic analysis, however, dynamic numerical instability might become a serious problem. As time marches, accumulations of numerical errors may deteriorate the accuracy of the solutions. Numerical stability is an important factor when dynamic problems are studied by using DQ method.

First in this chapter, a simple numerical example is given to demonstrate dynamic numerical instability associated with DQ discretization. It is revealed from the example that grid points near and on boundaries exert dominant influence on dynamic numerical instability, as confirmed by various researchers (Moradi and Taheri, 1998; Quan and Chang, 1989a, b; Bert and Malik, 1996a; Shu et al., 2001). This finding led to the proposal of variable order DQ method in which we distinguish two main classes of nodes (grid points), core nodes and cortical nodes according to their distance from boundaries (Zong, 2003b). Variable order DQ approximations are applied to core and cortical nodes. At core nodes, higher order DQ schemes are employed while at cortical nodes lower order DQ schemes are applied. This variable order approach turns out to be very effective in keeping the balance between the dynamic stability and accuracy. Numerical examples manifest that the approach introduced in this chapter is applicable to linear and highly nonlinear dynamic equations.

Numerical instability associated with dynamic analysis may be improved through use of variable DQ method for spatial discretization or use of a precision integration technique for temporal discretization. This is the content of section 5.3 in this chapter.

5.1 Direct DQ discretization and dynamic numerical instability

Consider a continuous function $f(x,t)$ defined in terms of time t and space x. Suppose on the following N discrete spatial grid points (or nodes)

$$x_1 < x_2 < \ldots < x_N \qquad (5.1)$$

the values of the function $f(x,t)$ are

$$f(x_1,t), f(x_2,t), \ldots, f(x_N,t) \qquad (5.2)$$

The DQ discretization of the r-th order partial derivative with respect to x is given by the following equation as repeatedly quoted

$$\frac{\partial^r f(x_i,t)}{\partial x^r} = \sum_{j=1}^{N} e_{ij}^{(r)} f(x_j,t), \quad i = 1,2,\ldots N, \quad r = 1,2,\ldots N-1 \qquad (5.3)$$

where $e_{ij}^{(r)}$ is the weighting coefficient for the r-th order derivative with respect to x as defined by Eq.(1.46). The derivatives are dependent on the function values on grid points and on spatial grid spacing.

For a dynamic equation, we also need to discretize time. There have been attempts to apply DQ approach for temporal discretization (see Chapter 1), but in this chapter we employ simpler temporal discretization method, fourth-order Runge–Kutta (RK) scheme (see Chapter 8).

To show dynamic numerical instability we first consider the following string vibration equation

$$\frac{\partial^2 u(x,t)}{\partial t^2} = \frac{\partial^2 u(x,t)}{\partial x^2}, \quad 0 \le x \le 1 \qquad (5.4)$$

subject to the following initial and boundary conditions

$$u(x,t=0) = \sin(\pi x), \quad u(x=0,t) = u(x=1,t) = 0 \qquad (5.5)$$

The analytical solution to the above equations is known to be

$$u(x,t) = \sin(\pi x)\cos(\pi t) \qquad (5.6)$$

To apply RK method, we rewrite Eq. (5.4) into the following form of a set of first-order ordinary differential equations

$$\begin{cases} \frac{du_i}{dt} = v_i \\ \frac{dv_i}{dt} = \frac{d^2 u_i}{dt^2} = \sum_{j=1}^{N} e_{ij}^{(2)} u_i \end{cases}, \quad i = 1,\ldots,N \qquad (5.7)$$

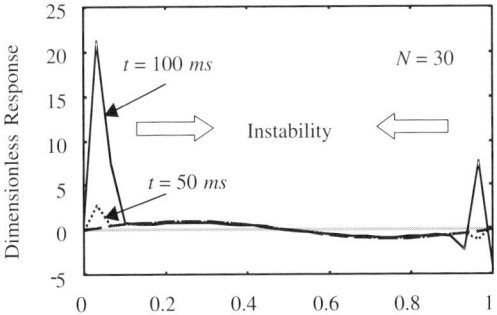

FIGURE 5.1: Dynamic instability in RK-DQ discretization of string vibration equation.

In Eq. (5.7), DQ approximation is applied for the second-order derivative with respect to x: see Eq. (5.3). For this problem we adopted $N = 30$ for spatial discretization. The time step adopted here is $\Delta t = 10^{-4}$. The numerical results obtained from Eq. (5.7) at three time steps ($t = 0$, 50 and 100 milliseconds) are shown in Fig. 5.1. The solid line and dotted line denote the displacement at $t = 100$ms and $t = 50$ms, respectively. It is observed from the figure that the solution becomes unstable very fast. At $t = 50$ms, the solution exhibits divergence at both ends. The instability spreads with time to the central part of the solution domain ($x = 0$) so fast that the solution at time step $t = 100$ ms is completely meaningless. This is a very simple dynamic equation, but it is clearly indicated that dynamic numerical instability may destroy the solution if DQ discretization of spatial variable is not properly treated.

Both spatial and temporal discretizations might contribute to the instability. We leave analysis of the instability due to temporal discretization to the next section. One comment is, however, made about the temporal discretization. We should note that in the above numerical example, the Courant-Friedriches-Levy (CFL) stability condition for time-step is satisfied. The CFL condition on the time-step Δt is a necessary condition for the convergence of an explicit numerical evolution algorithm (Smith, 1985). Suppose c is the wave velocity and Δx is grid spacing. From Eq. (5.4), the wave velocity is $c = 1$ and the grid spacing $\Delta x = 1/N = 1/30 \approx 3.3 \times 10^{-2}$. Then in this example, the CFL condition is satisfied. This demonstrates that temporal discretization-relevant instability, if any, may not be as significant as spatial discretization. Thus, we will restrict ourselves to the analysis of spatial discretization relevant instability.

$$\Delta t = 10^{-4} << 3.3 \times 10^{-2} = \frac{\Delta x}{c} \tag{5.8}$$

In the following section, a simple example is presented to demonstrate how numerical instability is produced.

5.2   Variable order approach

5.2.1   Boundary effect

For the sake of simplicity, we shall focus on DQ discretization only. Consider a cosine function defined on $[0, 2\pi]$

$$f(x) = \cos(2x), \quad 0 \leq x \leq 2\pi \tag{5.9a}$$

The second order derivative is

$$\frac{d^2 f(x)}{dx^2} = -4\cos(2x), \quad 0 \leq x \leq 2\pi \tag{5.9b}$$

The second-order derivative of the function in Eq. (5.9a) can be determined by DQ. Suppose $f(x)$ is approximated by its values f_i ($i = 1, \ldots, N$) on N equidistant grid points $0 = x_1 < \ldots < x_N = 2\pi$. The squared difference between the numerical value and analytical result at i-th node is given by

$$S_i^2 = \left[\sum_{j=1}^{n} a_{ij}^2 f(x_j) + 4\cos(2x_j) \right]^2 \tag{5.10a}$$

The results obtained from Eq. (5.10a) are shown in Fig. 5.2(a) for $N = 20$, 30, 40, 50 and 60. It can be readily seen from Fig. 5.2(a) that the errors are not uniformly distributed in the solution domain. The curves feature low accuracy near the two ends and high accuracy in the central part of the solution domain. This is expected. But such big difference of the errors on the points near boundaries and far away from boundaries is not expected. From Fig. 5.2(a), it is concluded that instability originates from the two ends and spreads fast to the rest of the solution domain. This finding agrees with our previous example (see Eq. (5.7)). Therefore, nodes near boundaries play the dominant role for numerical stability. In order to get a better insight into the instability, three errors are measured. They are defined as follows:

$$S_T^2 = \frac{1}{N} \sum_{i=1}^{N} \left[\sum_{j=1}^{N} a_{ij}^2 f(x_j) + 4\cos(2x_j) \right]^2 \qquad (5.10b)$$

$$S_C^2 = \left[\sum_{j=1}^{n} a_{n/2,j}^2 f(x_j) + 4\cos(2x_{n/2}) \right]^2 \qquad (5.10c)$$

$$S_E^2 = \left[\sum_{j=1}^{n} a_{1j}^2 f(x_j) + 4 \right]^2 \qquad (5.10d)$$

where S_T^2 is the average squared error, S_C^2 is squared error at the center point of x-coordinate and S_E^2 is squared error at point x_1 (left end point). S_E^2 and S_C^2 are components of S_T^2. The logarithmic errors versus the node number N are plotted in Fig. 5.2(b), in which S_T^2, S_E^2, and S_C^2 are denoted by dotted, solid and dashed lines, respectively. From Fig. 5.2(b), it is observed that the curves of S_T^2 and its component S_E^2 are coincident with each other with inappreciable difference, indicating that the error at end points (S_E^2) is the dominant part of the total error (S_T^2) whereas the contribution of the error at the center points (S_C^2) to (S_T^2) is insignificant. It is also clearly demonstrated in Fig. 5.2(b) that there is an optimal choice of node number N ($\approx$30) corresponding to the lowest errors of S_T^2 and S_E^2. The error increases along with N varying away from this optimal value at both sides. In other words, S_T^2 and S_E^2 are decreasing functions of N when $N < 30$ and increasing functions when $N > 30$. Once N is larger than 56, S_T^2 is so large that the numerical values of the derivatives are divergent. As for the insignificant error S_C^2 from center points, it is found that S_C^2 is decreased with increasing N when $N < 30$ and then it is approximately a constant independent of N as shown in Fig. 5.2(b). In summary, the main contribution to the total error S_T^2 comes from the errors at the end points (S_E^2) while the errors from center points S_E^2 are small and can be neglected.

In summary, we conclude from the above simple example that

(1) The accuracy near boundaries dominates the total accuracy;

(2) The accuracy near boundaries might become very poor if grid number N is large;

(3) Very good accuracy can be achieved at nodes far away from boundaries. The accuracy is independent on node number N if N is large enough.

It should be noted that the above conclusions are reached based on the example given in Eq. (5.9). A theoretical analysis for general cases is beyond the scope of the present book. But these conclusions are in agreement with

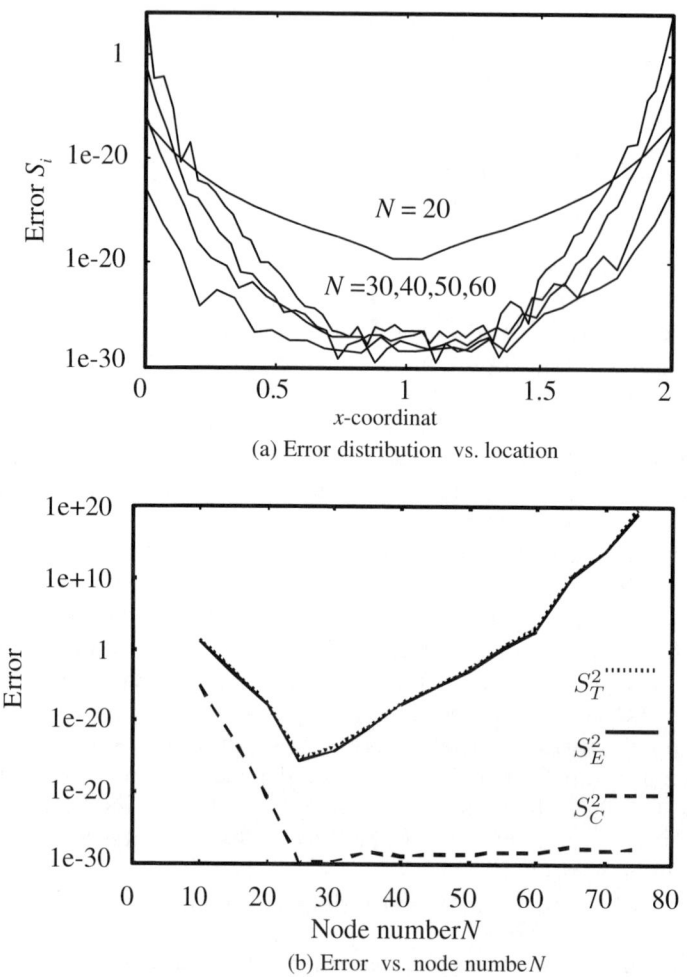

(a) Error distribution vs. location

(b) Error vs. node number N

FIGURE 5.2: Influence of node location and number on errors.

previous studies (Moradi and Taheri, 1998; Quan and Chang, 1989a,b; Bert and Malik, 1996; Shu et al, 2001). These studies and the results in Chapter 1 showed that a non-uniform grid distribution might produce better results when DQ discretization is applied. The commonly used Chebyshev nodes are

$$x_i = \frac{1}{2}\left[1 - \cos\left(\frac{i-1}{N-1}\pi\right)\right]^\alpha \Delta, \quad i = 1, \ldots, N \qquad (5.11)$$

where Δ is the interval length, and α is a positive constant near 1. Note that the above grid distribution is characterized by employing more nodes near boundaries and fewer nodes near the center of the solution domain. For nodes near boundaries the accuracy is low, and thus dense nodes are required. Far away from boundaries, the DQ is of high accuracy and thus lesser nodes are used.

5.2.2 Variable order DQ method

Due to the fact that boundary nodes dominate accuracy and that large grid number may lead to numerical instability, it is possible to distinguish two classes of nodes, *core* and *cortical* nodes. Cortex (cortical) is an anatomical word, which means outer shell or covering. Herein we adopted the word to represent the nodes near and on the boundaries. At a cortical node, the grid number used to approximate the derivatives at that point cannot be too large. Suppose only a small number of nodes, say $M \ll N$, are used to approximate the derivatives at a cortical node as shown in Fig. 5.3. At a core node, however, we still use N nodes because the accuracy at a core node can be well kept even if grid number is large. In Fig. 5.3, another parameter M_P is defined. It denotes transition from a core node to a cortical one. Generally, we have

$$M_P \ll M \ll N \qquad (5.12)$$

We use **B** to denote the set of cortical nodes and **C** to denote the set of core nodes. Then the derivatives are approximated by

$$\frac{\partial^r f(x_i, t)}{\partial x^r} = \sum_{j=1}^{M} e_{ij}^{(r)} f(x_j, t), \quad x_i \in \mathbf{B} \qquad (5.13a)$$

$$\frac{\partial^r f(x_i, t)}{\partial x^r} = \sum_{j=1}^{N} e_{ij}^{(r)} f(x_j, t), \quad x_i \in \mathbf{C} \qquad (5.13b)$$

Or in one-dimensional case, the set **B** and **C** are

$$B = \{x_1, \cdots, x_{M-M_P}\} \cup \{x_{N-M+M_P+1}, \cdots, x_N\} \qquad (5.14a)$$

$$C = \{x_{M-M_P+1}, \cdots, x_{N-M+M_P}\} \qquad (5.14b)$$

Similar formula may be easily established for two-dimensional cases in an analogous manner. Since DQ is equivalent to differentiation using Lagrange

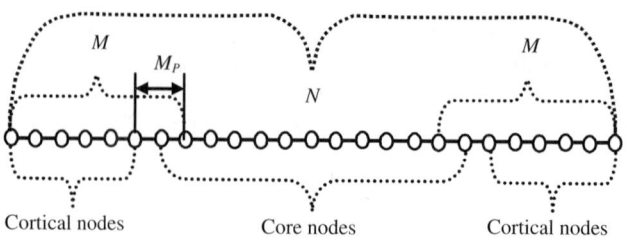

FIGURE 5.3:   Classification of cortical and core nodes.

polynomial, the above equations indicate that Lagrange polynomials of different orders are separately employed to find derivatives at a core node and a cortical node, resulting in a variable order DQ method as defined in Eq.(5.13).

This method is simple, but it does yield encouraging results. To show this point, we take the simple cosine function given by Eq. (5.9a) as an example. The parameters are taken as $N = 55$, $M = 10$ and $M_P = 1$. The results obtained from direct DQ method and from the present scheme are compared in Fig.5.4(a).

In Fig.5.4(a), the analytical results are denoted by dots while the numerical ones from direct DQ and the present variable order DQ (VODQ) are denoted by dotted and solid lines, respectively. At both ends, the numerical results obtained from direct DQ method are divergent while those obtained from the present method are stable. By using more nodes, say $N = 70$, it is found that the direct DQ gives very bad results as shown in Fig. 5.2. Using present variable order DQ, we obtained the second-order derivatives at each node as shown in Fig.5.4(b) denoted by solid lines. The results are stable and in very good agreement with the analytical results.

Figure 5.5 shows the dependence of the error in Eq.(5.10b) on the node number N. In Fig.5.5, the dotted line denotes the error from direct DQ approximation while the solid and dashed lines represent the results obtained by the present variable order DQ by using $M = 6$ and $M = 8$, respectively. It can be clearly seen from Fig.5.5 that the error is significantly reduced by using the variable order DQ method and further reduction can be achieved by increasing N. In other words, the present method developed from the direct DQ with minor manipulation can enhance the accuracy of the results considerably.

Next we investigate the effect of the cortical nodes number M on the stability of the results. The results obtained by adopting three different M are depicted in Fig. 5.6. It is readily seen from Fig. 5.6 that small M (such as $M = 3$ and 4) can give rise to instability of the results, especially at points near the boundaries. So far as there is a lack of theoretical value for M, we may obtain a rough estimate from Figs. 5.5 and 5.6. We suggest $M = 6 \sim 15$ and accordingly $M_P = 0 \sim 5$ based on Eq. (5.12).

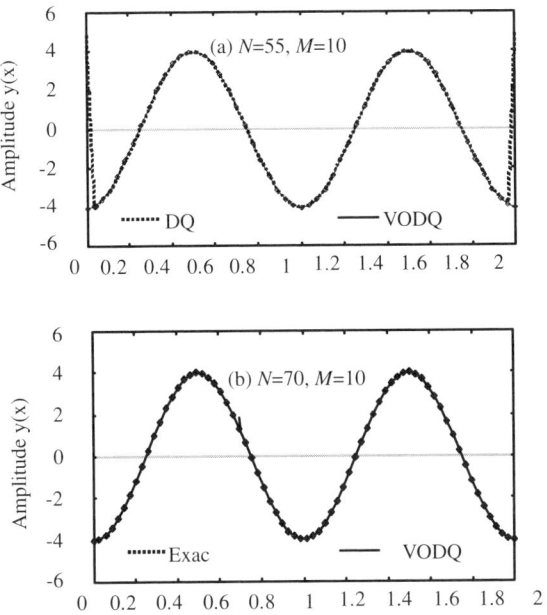

FIGURE 5.4:   Improvement by using variable order DQ (VODQ) schemes to cortical and core nodes.

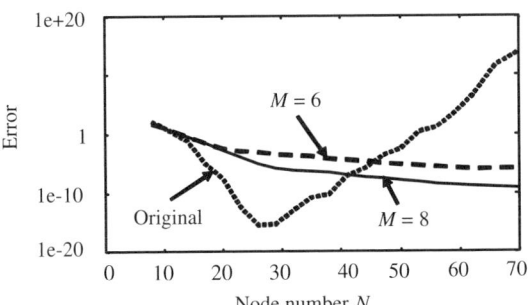

FIGURE 5.5:   Error dependence on N and M.

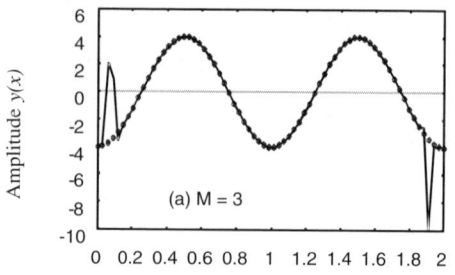

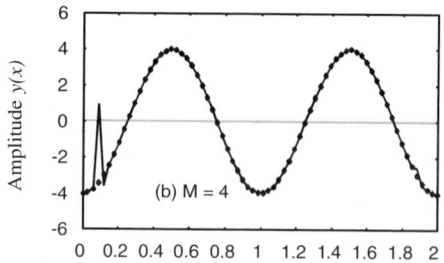

FIGURE 5.6: Dependence of the accuracy on the size M of cortical node set.

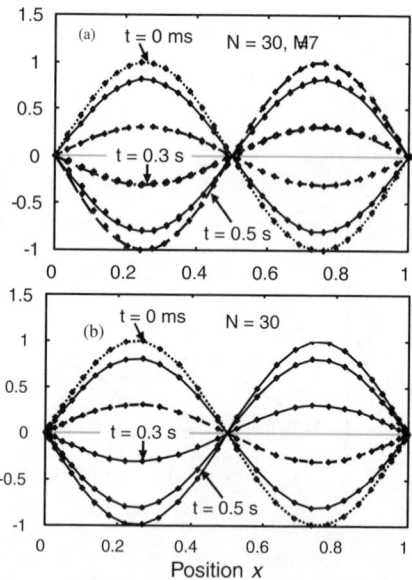

FIGURE 5.7: Position x string vibration by using (a) present approach and (b) direct DQ (non-uniformly spaced nodes).

5.2.3 Numerical examples

In this section, we present some numerical examples to validate the capability of the new variable order DQ method.

Example 5.1 *String Vibration*

First we reconsider the string vibration as expressed in Eq.(5.5). The parameters are taken as: total node number $N = 30$, cortical node $M = 7$, $M_P = 1$ and time step $\Delta t = 10^{-4}$. Both simulation results (solid lines) and analytical solutions (dots) are shown in Fig. 5.7(a) for comparison. It is observed that the numerical results agree well with the analytical ones. Recalling the example of Eq.(5.5), the numerical results obtained by the present method are very satisfactory.

For the sake of comparison, the direct DQ method with non-uniformly spaced grid points is also applied to this problem. The non-uniform grid distribution is given in Eq.(5.11) and the results are shown in Fig.5.7(b). As expected, the numerical results are also in very good agreement with the analytical ones.

Example 5.2 *Scalar Combustion Model*

Next a reaction-diffusion equation is considered. This is a highly nonlinear dynamic equation, in which a shock is formed. It is often used to verify feasibility of a new algorithm. The equation is described by Adjerid and Flaherty (1986) as a model of a single step reaction with diffusion. The equation is given by

$$\frac{\partial u(x,t)}{\partial t} = \frac{\partial^2 u(x,t)}{\partial x^2} + D[1 + a - u(x,t)] \exp(-d/u), -1 \le x \le 1, t > 0$$

$$(5.15a)$$

with the following boundary and initial conditions

$$u(-1,t) = u(1,t) = 1, \quad t > 0$$

$$(5.15b)$$

$$u(x,0) = 1, \quad -1 \le x \le 1$$

$$(5.15c)$$

where $D = Re^d/(ad)$ and R, d, a are constants. The solution represents the temperature $u(x,t)$ of a reactant in a chemical system. For short time, the temperature gradually increases from unity with a "hot spot" forming at $x = 0$. At a finite time ignition occurs, causing the temperature at $x = 0$ to jump to $1+a$. A sharp flame front then forms and propagates toward both ends (i.e., $x = -1$ and $x = 1$) with a speed proportional to $\exp[ad/2(1 + a)]$. In real problems, a is around unity and d is large; thus the flame front moves exponentially fast after ignition. The problem reaches a steady state once the flame propagates to $x = -1$ and $x = 1$.

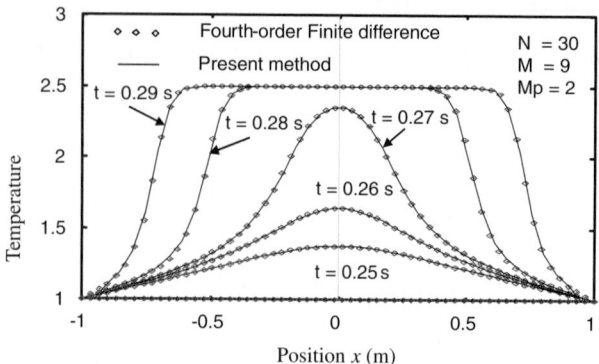

FIGURE 5.8:   Comparison of temperatures obtained by variable order DQ and fourth-order finite difference methods (Chapenter et al., 1994).

Equation (5.15) is solved for $a=1.5$, $d=15$ and $R=5$ by using the present variable order DQ method with $N=30$, $M=9$ and $M_P=2$. The temperatures obtained by using the present method and fourth-order finite difference at different time steps (Chapenter et al, 1994) are illustrated in Fig.5.8. It is readily seen from Fig. 5.8 that the numerical results are in excellent agreement with each other. This is a very difficult problem due to exponential non-linearity (Adjerid and Flaherty, 1986). However, variable order DQ method is capable of finding the solution with relative ease.

An attempt is also made to solve Eq.(5.15) by using direct DQ method with non-uniformly spaced grid points given in Eq.(5.11). The results by using $N=24$ and $N=25$ are depicted in Figs.5.9(a) and (b), respectively. From Fig.5.9(a), it is found that there is discrepancy between the two sets of results, particularly at the center $(x=0)$ at time step $t=0.27$s and at the flame front at $t=0.28$s and $t=0.29$s. The curves in Fig. 5.9(a) given by the direct DQ method are not as smooth as those by the fourth-order finite difference methods. Moreover, accuracy of the direct DQ method cannot be improved by using more nodes because instability would occur once node number N is greater than 25 as shown in Fig. 5.9(b). It is shown in Fig.5.9(b) that the solution by using $N=25$ becomes divergent as early as at time step $t=0.27$s. This example clearly indicates that little can be done to enhance the accuracy of the direct DQ method even if non-uniformly spaced grid points are applied.

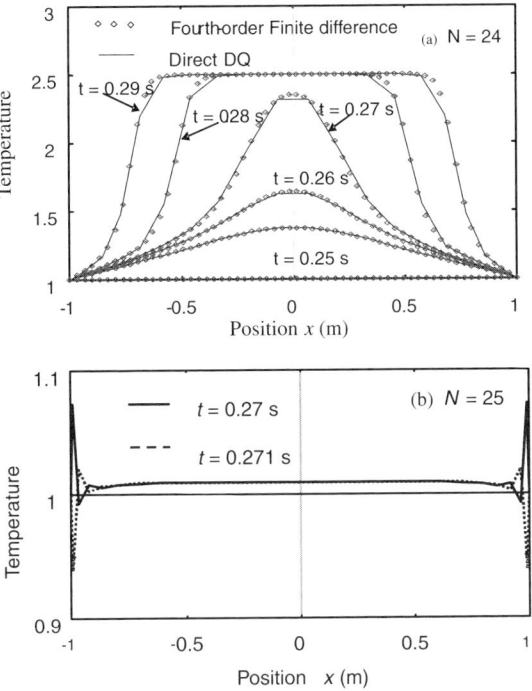

FIGURE 5.9: The scalar combustion model solved using direct DQ method with non-equally spaced grid points (a) $N = 24$ and (b) $N = 25$.

Example 5.3 *Two-Dimensional Scalar Combustion Model*

The two-dimensional scalar reaction model is described by the following equation

$$\frac{\partial u(x,y,t)}{\partial t} = \frac{\partial^2 u(x,y,t)}{\partial x^2} + \frac{\partial^2 u(x,y,t)}{\partial y^2} + D[1 + a - u(x,y,t)]\exp(-d/u),$$
$$-1 \leq x \leq 1, \quad -1 \leq y \leq 1, \quad t > 0$$
$$(5.16a)$$

with the following boundary and initial conditions

$$u(-1,y,t) = u(1,y,t) = 1, \quad u(x,-1,t) = u(x,1,t) = 1 \quad t > 0$$
$$(5.16b)$$

$$u(x,y,0) = 1, \quad -1 \leq x \leq 1, \quad -1 \leq y \leq 1$$
$$(5.16c)$$

The parameters used in the computation are $a=1$, $d=20$ and $R = 5$. It is straightforward to extend Eq.(5.13) to two-dimensional cases. Suppose the two-dimensional solution domain ($a \leq x \leq b, c \leq x \leq d$) are discretized by

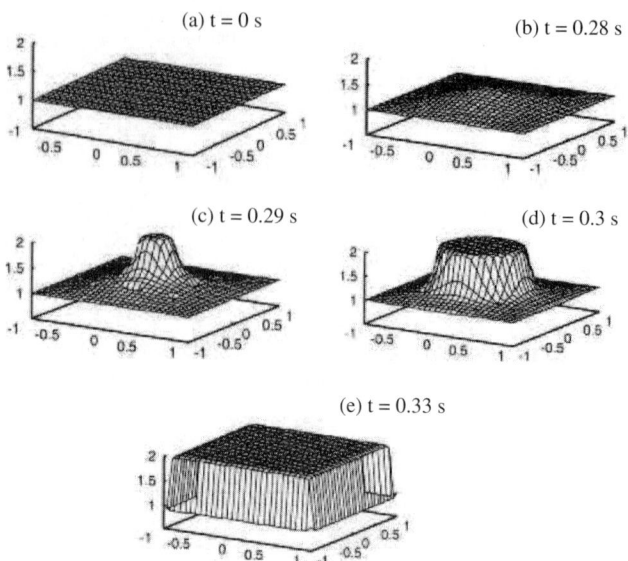

FIGURE 5.10: Temperature distribution at different time steps for two-dimensional reaction model.

$N_x \times N_y$ regular nodes. The direct application of Eq.(5.13) to x and y will yield the partial derivatives in both directions. The variable order DQ method is employed to solve Eq.(5.16) by using $N_x = N_y = 29$, $M = 9$ and $M_P = 2$. The computed temperatures with respect to the positions (x, y) are shown in Fig. 5.10 for several time steps. It can be observed from Fig. 5.10(b) and (c) that a sharp jump of temperature appears from $t = 0.28$ to $t = 0.29$s, similar to one-dimensional case. Later at $t = 0.3$s, a sharp flame front is formed as shown in Fig. 5.10(d). Figure 5.10(e) demonstrates the final steady state, in which the temperature in the whole domain is uniform with the magnitude of $1+a$.

Example 5.4 *Forced Vibration of a Simply-Supported Plate*

So far, the order of partial differential equations in the forgoing examples was not higher than two. In this example, a higher-order (fourth order) partial DQ is considered to verify the applicability of the present method. Consider the forced transverse vibration of a square thin plate. The governing equation is given by

$$\frac{\partial^2 W}{\partial t^2} + \frac{D}{m}\nabla^2\nabla^2 W(x, y; t) = \frac{f(x, y; t)}{m} = g(t)\sin(2\pi x)\sin(\pi y) \quad (5.17a)$$

where $W(x, y, z)$ is vertical displacement, D is stiffness and m is mass per unit area. If the plate is simply supported on the four sides, the boundary conditions are

$$\frac{\partial^2 W}{\partial x^2} = W = 0 \quad at \quad x = 0 \quad and \quad x = 1 \tag{5.17b}$$

$$\frac{\partial^2 W}{\partial y^2} = W = 0 \quad at \quad y = 0 \quad and \quad y = 1 \tag{5.17c}$$

This problem can be analytically solved by using Laplace transform and the analytical solution can be expressed in the form of convolution integral as

$$W(x, y; t) = A(t) \sin(2\pi x) \sin(\pi y) \tag{5.18a}$$

$$A(t) = \int_0^t g(\tau) \cos \beta(t - \tau) d\tau \tag{5.18b}$$

Suppose $g(t)$ is a sinusoidal function of time, i.e., $g(t) = G \sin(\omega t)$. Based on Eq. (5.18b), we have

$$A(t) = -\frac{G\omega \sin \beta t}{\beta(\beta^2 - \omega^2)} + \frac{G \sin \omega t}{\beta^2 - \omega^2} \tag{5.19}$$

The parameters adopted in the computation are: $D/m = 10^{-2}$, $G = 10$ and $\omega = 15$. The node number is $N_x = N_y = 20$, and $M = 7$, $M_P = 0$.

The solution of Eq.(5.17) obtained by using variable order DQ method is plotted in Fig.5.11. Figure 5.11 (a) illustrates the displacement field given by the variable order DQ method at time $t = 0.26$s. For the sake of comparison, the corresponding analytical results at $t = 0.26$s are shown in Fig.5.11(b). From Fig.5.11(a) and (b) it is observed that the numerical and analytical results agree well with each other. Figure 5.11(c) manifests the comparison of time history at the point $(0.25, 0.25)$. It is readily seen that the numerical results given by the present method are coincident with the analytical ones.

As demonstrated in Fig.5.11, the variable order DQ method can yield the vertical displacement for the vibrating plate with high accuracy. Once the displacement is known, the bending moments of the plate can be obtained by using the following relationship

$$M_x = -D\left(\frac{\partial^2 W}{\partial x^2} + \nu \frac{\partial^2 W}{\partial y^2}\right), \quad M_y = -D\left(\nu \frac{\partial^2 W}{\partial x^2} + \frac{\partial^2 W}{\partial y^2}\right) \tag{5.20}$$

where ν is Poisson's ratio of the plate, M_x and M_y are the bending moments about x and y axis, respectively.

Accordingly, the bending moments are calculated by twice differentiations. The variations of the bending moments M_x and M_y with time at point $(0.1, 0.45)$ are illustrated in Fig.5.12(a) and (b), respectively. From the figures,

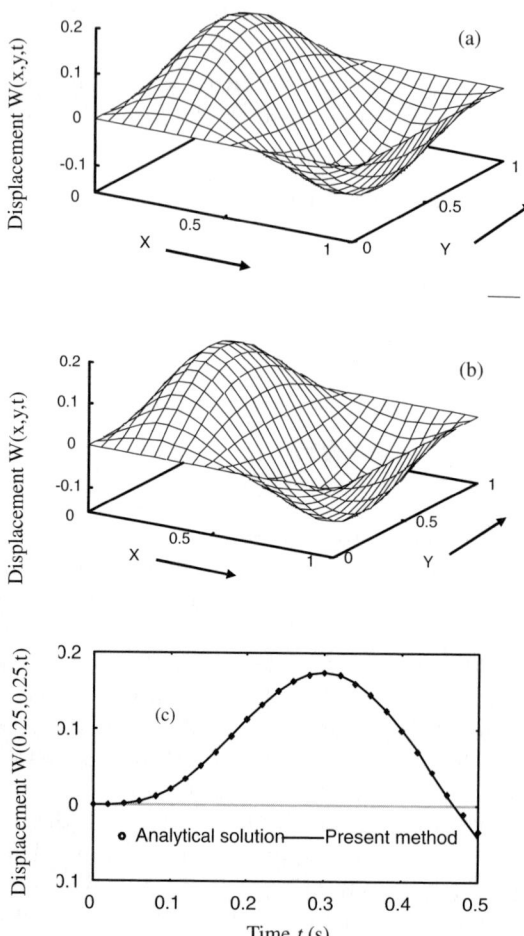

FIGURE 5.11: Vertical displacements at $t = 0.26$s (a) numerical results; (b) analytical solutions and (c) comparison of time history of displacement at point (0.25, 0.25).

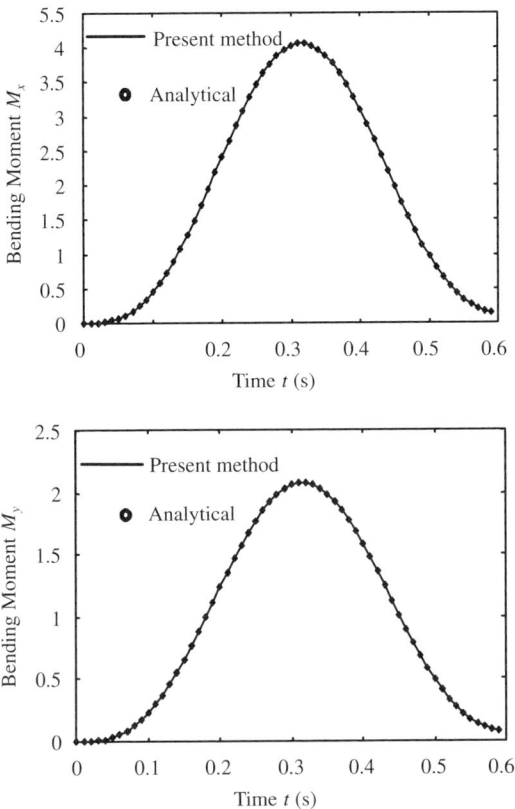

FIGURE 5.12: Time histories of bending moments at point (0.1, 0.45).

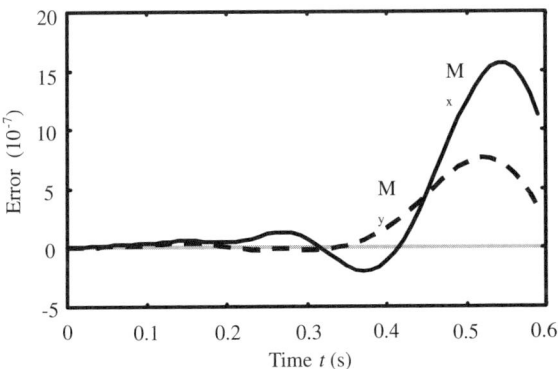

FIGURE 5.13: Error variation with time at point (0.1, 0.45).

the numerical results given by the present method are found to be close to the analytical ones with negligible difference. The time history of the errors, defined by the difference between numerical and analytical solutions, is plotted in Fig. 5.13. The errors grow with time initially and then decrease with time as time exceeds 0.55 seconds. In general, the errors are less than 15×10^{-7}, indicating high accuracy.

Again through this example we demonstrated the good accuracy of DQ method, as confirmed by other researchers in the previous studies.

5.2.4   Discussion

In this section, some potentially controversial issues are discussed. There exists an optimum grid distribution when direct DQ method is applied, which is often characterized by non-uniformity as given in Eq.(5.11). Instead of using a uniform grid distribution in the string vibration problem defined in Eq.(5.5), will a non-uniform distribution in Eq.(5.11) give rise to dynamic numerical instability, too? The answer is yes. Non-uniform distribution can increase accuracy, but cannot avoid or remove dynamic numerical instability completely if the grid number N is large. This can be explained as follows. Differential Quadrature scheme is equivalent to differentiation using Lagrange interpolation. In the theoretical framework of Lagrange interpolation, it can be shown that the divergence near the end of the interval will actually grow exponentially as the number of grid points is increased (Atkinson, 1989). Thus the easiest way is to use a small number of nodes to approximate the derivatives at cortical nodes to prevent the exponential growth.

Classification of core and cortical nodes is crucial in the application of the present variable order approach. So far, no theoretical works have been done to determine the magnitudes of M and M_P. But the rule of thumb given in the previous section (i.e., $M = 6 \sim 15$ and $M_P = 0 \sim 5$) will suffice practical applications. M_P plays the role of smooth transition from a cortical node to a core one. $M_P = 0$ means a hard transition while a larger M_P means a soft and smooth transition. It is not necessary to use large M_P. The suitable M_P is in the range of 0 and 5. We will explore this further in Chapter 7 where more powerful techniques will be developed based on such ideas.

It is worth mentioning that the numerical examples analyzed in this chapter are under the same type of boundary conditions, that is, the function values at the boundary points are given. This can be implemented with ease. At each time step in using Runge–Kutta scheme, the function values on boundaries are set to the given values. For general cases, however, more efforts are demanded to implement different boundary conditions. This problem has been introduced in Chapter 1.

5.3 Improvement of temporal integration

Another approach for improving the numerical accuracy of dynamic problems is to use an alternative temporal integration algorithm other than the Runge–Kutter approach, such as the block-marching with DQ discretization introduced in Chapter 1 and the precise integration introduced below (Ma and Qin 2005).

The dynamic responses of structures are usually governed by second-order differential equations, which are commonly analyzed by means of direct integration or step-by-step schemes such as the Newmark, Wilson-θ, RK or Houbolt schemes, and explicit methods such as the central difference scheme. Recently, a high-precision integration scheme was proposed by Zhong (1994) and Zhong and Williams (1994) for dynamics. Here "precision integration" is specially referred to as a class of time integration procedure (Lin et al., 1995, 1997) which uses a recurrence formula to reduce the computing effort and simplifies the exponential matrix method. It is a numerical rather than an analytical method with which the machine precision can be achieved for the solution of the homogeneous part on ordinary computers. The high precision integration is an unconditionally stable explicit method, which has been described in several research publications, exhibiting high precision and efficiency when compared with traditional methods, such as Newmark method (Lin et al., 1995). However, these high-precision integration schemes are based on a system of first-order differential equations, so that second-order differential equations need to be transformed into first-order equations before the numerical schemes can be performed. This would lead to a large system matrix, and as a consequence, a less efficient scheme. In the present chapter, precision integration is extended to direct solution of the second-order algebraic and differential equations, whereby efficiency and accuracy can be further improved. This is of significant importance especially for dynamic analysis with an emphasis on long-term dynamic evolution. The sine and cosine matrices involved in the second-order scheme are calculated using the so-called 2^N algorithm. The corresponding particular solution is also presented incorporated with the second-order scheme, where the excitation vector is approximated by the truncated Taylor series. Numerical tests show that both the efficiency and the accuracy of the homogeneous solution can be enhanced considerably by using the proposed second-order scheme.

The equation of motion of a discretized structural model can be written as a second-order algebraic and differential equation in matrix form as

$$\mathbf{M\ddot{u}} + \mathbf{B\dot{u}} + \mathbf{Ku} = \mathbf{f}(t) \tag{5.21}$$

with initial conditions

$$\mathbf{u}(t_0) = \mathbf{u}_0, \quad \dot{\mathbf{u}}(t_0) = \dot{\mathbf{u}}_0 \tag{5.22}$$

where $\mathbf{M}$, $\mathbf{B}$, and $\mathbf{K}$ represent time-invariant mass, damping and stiffness matrices with order n of the structure, respectively; $\mathbf{f}(t)$ is the excitation vector. $\mathbf{M}$ is assumed positively definite. The dot over a variable indicates differentiation of the state variable with respect to the time t.

5.3.1 The first-order scheme

By introducing the new variable $\mathbf{v} = \{\mathbf{u}^{\mathrm{T}}, (\mathbf{Mu} + \mathbf{B\dot{u}}/2)^{\mathrm{T}}\}^{\mathrm{T}}$, Eq.(5.21) is first transformed into a form of first-order algebraic and differential equations as

$$\dot{\mathbf{v}} = \mathbf{Hv} + \mathbf{r} \tag{5.23}$$

where

$$\mathbf{r} = \begin{Bmatrix} 0 \\ \mathbf{f} \end{Bmatrix} \qquad \mathbf{H} = \begin{bmatrix} \dfrac{-\mathbf{M}^{-1}\mathbf{B}}{2} & \mathbf{M}^{-1} \\ \dfrac{\mathbf{B}\mathbf{M}^{-1}\mathbf{B}}{4} - \mathbf{K} & \dfrac{-\mathbf{B}\mathbf{M}^{-1}}{2} \end{bmatrix} \tag{5.24}$$

Since H is a constant matrix, from the theory of ordinary differential equations, the homogeneous solution of Eq.(5.23) can be expressed as

$$\mathbf{v}_h = \exp\left[\mathbf{H}\left(t - t_0\right)\right]\mathbf{c_0} \tag{5.25}$$

where $\mathbf{c_0}$ is a constant vector to be determined by initial conditions. Let $\mathbf{v}_p$ be the particular integral of the inhomogeneous term, then the general solution of Eq. (5.23) has the form of

$$\mathbf{v} = \mathbf{v}_h + \mathbf{v}_p = \exp\left[\mathbf{H}\left(t - t_0\right)\right]\mathbf{c_0} + \mathbf{v}_p \tag{5.26}$$

The vector $\mathbf{c_0}$ can then be readily determined from Eq. (5.26) by taking $t = t_0$ as $\mathbf{c_0} = \mathbf{v}\left(t_0\right) - \mathbf{v}_p\left(t_0\right)$, therefore

$$\mathbf{v} = \exp\left[\mathbf{H}\left(t - t_0\right)\right]\left[\mathbf{v}\left(t_0\right) - \mathbf{v}_p\left(t_0\right)\right] + \mathbf{v}_p \tag{5.27}$$

Letting Δt be the constant time step size, then we can write the explicit recursive formula for the general solution at $(k+1)$-th time step, that is, $t = t_0 + (k+1)\Delta t$ as

$$\mathbf{v}_{k+1} = \exp\left(\mathbf{H}\Delta t\right)\left[\mathbf{v}_k - \mathbf{v}_{p,k}\right] + \mathbf{v}_{p,k+1}, \quad (k = 0, 1, ...) \tag{5.28}$$

The key point of the precision integration is the calculation of $\exp\left(\mathbf{H}\Delta t\right)$ as accurately as possible (Zhong and Williams, 1994) since the accuracy of Eq. (5.28) depends on how accurately the exponential matrix function $\exp\left(\mathbf{H}\Delta t\right)$ can be evaluated. Because of its wide application, the calculation of the matrix exponential has been discussed in the paper by Deif (1991). Zhong and Williams (1994) proposed an accurate and efficient method for the calculation of $\exp\left(\mathbf{H}\Delta t\right)$ also called 2^N algorithm, which is

$$\exp\left(\mathbf{H}\Delta t\right) = \left[\exp\left(\mathbf{H}r\right)\right]^m \tag{5.29}$$

where $r = \Delta t/m$, $m = 2^N$, and N is a positive integer. It has been argued (Zhong and Williams, 1994) that precision integration can achieve machine accuracy because r is, in general, very small, even for a small value of N. For example, if $N = 20$ then $r = \Delta t/1048576$. In view of structural dynamics, this value of r is much smaller than the highest modal period of any conventional idealized structure. However, this first-order scheme will make the system matrix relatively large. For example, if the size of M, B, and K in Eq. (5.21) is $n \times n$, then the size of H becomes $2n \times 2n$. Therefore, the efficiency of the method can be further improved, especially in the case of long-term integration, by introducing the following second-order scheme.

5.3.2 The second-order scheme

We now discuss in detail the second-order scheme for the homogeneous solution of Eq.(5.21) by assuming $\mathbf{f}\left(t\right) = 0$ as follows:

$$\mathbf{M}\ddot{\mathbf{u}}_h + \mathbf{B}\dot{\mathbf{u}}_h + \mathbf{K}\mathbf{u}_h = 0 \tag{5.30}$$

with the initial conditions

$$\mathbf{u}_h\left(t_0\right) = \mathbf{u}_0, \quad \dot{\mathbf{u}}_h\left(t_0\right) = \dot{\mathbf{u}}_0 \tag{5.31}$$

Suppose that the homogeneous solution has the form $\dot{\mathbf{u}}_h = \exp\left(\mathbf{A}t\right)\mathbf{a}$ and insert it into Eq.(5.30), where $\mathbf{a}$ denotes an arbitrary vector. We have the characteristic matrix equation as

$$\mathbf{M}\mathbf{A}^2 + \mathbf{B}\mathbf{A} + \mathbf{K} = 0 \tag{5.32}$$

The solution of Eq. (5.32) is

$$\mathbf{A} = \mathbf{D} \pm i\boldsymbol{\Omega} \tag{5.33}$$

where i is the imaginary number $(i = \sqrt{-1})$. Matrices $\mathbf{D}$ and $\boldsymbol{\Omega}$ are expressed as follows, respectively:

$$\mathbf{D} = \frac{1}{2}\mathbf{M}^{-1}\mathbf{B} \tag{5.34}$$

$$\boldsymbol{\Omega} = \sqrt{\mathbf{J}} = \sqrt{\frac{\mathbf{M}^{-2}\mathbf{B}}{4} + \mathbf{M}^{-1}\mathbf{K}} \tag{5.35}$$

Matrix $\boldsymbol{\Omega}$ is the square root of matrix $\mathbf{J}$, which can be computed by the technique of matrix decomposition

$$\mathbf{J} = \mathbf{Q}\mathbf{\Lambda}\mathbf{Q}^{-1} = -\mathbf{Q}\left(i\sqrt{\mathbf{\Lambda}}\right)\mathbf{Q}^{-1}\mathbf{Q}\left(i\sqrt{\mathbf{\Lambda}}\right)\mathbf{Q}^{-1} = -\left(i\mathbf{\Omega}\right)^2 = \mathbf{\Omega}^2, \ (\lambda_k < 0)$$
$$\mathbf{J} = \mathbf{Q}\mathbf{\Lambda}\mathbf{Q}^{-1} = \mathbf{Q}\sqrt{\mathbf{\Lambda}}\mathbf{Q}^{-1}\mathbf{Q}\sqrt{\mathbf{\Lambda}}\mathbf{Q}^{-1} = \mathbf{\Omega}^2. \ (\lambda_k > 0)$$

$$(5.36)$$

That is

$$\mathbf{\Omega} = \mathbf{Q}\sqrt{\mathbf{\Lambda}}\mathbf{Q}^{-1} \tag{5.37}$$

where $\mathbf{\Lambda}$ stands for the diagonal matrix which is composed of the eigenvalues of J as follows:

$$\mathbf{\Lambda} = \begin{bmatrix} |\lambda_1| & & \\ & \ddots & \\ & & |\lambda_n| \end{bmatrix} \tag{5.38}$$

and the kth column in $\mathbf{Q}$ represents the eigenvector corresponding to the eigenvalue λ_k. The eigenvalues and eigenvectors of matrix $\mathbf{J}$ can be computed numerically using the subroutines described in Smith et al., (1974). Here an assumption has to be made that all the eigenvalues λ_k ($k = 1, \ldots, n$) are distinct, real, and nonzero ($\lambda_k \neq 0$). With the theory of ordinary differential equations (Deif, 1991; Bugl, 1995), the homogeneous solution of Eq. (5.30) can be expressed as

$$\mathbf{u}_h = \exp\left[-\mathbf{D}\left(t - t_0\right)\right]\left[\cos\mathbf{\Omega}\left(t - t_0\right)\mathbf{c}_1 + \sin\mathbf{\Omega}\left(t - t_0\right)\mathbf{c}_2\right] \tag{5.39}$$

where $\mathbf{c}_1$ and $\mathbf{c}_2$ represent constant vectors to be determined by initial conditions in Eq. (5.31). Therefore, we have

$$\begin{aligned}\mathbf{u}_h = \exp\left[-\mathbf{D}\left(t - t_0\right)\right]&\left\{\left[\cos\mathbf{\Omega}\left(t - t_0\right) + \mathbf{\Omega}^{-1}\mathbf{D}\sin\mathbf{\Omega}\left(t - t_0\right)\right]\mathbf{u}_0 \right.\\ &\left. + \mathbf{\Omega}^{-1}\sin\mathbf{\Omega}\left(t - t_0\right)\dot{\mathbf{u}}_0\right\}\end{aligned} \tag{5.40}$$

$$\begin{aligned}\dot{\mathbf{u}}_h = -\exp\left[-\mathbf{D}\left(t - t_0\right)\right]&\left\{\left(\mathbf{\Omega} + \mathbf{\Omega}^{-1}\mathbf{D}^2\right)\sin\mathbf{\Omega}\left(t - t_0\right)\mathbf{u}_0 \right.\\ &\left. + \mathbf{\Omega}^{-1}\mathbf{D}\sin\mathbf{\Omega}\left(t - t_0\right) - \cos\mathbf{\Omega}\left(t - t_0\right)\dot{\mathbf{u}}_0\right\}\end{aligned} \tag{5.41}$$

Taking the notations $\mathbf{C} = \cos\left(\mathbf{\Omega}\Delta t\right)$, $\mathbf{S} = \sin\left(\mathbf{\Omega}\Delta t\right)$, and $\mathbf{E} = \exp\left(-\mathbf{D}\Delta t\right)$ and letting Δt be the length of a constant time step, then the explicit recursive formula for the homogeneous solution at the $(k + 1)^{th}$ time step, i.e., $t = t_0 + (k + 1)\Delta t$, can be written as

$$u_{h,k+1} = E\left[\left(C + \mathbf{\Omega}^{-1}DS\right)u_{h,k} + \mathbf{\Omega}^{-1}S\dot{u}_{h,k}\right] \tag{5.42}$$

$$\dot{u}_{h,k+1} = -E\left[\left(\mathbf{\Omega} + \mathbf{\Omega}^{-1}D^2\right)Su_{h,k} + \left(\mathbf{\Omega}^{-1}DS - C\right)\dot{u}_{h,k}\right] \tag{5.43}$$

The method of precise calculation of the exponential matrix E was proposed by Zhong and Williams (1994). Here we give the method for the precise calculation of the matrix cosine and sine functions C and S. This is obviously the key point of the scheme. We start from the following exponential relation by letting $r = \Delta t/m$ and $m = 2^N$

$$\exp(i\mathbf{\Omega}\Delta t) = \{\exp(i\mathbf{\Omega}r)\}^m \tag{5.44}$$

Expand the right-hand side of Eq. (5.44) to the form of Taylor series (Deif, 1991)

$$\exp(i\mathbf{\Omega}r) = \sum_{k=0}^{\infty} \frac{1}{k!}(i\mathbf{\Omega}r)^k \tag{5.45}$$

As mentioned in the previous section, in general r is extremely small, and so the series can be truncated to retain limited terms as follows:

$$\exp(i\mathbf{\Omega}r) \approx \mathbf{I} + \mathbf{T}_0 + i\mathbf{F}_0 \tag{5.46}$$

where $\mathbf{I}$ refers the identity matrix of order n and

$$\mathbf{T}_0 = -\frac{1}{2}(\mathbf{\Omega}r)^2 + \frac{1}{24}(\mathbf{\Omega}r)^4 - \frac{1}{720}(\mathbf{\Omega}r)^6 \tag{5.47}$$

$$\mathbf{F}_0 = \mathbf{\Omega}r - \frac{1}{6}(\mathbf{\Omega}r)^3 + \frac{1}{120}(\mathbf{\Omega}r)^5 \tag{5.48}$$

With the following relations

$$(\mathbf{I} + \mathbf{T}_k + i\mathbf{F}_k)^2 = \mathbf{I} + (\mathbf{T}_k^2 - \mathbf{F}_k^2 + 2\mathbf{T}_k) + 2i\mathbf{F}_k(\mathbf{I} + \mathbf{T}_k) = \mathbf{I} + \mathbf{T}_{k+1} + i\mathbf{F}_{k+1},$$

$$(k = 0, 1, ..., N) \tag{5.49}$$

we can compute matrices $\mathbf{T}_N$ and $\mathbf{F}_N$ recursively. In FORTRAN language, the loop statement for the computation looks like this:

DO $k = 0$, $N - 1$

$$T_{k+1} = T_k^2 - F_k^2 = 2T_k$$

$$F_{k+1} = 2F_k(I + T_k)$$

ENDDO

Invoking the Euler formula $\exp(i\mathbf{\Omega}\Delta t) = \cos(\mathbf{\Omega}\Delta t) + i\sin(\mathbf{\Omega}\Delta t)$, we can write

$$\{\exp(i\mathbf{\Omega}r)\}^m = \mathbf{C} + i\mathbf{S} \approx (\mathbf{I} + \mathbf{T}_0 + i\mathbf{F}_0)^m = (\mathbf{I} + \mathbf{T}_N + i\mathbf{F}_N) \tag{5.50}$$

That is, $\mathbf{C} = \cos\left(\mathbf{\Omega}\Delta t\right) \approx \mathbf{I} + \mathbf{T}_N$ and $\mathbf{S} = \sin\left(\mathbf{\Omega}\Delta t\right) \approx \mathbf{F}_N$. The truncated error of the expansion equation (5.46) can be estimated as $\left(\mathbf{\Omega}r\right)^7 \big/ 7!$. By substituting Eq.(5.37) into Eq.(5.45) then taking the first seven terms, we have

$$\exp\left(i\mathbf{\Omega}r\right) \approx \mathbf{Q} \left[\sum_{k=0}^{6} \frac{1}{K!} \left(i\sqrt{\mathbf{\Lambda}}r\right)^k\right] \mathbf{Q}^{-1} \tag{5.51}$$

Thus the truncated error from the kth eigensolution corresponding to λ_k would be $\left(r\sqrt{|\lambda_k|}\right)^7 = 7!$. Suppose ε is the allowed truncation error, then we have $\Delta t\sqrt{|\lambda_k|} < 2^N \left(7!\varepsilon\right)^{1/7}$. If we neglect the inherent damping, the eigenvalue $|\lambda_k|$ would in fact be the kth angular frequency of the structure, that is, $|\lambda_k| = \omega_k$. It follows, by substituting $\omega_k = 2\pi/T_k$, where $\mathbf{T}_k$ is the kth natural period of the structure, that

$$\frac{\Delta t}{\sqrt{T_k}} < \frac{2^N \left(7!\varepsilon\right)^{1/7}}{\sqrt{2\pi}} \tag{5.52}$$

Since 2^N is, in general, a *very* large number, the time step size Δt can be assumed to be relatively large in numerical computation, even for an extremely small value of ε. If N is a moderate number which is not too small, the accuracy of the algorithm would not be dominated by the step size Δt in the sense of numerical computation. For example, suppose $\varepsilon = 10^{-17}$, which reaches or exceeds computer precision, and $N = 20$, then we have from Eq.(5.52) that $\Delta t/\sqrt{T_k} < 2100$. This means that there would be no significant truncation error induced by using Eq.(5.46), even if the time step size is chosen to be 2100 times the square root of the kth natural period of the structure. In practice, taking into consideration the effect of higher modes, the contribution of high modes to the solution would be damped out because of the inherent damping effect of the structure. It follows that if $N = 20$ the precision integration stated above will give essentially the exact homogeneous solution. In other words, matrices C and S, thus computed reflect the characteristics of the structure, including those of its higher mode. As a result of this, the accuracy of the general solution of Eq.(5.21) would be dominantly controlled by the accuracy of the particular solution.

5.3.3   The particular solution

If $\mathbf{f}(t) \neq 0$, the general solution of Eq. (5.21) consists of the homogeneous and the inhomogeneous or the particular solution which satisfies

$$\mathbf{M}\ddot{\mathbf{u}}_p + \mathbf{B}\dot{\mathbf{u}}_p + \mathbf{K}\mathbf{u}_p = \mathbf{f}(t) \tag{5.53}$$

The particular solution incorporated with the second-order scheme can be derived (Bugl, 1995) with the aid of the homogeneous solution in Eq. (5.39). Suppose

$$\mathbf{u}_p = \mathbf{U}_1 \mathbf{k}_1 + \mathbf{U}_2 \mathbf{k}_2 \tag{5.54}$$

where $\mathbf{k}_1$ and $\mathbf{k}_2$ are vectors to be determined and

$$\mathbf{U}_1 = \exp\left[-\mathbf{D}\left(t - t_0\right)\right] \cos\left[\mathbf{\Omega}\left(t - t_0\right)\right],$$

$$\mathbf{U}_2 = \exp\left[-\mathbf{D}\left(t - t_0\right)\right] \sin\left[\mathbf{\Omega}\left(t - t_0\right)\right]. \tag{5.55}$$

By solving the following matrix differential equations:

$$\begin{vmatrix} \mathbf{U}_1 & \mathbf{U}_2 \\ \dot{\mathbf{U}}_1 & \dot{\mathbf{U}}_2 \end{vmatrix} \left\{ \begin{matrix} \dot{\mathbf{k}}_1 \\ \dot{\mathbf{k}}_2 \end{matrix} \right\} = \left\{ \begin{matrix} 0 \\ \mathbf{M}^{-1}\mathbf{f} \end{matrix} \right\} \tag{5.56}$$

we can derive

$$\mathbf{k}_1 = \int_{t_0}^t \Delta^{-1} \begin{vmatrix} \mathbf{0} & \mathbf{U}_2 \\ \mathbf{M}^{-1}\mathbf{f} & \dot{\mathbf{U}}_2 \end{vmatrix} \mathbf{f}\left(\eta\right) d\eta, \mathbf{k}_2 = \int_{t_0}^t \Delta^{-1} \begin{vmatrix} U_1 & 0 \\ \dot{U}_2 & M^{-1}f \end{vmatrix} \mathbf{f}\left(\eta\right) d\eta \tag{5.57}$$

where

$$\Delta = \begin{vmatrix} \mathbf{U}_1 & \mathbf{U}_2 \\ \dot{\mathbf{U}}_1 & \dot{\mathbf{U}}_2 \end{vmatrix} = \mathbf{\Omega} \exp\left[-2\mathbf{D}\left(t - t_0\right)\right] \tag{5.58}$$

Combining Eqs.(5.54) to (5.58), we can write

$$\mathbf{u}_p = \mathbf{M}^{-1}\mathbf{\Omega}^{-1} \int_{t_0}^t \exp\left[-\mathbf{D}\left(t - \eta\right)\right] \sin\mathbf{\Omega}\left(t - \eta\right) \mathbf{f}\left(\eta\right) d\eta \tag{5.59}$$

$$\dot{\mathbf{u}}_p = \mathbf{M}^{-1} \int_{t_0}^t \exp[-\mathbf{D}(t - \eta)][\cos\mathbf{\Omega}(t - \eta) - \mathbf{\Omega}^{-1}\sin\mathbf{\Omega}(t - \eta)]\mathbf{f}(\eta)d\eta \tag{5.60}$$

By substituting the integral variable with $r = t - \eta$, the particular solution at the $(k + 1)^{th}$ time step, i.e., $t = t_0 + (k + 1)\Delta t$, can be written as

$$\mathbf{u}_{p,k+1} = \mathbf{M}^{-1}\mathbf{\Omega}^{-1} \int_0^{\Delta t} \exp\left(-\mathbf{D}r\right) \sin\left(\mathbf{\Omega}r\right) \mathbf{f}\left(t_k + \Delta t - r\right) dr \tag{5.61}$$

$$\dot{\mathbf{u}}_{p,k+1} = \mathbf{M}^{-1} \int_0^{\Delta t} \exp(-\mathbf{D}r)[\cos(\mathbf{\Omega}r) - \mathbf{\Omega}^{-1}\sin(\mathbf{\Omega}r)]\mathbf{f}(t_k + \Delta t - r)dr$$
$$\tag{5.62}$$

Methods have been proposed (Lin et al., 1995) for the numerical treatment of particular solutions according to the properties of $\mathbf{f}(t)$. Here, suppose that $\mathbf{f}(t)$ are smooth functions within one time step, so we can compute the particular solution with the aid of Taylor expansion as

$$\mathbf{f}\left(t_k + \Delta t - r\right) = \sum_{m=0}^{\infty} \frac{\mathbf{c}_m^k}{m!} \left(\Delta t - r\right)^m \tag{5.63}$$

where

$$\mathbf{c}_m^k = \mathbf{f}^{(m)}\left(t_k\right) \tag{5.64}$$

Define the integrals including the term $(\Delta t - r)^m$ in the Taylor series as

$$\mathbf{I}_S^m = \int_0^{\Delta t} \exp\left(-\mathbf{D}r\right) \sin\left(\mathbf{\Omega}r\right) \left(\Delta t - r\right)^m dr \tag{5.65}$$

$$\mathbf{I}_C^m = \int_0^{\Delta t} \exp\left(-\mathbf{D}r\right) \cos\left(\mathbf{\Omega}r\right) \left(\Delta t - r\right)^m dr \tag{5.66}$$

We can deduce the recurrence formula for the calculation of the above integrals as follows:

$$\mathbf{I}_S^0 = \left(\mathbf{D}^2 + \mathbf{\Omega}^2\right)^{-1} \left(\mathbf{\Omega} - \mathbf{DES} - \mathbf{\Omega EC}\right), \tag{5.67}$$

$$\mathbf{I}_C^0 = \left(\mathbf{D}^2 + \mathbf{\Omega}^2\right)^{-1} \left(\mathbf{D} + \mathbf{\Omega ES} - \mathbf{DEC}\right), \tag{5.68}$$

$$\mathbf{I}_S^m = m\left(\mathbf{D}^2 + \mathbf{\Omega}^2\right)^{-1} \left(\frac{1}{m}\mathbf{\Omega}\Delta t^m - \mathbf{DI}_S^{m-1} - \mathbf{\Omega I}_C^{m-1}\right), \tag{5.69}$$

$$\mathbf{I}_C^m = m\left(\mathbf{D}^2 + \mathbf{\Omega}^2\right)^{-1} \left(\frac{1}{m}\mathbf{D}\Delta t^m + \mathbf{\Omega I}_S^{m-1} - \mathbf{DI}_C^{m-1}\right), \quad (m = 1, 2, \ldots). \tag{5.70}$$

By combining the homogeneous solutions in Eqs.(5.42) to (5.43) with the particular solutions in Eqs.(5.61) to (5.62), we have the general solutions of Eq.(5.21) using the second-order scheme as follows:

$$\begin{aligned} \mathbf{u}_{k+1} &= \mathbf{E}\left[\mathbf{C} + \mathbf{\Omega}^{-1}\mathbf{S}\left(\mathbf{u}_k + \dot{\mathbf{u}}_k\right)\right] \\ &+ \mathbf{\Omega}^{-1}\mathbf{M}^{-1}\int_0^{\Delta t} \exp\left(-\mathbf{D}r\right) \sin\left(\mathbf{\Omega}r\right) \mathbf{f}\left(t_k + \Delta t - r\right) dr \end{aligned} \tag{5.71}$$

$$\begin{aligned} \dot{\mathbf{u}}_{k+1} &= -\mathbf{E}\left[(\mathbf{\Omega} + \mathbf{\Omega}^{-1}\mathbf{D}^2)\mathbf{Su}_k + (\mathbf{\Omega}^{-1}\mathbf{DS} - \mathbf{C})\dot{\mathbf{u}}_k\right] \\ &+ \mathbf{M}^{-1}\int_0^{\Delta t} \exp(-\mathbf{D}r)[\cos(\mathbf{\Omega}r) - \mathbf{\Omega}^{-1}\mathbf{D}\sin(\mathbf{\Omega}r)]\mathbf{f}(t_k + \Delta t - r)dr \end{aligned} \tag{5.72}$$

5.3.4 Numerical examples

In this section, several simple numerical examples of one-dimensional problem are presented to test the proposed second-order scheme. The eigenvalues and the corresponding eigenvectors of the system matrix are computed using the FORTRAN packages in reference (Smith et al., 1976). $N = 20$ is used in the 2^N algorithm. The first example serves as a basic test using the second-order algorithm for the solution of a homogeneous equation, which is compared with the first-order algorithm. The next two examples are for the test of the effect of the inhomogeneous part and the damping term of the differential equation on the solutions with the proposed algorithm. The numerical results of the first three examples are presented by the maximum errors, which are compared with the theoretical solutions. The last two numerical examples are wave propagation and impact problem, respectively. Their solution has some spatial and temporal non-smoothness which is for further verification of the algorithm.

Example 5.5 *One-Dimensional Wave Equation*

Consider the vibration of a string clamped at both ends. The governing equation is a one-dimensional wave equation

$$\ddot{u} = \frac{\partial^2 u}{\partial x^2}, \quad x \in [0, 2\pi] \tag{5.73}$$

with the boundary and initial conditions

$$u(0, t) = u(2\pi, t) = 0, \quad u(x, 0) = \sin(x), \quad \dot{u}(x, 0) = 0 \tag{5.74}$$

where $u(x, t)$ is the transverse displacement of the string with unit amplitude of initial displacement. In Eq. (5.74), the first two relations denote the fixed boundary condition and the last two relations represent the initial condition. The solution to Eq. (5.73) is given by the D'Alembert integral as

$$u(x, t) = \sin(x)\cos(t) \tag{5.75}$$

We now discretize the spatial coordinate in Eq. (5.73) with the direct DQ method. In this method, the derivatives of $u(x, t)$ with respect to the spatial coordinate x are approximated by a weighted sum of function values at all discrete points within the interval $x \in [0, 2\pi]$ under consideration, i.e.,

$$u^{(r)}(x_j, t) = \left. \frac{d^r}{dx^r} u(x, t) \right|_{x = x_j} = \sum_{k=0}^{n} A_{jk}^{(r)} u(x_k, t), \quad (j = 0, 1, ..., n+1) \tag{5.76}$$

The details of computation of weighting coefficients $A_{jk}^{(r)}$ $(r = 2)$ are presented in Chapter 1. In the quadrature formula in Eq. (5.76), $j = 0$ and

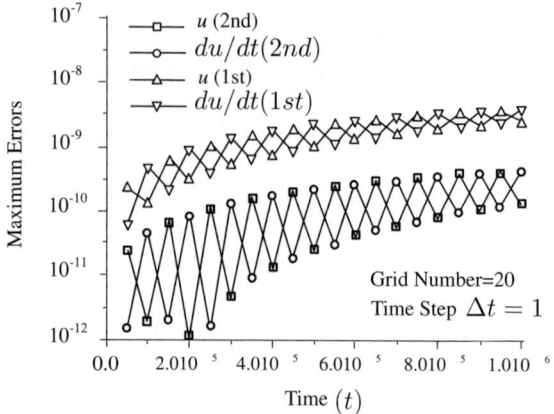

FIGURE 5.14:   Comparison between the first-and second-order algorithms.

$j = n + 1$ correspond to the boundary nodes, respectively, at $x = 0$ and $x = 2\pi$ so that n is the number of the interior nodes. The positions of the interior grid points are placed at Gaussian points, which are computed by the subroutine presented in reference (Press et al., 1986).

After discretization incorporated with the boundary condition, Eq. (5.73) is transformed to the following form of algebraic and differential equation as

$$\ddot{u} + Au = 0 \quad \text{or} \quad \ddot{u} + \Omega^2 u = 0 \tag{5.77}$$

where $\Omega = \sqrt{A}$. The numerical solution of the wave equation with the second-order precision integration algorithm at $(k + 1)^{th}$ time step, i.e., $t = t_0 + (k + 1)\Delta t$, is written as follows:

$$u_{k+1} = Cu_k + \Omega^{-1}S\dot{u}_k, \tag{5.78}$$

$$\dot{u}_{k+1} = -\Omega Su_k + C\dot{u}_k. \tag{5.79}$$

The maximum solution errors using both the first- and second-order algorithms are compared in Fig. 5.14. Twenty interior grid points and the time step $\Delta t = 1$ were used in the computation.

It can be seen from Fig. 5.14 that both the algorithms are very stable, but better accuracy is achieved by using the second-order algorithm. The CPU time used for the second-order algorithm is only about 37.5% of that for the first-order algorithm for the same amount of computation. This is mainly due to the small size of the system matrix in the second-order algorithm during long-term computation since the major computational effort comes from the step-by-step integration realized by matrix multiplications.

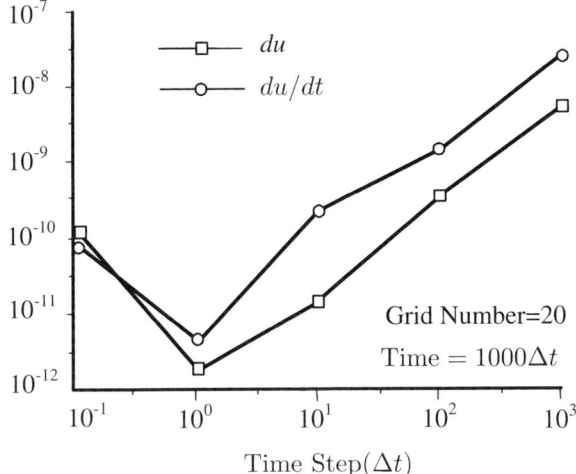

FIGURE 5.15: Errors as function of time step Δt with the second-order algorithm.

The effect of the time step Δt on the accuracy of the second-order algorithm is shown in Fig. 5.15. It can be seen from Fig. 5.15 that very large time steps can be employed for the solution of homogeneous equations with the precision integration scheme which also verifies the error analysis presented in Section 5.4.2. The effect of the interior grid number on the accuracy is shown in Fig. 5.16. In the work of Zong and Lam (2002), the wave equation was solved by using the DQ discretization together with the fourth-order Runge–Kutta method with a time step $\Delta t = 10^{-4}$. The solution becomes unstable at $t = 0.45$ even though the grid number is not too large ($N = 15$) in global discretization. In contrast, the results obtained by the proposed algorithm show that no instability phenomenon occurs with the increase of the grid number after long-term integration ($t = 1000$) as shown in Fig. 5.16. This finding demonstrates the superior feature of the precision integral when compared with the traditional finite-difference-based time-marching method.

Example 5.6 *Forced One-Dimensional Wave Equation*

Let us consider the forced vibration of a string clamped at both ends. The governing equation is the following inhomogeneous one-dimensional wave equation

$$\ddot{u} = \frac{\partial^2 u}{\partial x^2} + f(x, t) \tag{5.80}$$

where

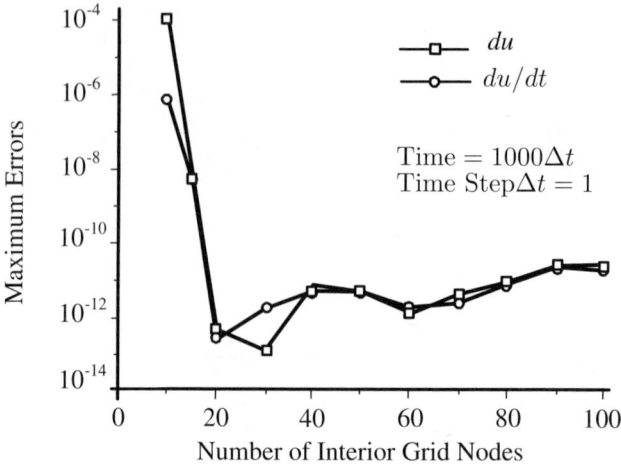

FIGURE 5.16: Errors as function of interior grid numbers with the second-order algorithm.

$$f(x,t) = -3\sin(x)\sin(2t) \tag{5.81}$$

subject to the boundary and the initial conditions

$$u(0,t) = u(2\pi,t) = 0, \quad u(x,0) = \sin(x), \quad \dot{u}(x,0) = 2\sin(x) \tag{5.82}$$

The solution to Eq. (5.80) is as follows:

$$u(x,t) = \sin(x)[\cos(t) + \sin(2t)] \tag{5.83}$$

After spatial discretization by the direct DQ method, Eq. (5.80) becomes

$$\ddot{u} + \Omega^2 u = f(t) \tag{5.84}$$

The numerical solution of the forced wave equation with the second-order precision integration algorithm at the $(k+1)^{th}$ time step, i.e., $t = t_0 + (k+1)\Delta t$, is written as follows:

$$u_{k+1} = Cu_k + \Omega^{-1}S\dot{u}_k + \Omega^{-1}\int_0^{\Delta t} \sin(\Omega r)f(t_k + \Delta t - r)\,dr \tag{5.85}$$

$$u_{k+1} = -S\Omega u_k + C\dot{u}_k + \int_0^{\Delta t} \cos(\Omega r)f(t_k + \Delta t - r)\,dr \tag{5.86}$$

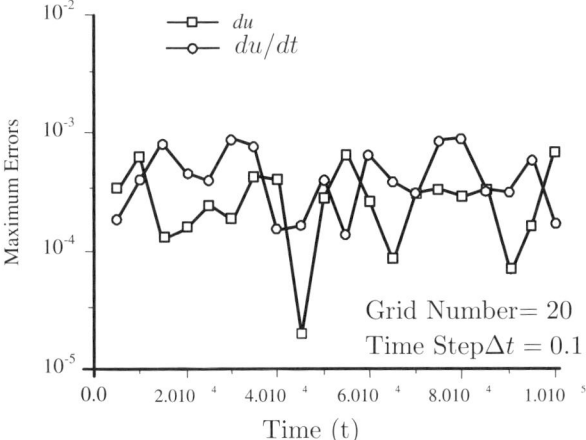

FIGURE 5.17: Errors as function of time for the forced wave equation.

The integrals for the terms in the Taylor expansion of $f(t)$ for the forced wave equation can be computed recursively by the following formula

$$I_S^0 = \int_0^{\Delta t} \sin(\Omega r)\, dr = \Omega^{-1}(I - C),$$ (5.87)

$$I_C^0 = \int_0^{\Delta t} \cos(\Omega r)\, dr = \Omega^{-1} S,$$ (5.88)

$$I_S^m = \int_0^{\Delta t} \sin(\Omega r)(\Delta t - r)^m\, dr = m\Omega^{-1}\left(\frac{1}{m} I \Delta t^m - I_C^{m-1}\right),$$ (5.89)

$$I_C^m = \int_0^{\Delta t} \cos(\Omega r)(\Delta t - r)^m\, dr = m\Omega^{-1} I_S^{m-1}, \quad (m = 1, 2, ...).$$ (5.90)

The maximum errors as a function of time for the forced wave equation are shown in Fig. 5.17. In the computation, only three terms in the Taylor expansion are sufficient to keep reasonable accuracy, and are retained. Figure 5.17 shows that the algorithm is long-term stable although the time step chosen should not be too large ($\Delta t = 0.1$) owing to the existence of the inhomogeneous term. By comparing Fig. 5.17 with Figs. 5.14 to 5.16, it is evident that the accuracy of the general solution is dominantly controlled by the accuracy of the particular solution.

5.4   Conclusions

When direct DQ method is applied, large errors may occur if the grid number is large. The errors exponentially grow as the grid number increases. The accuracy of direct DQ method is dominated by nodes near and on boundaries. These nodes lead to unavoidable dynamic numerical instability.

The variable order DQ method presented in this chapter is used to eliminate the divergence near boundaries with ease by applying a small number of points for cortical nodes. It is verified that the variable order DQ method yields very good results for both linear and strongly nonlinear dynamic problems. The accuracy of direct DQ method is significantly improved by adopting different order DQ schemes at the core and cortical nodes.

Precision integration technique is also introduced in this chapter as a means to improve numerical differentiation accuracy. Variable order DQ may improve spatial accuracy and precision integration may improve temporal accuracy. Both of them are effective approaches.

Chapter 6

Multi-Domain Differential Quadrature Method

There arise difficulties when we apply DQ method to the elastic problems where discontinuities are present. If an elastic structure under consideration is made of two or more materials, at least one of stress components at the interface of the two different materials is not continuous, leading to a finite jump at the interface. It is mathematically clear that a finite discontinuity can only be described by a first-order numerical scheme (Sod, 1978; Zong et al., 2005). Any high-order numerical scheme must change to first-order scheme at the discontinuity (but remains a high–order scheme elsewhere).

DQ method is equivalent to Lagrange interpolation and differentiation using Lagrange polynomial, and thus it is a high-order numerical scheme, being differentiable to many orders. However, it fails at the discontinuity where a first-order scheme is required. In these cases, direct application of DQ method would yield very bad results due to the nature of the method.

A multi-domain DQ method is formulated in this chapter to solve plane elastic problems in the presence of material discontinuity. By putting the boundary of each sub-domain on the interface of two different materials, discontinuity is transformed to proper imposition of compatibility conditions at the interface. It turns out that the compatibility conditions at the interfaces have dominant influence on the numerical accuracy. Numerical examples show that the proposed method, which is first-order accurate at the interfaces and high-order accurate elsewhere, can properly capture the material discontinuity and yield good results when compared with those given by FEM.

6.1 Linear plane elastic problems with material discontinuity

First of all, let us show the incapability of the direct DQ method for the elastic problem with material discontinuity and highlight the importance of the development of a new multi-domain DQ method.

Consider a plane elastic problem with material continuity as shown in Fig. 6.1. The domain under consideration is divided into two parts at the middle.

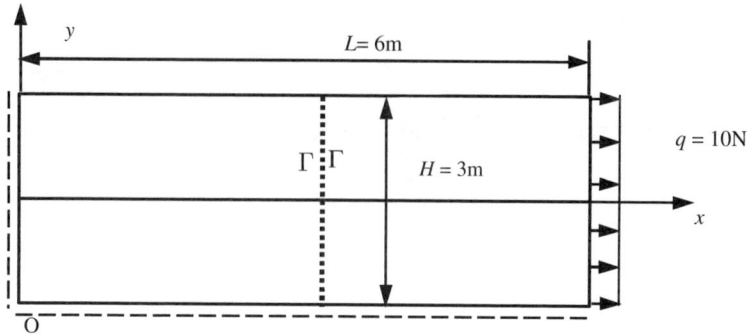

FIGURE 6.1: Two-dimensional elastic structure made of two materials, simply supported boundary condition.

The left half is made of material 1 and the right half is made of material 2. There is a material discontinuity at the middle. If a plane stress problem is considered, the governing equation is (Timoshenko and Goodier, 1970)

$$\frac{E}{1-\mu^2}\left(\frac{\partial^2 u}{\partial x^2} + \frac{1-\mu}{2}\frac{\partial^2 u}{\partial y^2} + \frac{1+\mu}{2}\frac{\partial^2 v}{\partial x \partial y}\right) = P^x \qquad (6.1a)$$

$$\frac{E}{1-\mu^2}\left(\frac{\partial^2 v}{\partial y^2} + \frac{1-\mu}{2}\frac{\partial^2 v}{\partial x^2} + \frac{1+\mu}{2}\frac{\partial^2 u}{\partial x \partial y}\right) = P^y \qquad (6.1b)$$

where u and v are the displacement components in x- and y-directions, respectively; E and μ are Young's modulus and Poisson's ratio. P^x and P^y are the components of the body force in x- and y-directions, respectively. Note that E and μ take different values in the left and right halves.

Discretize the domain using N_x nodes in the x-direction and N_y nodes in the y-direction. So the nodes along the x-direction are $x_1, x_2, \cdots, x_{N_x}$, and the nodes along the y-direction are $y_1, y_2, \cdots, y_{N_y}$. The displacements at each point (x_i, y_j) are u_{ij} and v_{ij}, $i = 1, \cdots N_x$, $j = 1, \cdots N_y$. By using the DQ procedure, we obtain the following discrete form of Eq. (6.1)

$$\frac{E}{1-\mu^2}\left(\sum_{k=1}^{N_x} b_{ik}^x u_{kj} + \frac{1-\mu}{2}\sum_{\ell=1}^{N_y} b_{j\ell}^y u_{i\ell} + \frac{1+\mu}{2}\sum_{k=1}^{N_x}\sum_{\ell=1}^{N_y} a_{ik}^x a_{j\ell}^y v_{k\ell}\right) = P_{ij}^x$$

$$(6.2a)$$

$$\frac{E}{1-\mu^2}\left(\sum_{\ell=1}^{N_y} b_{j\ell}^y v_{i\ell} + \frac{1-\mu}{2}\sum_{k=1}^{N_x} b_{ik}^x v_{kj} + \frac{1+\mu}{2}\sum_{k=1}^{N_x}\sum_{\ell=1}^{N_y} a_{ik}^x a_{j\ell}^y u_{k\ell}\right) = P_{ij}^y$$

$$(6.2b)$$

The boundary conditions are

$$u(x = 0) = 0, \quad \tau(x = 0) = \tau_{xy}(x = 0) = 0 \tag{6.3a}$$

$$\sigma_x(x = 6) = 10, \quad \tau(x = 6) = 0 \tag{6.3b}$$

$$v(y = 0) = 0, \quad \tau(y = 0) = 0 \tag{6.3c}$$

$$\sigma_y(y = 3) = 0, \quad \tau(y = 3) = 0 \tag{6.3d}$$

Based on the stress-displacement relationship in elasticity, we have the following discrete forms of stress components

$$(\sigma_x)_{ij} = \frac{E}{1 - \mu^2}(\sum_{k=1}^{N_x} a_{ik}^x u_{kj} + \mu \sum_{l=1}^{N_y} a_{jl}^y v_{il}) \tag{6.4a}$$

$$(\sigma_y)_{ij} = \frac{E}{1 - \mu^2}(\mu \sum_{k=1}^{N_x} a_{ik}^x u_{kj} + \sum_{l=1}^{N_y} a_{jl}^y v_{il}) \tag{6.4b}$$

$$\tau = G(\sum_{k=1}^{N_x} a_{ik}^x v_{kj} + \sum_{\ell=1}^{N_y} a_{j\ell}^y u_{i\ell}) \tag{6.4c}$$

where $G = E/2(1 + \mu)$ is shear modulus. Substituting the above equations into Eq. (6.3), we obtain the discrete form of boundary conditions along x-axis

$$u_{ij} = 0, \quad \frac{E}{2(1 + \mu)}(\sum_{k=1}^{N_x} a_{ik}^x v_{kj} + \sum_{\ell=1}^{N_y} a_{j\ell}^y u_{i\ell}) = 0, \quad i = 1 \tag{6.5a}$$

$$\frac{E}{1 - \mu^2}(\sum_{k=1}^{N_x} a_{ik}^x u_{kj} + \mu \sum_{l=1}^{N_y} a_{jl}^y v_{il}) = 10, \quad i = N_x \tag{6.5b}$$

Note that shear stress is zero on all boundaries, and thus we omitted the equation for shear stress in Eq. (6.41) and in the following equations. The boundary conditions along y-axis are

$$v_{ij} = 0, \quad j = 1 \tag{6.5c}$$

$$\frac{E}{1 - \mu^2}(\mu \sum_{k=1}^{N_x} a_{ik}^x u_{kj} + \sum_{l=1}^{N_y} a_{jl}^y v_{il}) = 0, \quad j = N_y \tag{6.5d}$$

To apply Gauss elimination for the solution of the elastic problem, we introduce the following indices

$$\alpha = (i - 1)N_y + j, \quad N = N_x \times N_y, \quad i = 1, \cdots N_x \tag{6.6}$$

Then Eq. (6.2) can be recast in the following form

$$\sum_{\beta=1}^{N} d_{\alpha\beta}^{11} u_\beta + \sum_{\beta=1}^{N} d_{\alpha\beta}^{12} v_\beta = P_\alpha^x \tag{6.7a}$$

$$\sum_{\beta=1}^{N} d_{\alpha\beta}^{21} u_\beta + \sum_{\beta=1}^{N} d_{\alpha\beta}^{22} v_\beta = P_\alpha^y \tag{6.7b}$$

where the coefficient matrices are obtained from Eq. (6.2) through simple manipulation. They are

$$d_{\alpha\beta}^{11} = \frac{E}{1-\mu^2} b_{ik}^x \delta_{j\ell} + \frac{E}{2(1+\mu)} b_{j\ell}^y \delta_{ik}, \quad d_{\alpha\beta}^{12} = \frac{E}{2(1-\mu)} a_{ik}^x a_{j\ell}^y \tag{6.8a}$$

$$d_{\alpha\beta}^{21} = \frac{E}{2(1-\mu)} a_{ik}^x a_{jl}^y, \quad d_{\alpha\beta}^{22} = \frac{E}{1-\mu^2} b_{jl}^y \delta_{ik} + \frac{E}{2(1+\mu)} b_{ik}^x \delta_{jl} \tag{6.8b}$$

where the delta function $\delta_{ij} = 1$ if $i = j$ and it is zero if $i \neq j$. For nodes on boundaries, we may similarly write out the equations. Combining them together and with aid of Eq. (6.4), the unknowns satisfy

$$\begin{bmatrix} \mathbf{d}^{11} & \mathbf{d}^{12} \\ \mathbf{d}^{21} & \mathbf{d}^{22} \end{bmatrix} \begin{pmatrix} \mathbf{u} \\ \mathbf{v} \end{pmatrix} = \begin{pmatrix} \mathbf{P}^x \\ \mathbf{P}^y \end{pmatrix} \tag{6.9}$$

with suitable boundary conditions. This is a system of linear algebraic equations with $2N$ unknowns. In compact form, Eq. (6.49) becomes

$$\mathbf{Dw} = \mathbf{Q} \tag{6.10}$$

with

$$\mathbf{D} = \begin{bmatrix} \mathbf{d}^{11} & \mathbf{d}^{12} \\ \mathbf{d}^{21} & \mathbf{d}^{22} \end{bmatrix}, \quad \mathbf{w} = \begin{pmatrix} \mathbf{u} \\ \mathbf{v} \end{pmatrix}, \quad \mathbf{Q} = \begin{pmatrix} \mathbf{P}^x \\ \mathbf{P}^y \end{pmatrix} \tag{6.11}$$

This system of equations can be solved by Gauss elimination. After the unknowns are produced, the stresses at each node can be estimated by using Eq. (6.4).

Based on the above formulations, a FORTRAN code was developed by Zong et al. (2005) and a numerical example as shown in Fig. 6.1 is solved. A two-dimensional elastic structure, made of two different materials is considered. The Young's modulus in the left half is $E_1 = 3.0 \times 10^7 N/m^2$ while the Young's modulus in the right half is $E_2 = 3.0 \times 10^6 N/m^2$. Poisson's ratios for both materials are $\mu = 0.25$. The domain is $L = 6m$ long, and $H = 3m$ high. The uniform distributed force $q = 10$ N is applied on the right side of the domain. The domain is discretized by 13×11 nodes. The stress components at the central section $y = 1.5m$ calculated from direct DQ are shown in Fig. 6.2. They are represented by solid lines. In the same figure, the results obtained from the commercial FEM software ABAQUS (ABAQUS User Guide, 2002) are also presented for the sake of comparison.

The result for stress component σ_y obtained from ABAQUS shows a discontinuous finite jump at the interface of the two materials while those obtained by direct DQ are nearly equal to zero as shown in Fig. 6.2(a). The physically finite jump is "smeared out" by direct DQ method. Such "smearing out" phenomenon is in fact not rare. It also occurs in some finite difference schemes when used to simulate shock waves (Sod, 1978). It is clearly demonstrated that discontinuity or finite jump must be carefully treated. At a discontinuity, both left and right derivatives exist, but they are not equal. So the second-order or higher order derivatives do not exist physically. Any numerical schemes which are capable of simulating such finite discontinuity must also be first-order accurate at the interface. It cannot be second-order accurate at the interface, which is against physical observations. This is an interesting topic and a detailed discussion would be beyond the scope of this chapter. The interested reader may refer to the paper by Sod (1978) and the numerous papers on this topic in the literature.

As confirmed by Shu (2000a), DQ is a high-order numerical scheme with order up to min $\{N_x\text{-}1, N_y\text{-}1\}$. If ten nodes are used in each direction, DQ is a 9-th order scheme. So it is not surprising that it cannot capture the discontinuity at the interface of two different materials.

Figure 6.2(b) shows the shear stress distributions obtained by direct DQ method and ABAQUS. It is found that the curve of shear stress given by ABAQUS is continuous, but not smooth. There exists a sharp turning point at the interface. Direct DQ procedure fails again to reproduce the expected phenomenon, yielding an almost zero shear stress distribution across the middle section, incomparable with those given by FEM.

It should be noted that the ordinates and abscissas are not proportionally drawn in Fig. 6.2. To see the errors the direct DQ method produces, the ordinates are deliberately prolonged in Fig. 6.2(a) and (b).

6.2 A multi-domain approach for numerical treatment of material discontinuity

In order to capture the discontinuity correctly, the order of the scheme used must be reduced to one at the discontinuity as mentioned in the previous section. To do so, the two-dimensional domain Ω is decomposed into several sub-domains: Ω_s, $s= 1, 2\ldots, M$ in such a way that any two of sub-domains are disjoint, that is, $\Omega_s \cap \Omega_t = \emptyset$ if $s \neq t$. Here $\emptyset$ is the empty set. Then, we have $\Omega = \Omega_1 \cap \Omega_2 \cap \cdots \cap \Omega_M$.

The decomposition of the domain into several sub-domains should follow

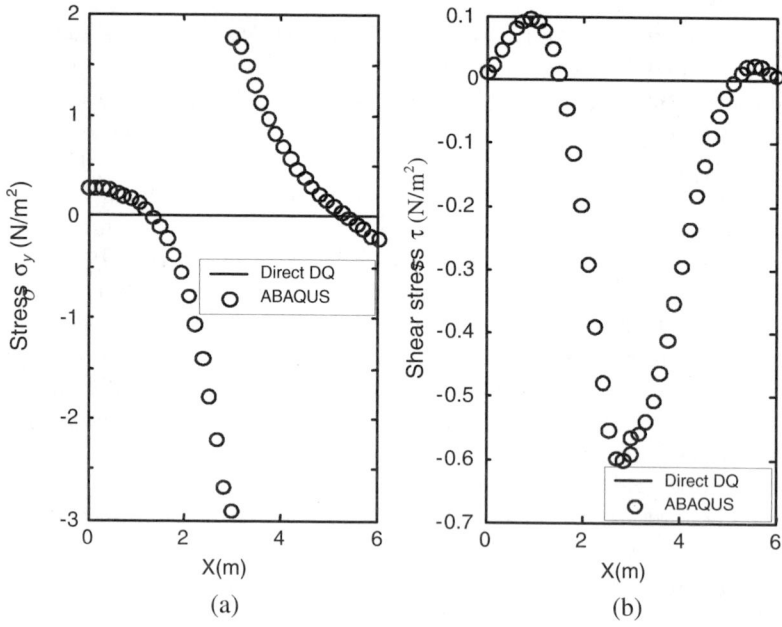

FIGURE 6.2:   Stresses obtained by direct DQ and ABAQUS at the section $y = H/2$ for the bi-material elastic structure with interface at $x = L/2$.

the general guideline that the physical constants in each sub-domain such as Young's modulus and Poisson's ratio are constant. The boundaries of sub-domains coincide with the interfaces of different materials. Take Fig. 6.1 as an example. The interface of the two different materials defines the common boundaries of sub-domain Ω_1 and Ω_2. In this way, each sub-domain Ω_s has two types of boundaries: boundary Γ_s^o, which is a part of the whole domain Ω, and boundary Γ_{st}^i, which is the interface of sub-domain Ω_s and Ω_t; see Fig. 6.3. We use superscript "s" or "t" to denote displacements and stresses in each sub-domain. So $\mathbf{w}^{(s)}$ is the displacement vector in the s-th sub-domain, which has the structure as shown in Eq. (6.10).

The boundary conditions on Γ_s^o can be treated in the same way as before. On the interface Γ_{st}^i, however, special treatment is required. There are two ways to impose the boundary conditions on the interfaces (namely compatibility condition). In the first way, the displacements and their normal derivatives calculated from the two neighboring domains are set equal. Numerical tests have shown that this procedure furnishes correct results for the homogeneous problems but wrong results for the problems in the presence of material discontinuity (see Example 6.4). Therefore, compatibility conditions are strongly dependent on the material discontinuity. Instead, in the second procedure, the normal and shear stresses as well as the displacements on the interface

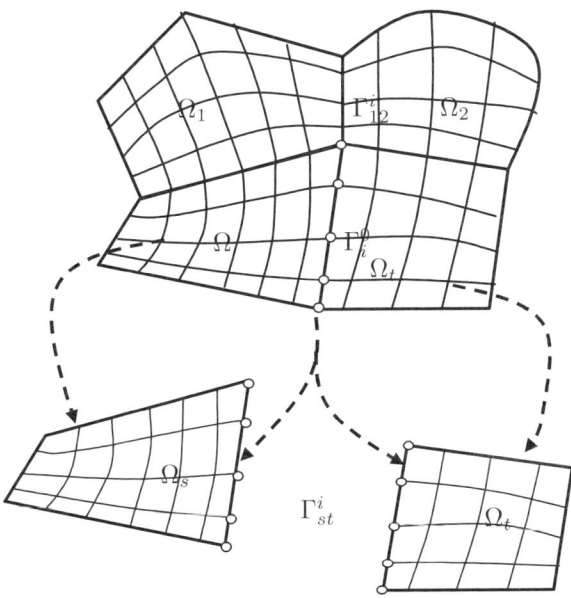

FIGURE 6.3: Multi-domain decomposition with interface boundaries Γ^i_{st} and boundaries of the original domain Γ^o_s.

are set equal. But the stress component along the interface is not fixed. This procedure works very well for the present discontinuous elastic problems as shown in subsequent sections.

The second procedure can be expressed in a mathematical form as

$$\mathbf{u}^{(s)} = \mathbf{u}^{(t)}, \quad \mathbf{v}^{(s)} = \mathbf{v}^{(t)} \quad \text{on } \Gamma_{st}^{i} \tag{6.12a}$$

$$\sigma_n^{(s)} = \sigma_n^{(t)}, \quad \tau_n^{(s)} = \tau_n^{(t)} \quad \text{on } \Gamma_{st}^{i} \tag{6.12b}$$

where $\sigma_n^{(s)}$ denotes the stress component perpendicular to the interface, and $\tau_n^{(s)}$ denotes the shear stress along the interface.

Discretize each sub-domain $\Omega^{(s)}$ using $N_x^{(s)} \times N_y^{(s)}$ nodes. In each sub-domain, the discrete governing equations have the form of Eq. (6.10), that is,

$$\mathbf{D}^{(s)} \mathbf{w}^{(s)} = \mathbf{Q}^{(s)}, \quad s = 1, 2, \cdots, M \tag{6.13}$$

which is valid for nodes inside the sub-domain $\Omega^{(s)}$ and the nodes on boundary Γ_s^o. For nodes on the interface Γ_{st}^i, Eq. (6.12) is applied instead as

$$\mathbf{H}^{(s)} \mathbf{w}^{(s)} = \mathbf{H}^{(t)} \mathbf{w}^{(t)} \quad \text{on } \Gamma_{st}^{i} \tag{6.14}$$

By assembling Eq. (6.13) with (6.14), we obtain a system of linear algebraic equations

$$\begin{bmatrix} \mathbf{D}^{(1)} & & & & \\ & \mathbf{D}^{(2)} & & & \\ & & \ddots & & \\ & & & \mathbf{D}^{(M)} & \\ \mathbf{H}^{(1)} & & & & \\ \vdots & & & & \\ \mathbf{H}^{(M)} & & & & \end{bmatrix} \begin{pmatrix} \mathbf{w}^{(1)} \\ \mathbf{w}^{(2)} \\ \vdots \\ \mathbf{w}^{(M)} \end{pmatrix} = \begin{pmatrix} \mathbf{Q}^{(1)} \\ \mathbf{Q}^{(2)} \\ \vdots \\ \mathbf{Q}^{(M)} \end{pmatrix} \tag{6.15}$$

There are in total $\prod_{s=1}^{M} \left(N_x^{(s)} N_y^{(s)} \right)$ unknowns, which can be computed from solving Eq. (6.15) by Gauss elimination. It is worth pointing out that local DQ scheme is applied to each regular sub-domain in the same manner as in the single domain. Since the coordinates of each sub-domain in the Cartesian plane vary in different sub-domains, therefore, the weighting coefficients for DQ discretization are location-dependent. This means the weighting coefficients vary from sub-domain to sub-domain.

In each sub-domain, DQ is employed locally and thus the numerical scheme is high-order accurate. On the interface, the normal stress perpendicular to the interface and shear stress along the interface are set equal. Because stresses are the combinations of the first-order derivatives of displacements, the multi-domain DQ scheme at the interface is reduced to first-order accurate.

This is qualitatively in agreement with what we proposed in the previous section. This will also be validated by the numerical examples given in the following section.

6.2.1 Numerical examples

Numerical examples of discontinuous plane elastic problems made of different materials, in the absence of body forces, are presented in this section to show the validity and efficiency of the present method. The direct DQ method can not be employed to solve all of the problems directly but multi-domain DQ can. All of the numerical results obtained by the present method are compared with the results from the FEM commercial software, ABAQUS (ABAQUS User Guide, 2002).

Example 6.1 *Two-dimensional elastic structure made of two different materials*

We first return to the example we have studied in Section 6.1, where it is found that direct DQ cannot capture the finite discontinuity in stress component σ_y at all. It also fails to capture the sharp turning point occurring in shear stress distribution.

The multi-domain DQ is employed to this problem instead. The stress distributions along the section $y = H/2$ obtained by multi-domain DQ and ABAQUS are shown in Fig. 6.4. It is demonstrated in the figure that the present multi-domain approach characterizes the finite jump in σ_y and the sharp turning point in the shear stress τ results obtained from both methods are quantitatively in excellent agreement.

A convergence study has been performed by using six sets of different node numbers. The results at two points A and B (point A is at the center of the domain and point B is at the right top corner) are presented in Table 6.1. It is observed from Table 6.1 that the results are slightly dependent on the node numbers. In other words, the results are convergent with respect to the node numbers, and the convergent rate is very fast. It is clear from the observation that very accurate results can be obtained by using small number of nodes.

TABLE 6.1: The convergence study for Example 6.1

Mesh	u_B $(\times 10^{-5})$	v_B $(\times 10^{-6})$	σ_{xA}	σ_{yA} Domain 1	σ_{yA} Domain 2	τ_A
5x5	1.1393	-2.7528	9.5378	-3.0342	1.8755	-0.6151
7x7	1.1321	-2.6853	9.5612	-3.2745	1.8672	-0.5534
9x9	1.1312	-2.6837	9.4914	-3.1060	1.8771	-0.5778
11x11	1.1311	-2.6866	9.4918	-3.2291	1.8704	-0.5643
13x13	1.1310	-2.6888	9.4491	-3.1172	1.8782	-0.5725
15x15	1.1308	-2.6901	9.4458	-3.2054	1.8727	-0.5671

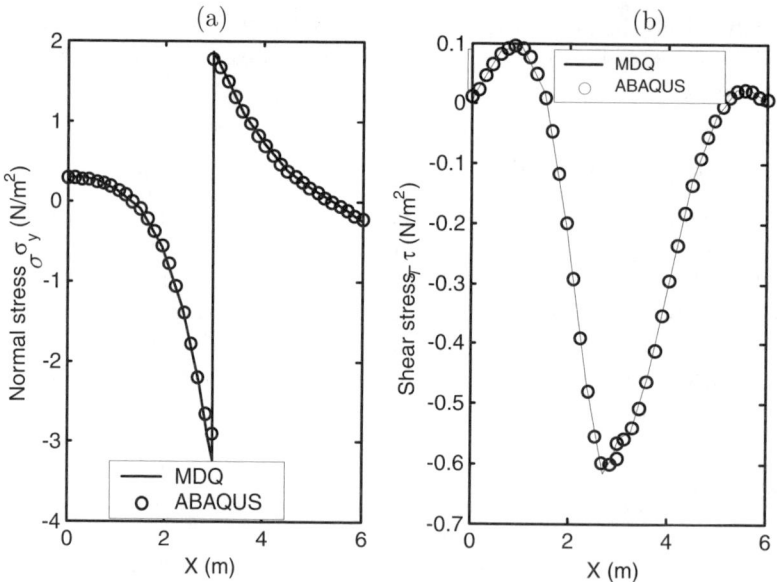

FIGURE 6.4: Stress distribution at the section $y = H/2$ for the bi-material beam with interface in x-direction: (a) Normal stress σ_y; (b) shear stress τ.

Example 6.2 *Sandwich composite plate*

A sandwich composite cantilever plate consisting of three layers of two materials is analyzed (Zong et al., 2005) shown in Fig. 6.5. The upper and lower surface layers are made of the same materials. The core layer, on the other hand, is made of a material different from the surface layers. According to the geometry and materials used in the cantilever plate, the problem is decomposed into three sub-domains that are connected by the interfaces along the y-direction. Young's modulus for the surface layers is $1.67 \times 10^9 \text{N/m}^2$ and $1.67 \times 10^8 \text{ N/m}^2$ for core layer. Poisson's ratio is assumed to be 0.3 for both materials. Other parameters are $L = 4.8m$, $H = 1.2m$, $t_c = 0.8m$, $t_f = 0.2m$ and $p = 100N$. The mesh used in this example is 15×7 for each sub-domain.

The boundary conditions for the sandwich cantilever plate are given as follows:

$$u(x = 0) = 0, \qquad v(x = 0) = 0$$

$$\sigma_x(x = 4.8) = 0, \qquad \tau(x = 4.8) = 0$$

$$\sigma_y(y = 0) = 0, \qquad \tau(y = 0) = 0$$

$$\sigma_y(y = 1.2) = -100, \qquad \tau(y = 1.2) = 0$$

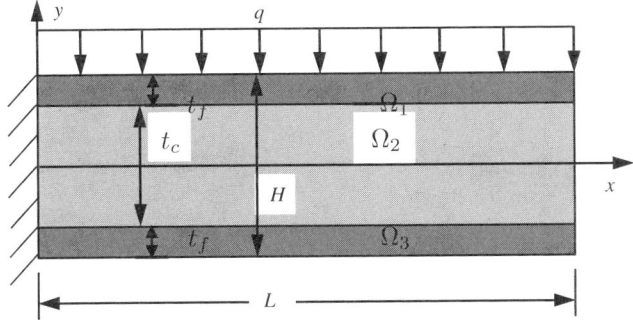

FIGURE 6.5: Sandwich composite cantilever plate made of two different materials.

The solutions for the displacements and the stresses at the middle section $x = L/2$ are given in Figs. 6.6 and 6.7, respectively. Displacements are continuous as required by the compatibility condition. Again discontinuous finite jumps are observed at the material interfaces in σ_x and sharp turning point in shear stress τ. It is also found from Figs. 6.6 and 6.7 that the numerical results produced by the multi-domain DQ approach agree well with those given by ABAQUS, verifying the accuracy of the present approach.

Example 6.3 *L-Shaped plane elastic problem*

The multi-domain DQ approach is applied to solve a plane elastic problem in an *L*-shaped domain, consisting of three parts made of two materials as shown in Fig. 6.8. Based on its geometry and materials, the physical domain is divided into three sub-domains denoted by Ω_1, Ω_2 and Ω_3, respectively, as shown in Fig. 6.8. The sub-domains Ω_1 and Ω_3 have the same material properties with Young's modulus $E_1 = E_3 = 3.0 \times 10^7 N/m^2$, while Ω_2 is made of a different material with Young's modulus $E_2 = 3.0 \times 10^6 N/m^2$. Poisson's ratios for two materials are taken as $\mu = 0.3$. The other parameters are indicated in Fig. 6.8. The mesh adopted in this example is 11×11 for each sub-domain. Figures 6.9 and 6.10 show the comparison between the stresses obtained by multi-domain DQ and ABAQUS at the section of $x = 3.75m$ and $y = 3.75m$, respectively. Again the results are in excellent agreement, validating the capability of the present approach and its applicability.

Example 6.4 *Compatibility conditions*

As mentioned in the previous section, the compatibility conditions at the interfaces have a significant influence on the solutions. Two types of com-

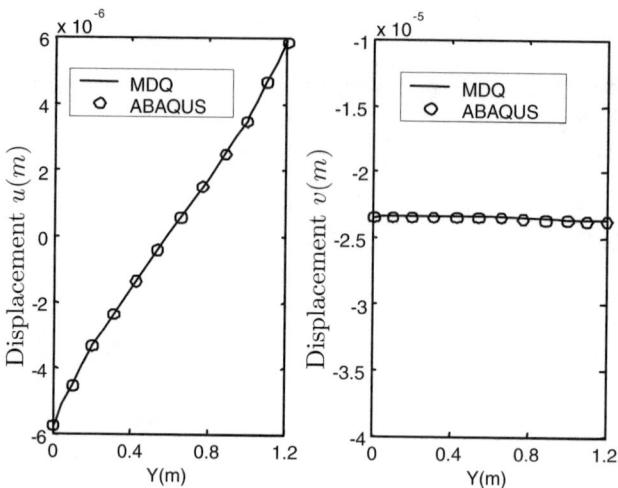

FIGURE 6.6: Displacements of sandwich cantilever plate at the section $x = L/2$.

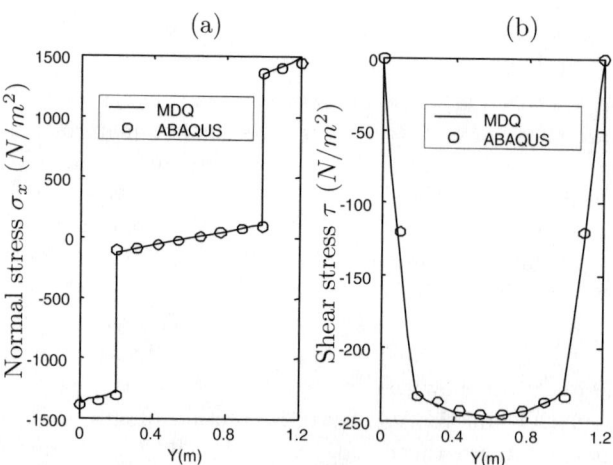

FIGURE 6.7: Stress distribution of sandwich cantilever at the section $x = L/2$: (a) stress σ_x; (b) shear stress τ.

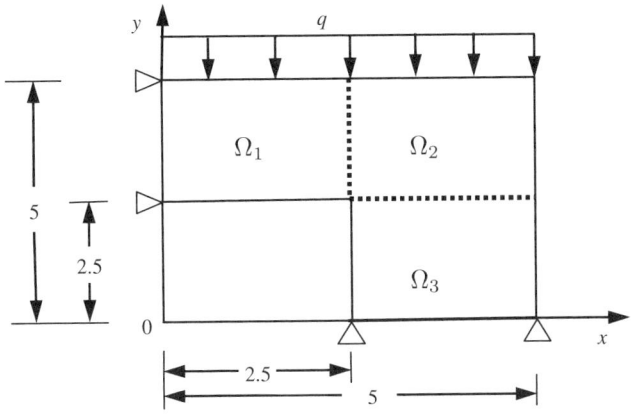

FIGURE 6.8: *L*-shaped plane elastic problem.

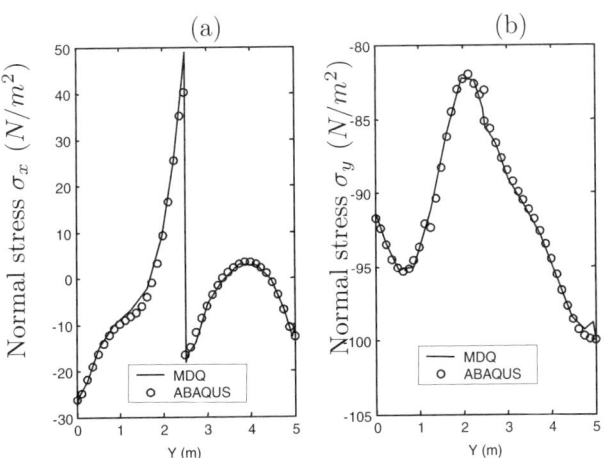

FIGURE 6.9: Stress distribution of the L-shaped problem on the line at $x = 3.75m$: (a) normal stress σ_x; (b) normal stress σ_y.

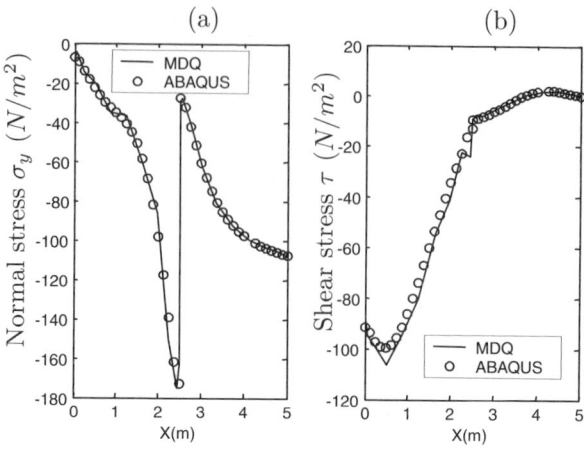

FIGURE 6.10:   Stress distribution of the L-shaped problem on the line at $y = 3.75m$: (a) normal stress; (b) shear stress.

patibility conditions can be imposed on the interfaces. Both assume that the displacements on the interfaces calculated from two neighboring sub-domains must be equal as expressed in Eq. (6.12a). The difference lies in that one type assumes the equivalence of normal derivatives whereas the other type assumes the equivalence of stress components instead; see Eq. (6.12b). For the sake of brevity, we shall refer the former type as normal derivative boundary condition (BC) and the latter type as stress BC. Both types of compatibility conditions were tested to the problem as shown in Fig. 6.11. A two-dimensional elastic medium made of two different materials separated at the section $y = 0$ is considered. The results for stress distribution along the section $x = L/2$ obtained by implementing the two different compatibility conditions are illustrated in Fig. 6.12, together with those obtained by ABAQUS. Figure 6.12 shows that the finite jump at the interface can be captured by applying normal derivative BC but the resultant stresses are not comparable with those given by ABAQUS. In contrast, the stresses given by adopting stress BC are close to those given by ABAQUS. Therefore, it is confirmed from the comparison in Fig. 6.12 that stress compatibility conditions are suitable for elastic problems in the presence of material discontinuity.

6.3   Multi-domain DQ method for irregular domain

The multi-domain differential quadrature approach, with the character of being first-order accurate at the discontinuity while remaining high-order ac-

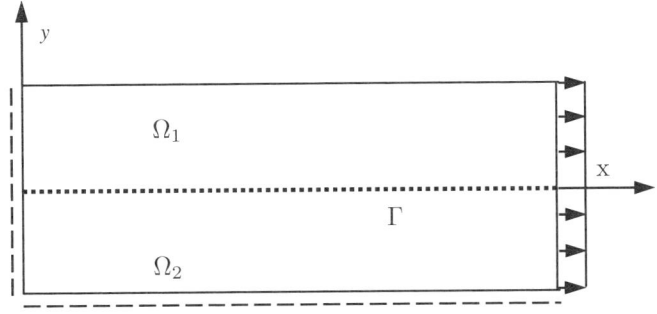

FIGURE 6.11: Two-dimensional elastic medium made up of two different materials separated at y = 0, - - - -, simply supported boundary condition.

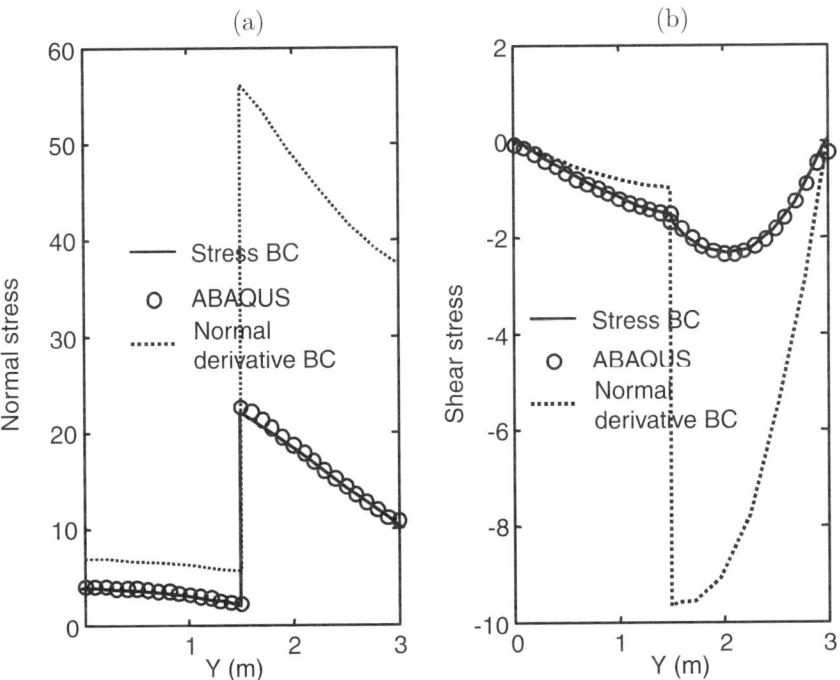

FIGURE 6.12: Stress distribution at the section $x = L/2$ for the plane elastic problem in Fig. 6.11: (a) normal stress σ_x; (b) shear stress τ.

curate elsewhere, is developed in the previous sections for the analysis of plane elastic problems made of different materials. By setting the boundary of each sub-domain on the interface of two different materials, discontinuity is transformed to proper imposition of first-order compatibility conditions at the interface. The problems considered in the previous sections are the combination of regular sub-domains. That means the boundaries of sub-domains are parallel to the coordinate axes. In view of this, local DQ scheme can be applied directly to each sub-domain in the same manner as the single domain.

In multi-domain DQ approach, the analysis domain is firstly divided into a certain number of sub-domains according to the domain's geometry, boundary condition or material discontinuity as shown in Fig. 6.3. If the problems are assemblies of regular sub-domains, same as the problems studied in the previous section, DQ scheme is carried out locally in each sub-domain. However, multi-domain DQ method is inapplicable to general irregular domains with curvilinear sides which are not parallel to the coordinate axes and which cannot be divided into regular sub-domains. In this case, coordinate transformation, the common technique in FEM, is employed to map the irregular sub-domain into a square computational domain in the natural coordinate. The governing equations and boundary conditions, including the first-order compatibility conditions of two adjacent sub-domains, are also transformed into the natural coordinate. Thus all the computations are performed in the rectangular computational domain.

It is well-known that inhomogeneities play an important role and may even dictate the mechanical behavior of advanced composite materials. In recent years, inhomogeneous problems have attracted considerable attentions and have triggered extensive investigations. For example, Eshelby (1957), Jaswon and Bhargava (1961), Honein and Hermann (1990), Gong and Meguid (1992, 1993) have studied the elastostatic behavior of one single inclusion in the elastic medium under particular loading. In addition, the problem of an elastic medium containing a finite number of arbitrary located inhomogeneous inclusions under simplified loading has been investigated by using the complex potentials of Mushelishvili (1953), Zhang and Katsube (1995). In the last decade, a lot of numerical methods, such as finite element method, boundary integral equation approach have been developed to analyze the elastic and thermoelastic behaviors of various kinds of inclusion problems (Nakasone et al., 2000; Zhu and Meguid, 2000; Dong et al., 2002, 2003).

Herein, attempt is made to apply multi-domain DQ method in conjunction with coordinate transformation to the plane elastic inhomogeneity problems. The numerical examples, a square plate with a square inclusion, a square plate with one and two circular inclusions under uniformly distributed loads, are presented to demonstrate the feasibility and accuracy of the method in the inhomogeneity problems.

6.3.1   Coordinate transformation for irregular domain

Coordinate transformation is employed to transform the irregular sub-domain in the Cartesian $x - y$ plane to a square computational domain in the natural coordinate $\xi - \eta$ by using the following equations

$$\begin{cases} x = x(\xi, \eta) \\ y = y(\xi, \eta) \end{cases} \tag{6.16}$$

All the variables, including their derivatives with respect to the space variable x or y, should be mapped to the new coordinate system $\xi - \eta$. The first-order derivatives of an arbitrary function defined in the Cartesian $x - y$ plane with respect to x and y, are given by

$$\frac{\partial}{\partial x} = \frac{\partial}{\partial \xi} \xi_x + \frac{\partial}{\partial \eta} \eta_x \tag{6.17a}$$

$$\frac{\partial}{\partial y} = \frac{\partial}{\partial \xi} \xi_y + \frac{\partial}{\partial \eta} \eta_y \tag{6.17b}$$

where ξ_x and η_x are the first-order derivatives of ξ and η with respect to x, respectively. ξ_y and η_y are the first-order derivatives of ξ and η with respect to y, respectively.

The second-order derivatives of a function could be derived from Eq. (6.17) as

$$\frac{\partial^2}{\partial x^2} = \frac{\partial^2}{\partial \xi^2} \xi_x^2 + \frac{\partial^2}{\partial \eta^2} \eta_x^2 + 2\xi_x \eta_x \frac{\partial^2}{\partial \xi \partial \eta} + \xi_{xx} \frac{\partial}{\partial \xi} + \eta_{xx} \frac{\partial}{\partial \eta} \tag{6.18a}$$

$$\frac{\partial^2}{\partial y^2} = \frac{\partial^2}{\partial \xi^2} \xi_y^2 + \frac{\partial^2}{\partial \eta^2} \eta_y^2 + 2\xi_y \eta_y \frac{\partial^2}{\partial \xi \partial \eta} + \xi_{yy} \frac{\partial}{\partial \xi} + \eta_{yy} \frac{\partial}{\partial \eta} \tag{6.18b}$$

$$\frac{\partial^2}{\partial xy} = \frac{\partial^2}{\partial \xi^2} \xi_x \xi_y + \frac{\partial^2}{\partial \eta^2} \eta_x \eta_y + (\xi_x \eta_y + \xi_y \eta_x) \frac{\partial^2}{\partial \xi \partial \eta} + \xi_{xy} \frac{\partial}{\partial \xi} + \eta_{xy} \frac{\partial}{\partial \eta} \tag{6.18c}$$

Equation (6.17) can be expressed in the matrix form as follows

$$\left\{ \begin{matrix} \frac{\partial}{\partial x} \\ \frac{\partial}{\partial y} \end{matrix} \right\} = \begin{bmatrix} \xi_x & \eta_x \\ \xi_y & \eta_y \end{bmatrix} \left\{ \begin{matrix} \frac{\partial}{\partial \xi} \\ \frac{\partial}{\partial \eta} \end{matrix} \right\} = [J]^{-1} \left\{ \begin{matrix} \frac{\partial}{\partial \xi} \\ \frac{\partial}{\partial \eta} \end{matrix} \right\} \tag{6.19}$$

The above 2×2 matrix denoted by $[J]^{-1}$ is the inverse of Jacobian matrix of the transformation $[J]$ defined as

$$[J] = \begin{bmatrix} x_\xi & y_\xi \\ x_\eta & y_\eta \end{bmatrix} \tag{6.20}$$

By use of Eq. (6.33b), the inverse matrix of Jacobian is obtained

$$[J]^{-1} = \frac{1}{|J|} \begin{bmatrix} y_\eta & -y_\xi \\ -x_\eta & x_\xi \end{bmatrix} \qquad |J| = x_\xi y_\eta - x_\eta y_\xi \tag{6.21}$$

where $|J|$ is the determinant of the Jacobian.

Comparing the inverse matrix of Jacobian in Eq. (6.33a) with that in Eq. (6.34), we have the following relationship

$$\xi_x = y_\eta/|J|, \quad \eta_x = -y_\xi/|J|, \quad \xi_y = -x_\eta/|J|, \quad \eta_y = x_\xi/|J| \qquad (6.22)$$

The substitution of Eq. (6.35) into Eqs. (6.17) yields

$$\frac{\partial}{\partial x} = \frac{1}{|J|}\left(y_\eta\frac{\partial}{\partial\xi} - y_\xi\frac{\partial}{\partial\eta}\right) \qquad (6.23a)$$

$$\frac{\partial}{\partial y} = \frac{1}{|J|}\left(-x_\eta\frac{\partial}{\partial\xi} + x_\xi\frac{\partial}{\partial\eta}\right) \qquad (6.23b)$$

Therefore, the second-order derivatives of ξ with respect to x and y can be expressed as

$$
\begin{aligned}
\frac{\partial^2\xi}{\partial x^2} &= \frac{\partial}{\partial x}(\xi_x) = \frac{1}{|J|}\left(y_\eta\frac{\partial}{\partial\xi} - y_\xi\frac{\partial}{\partial\eta}\right)\left(\frac{y_\eta}{|J|}\right)\\
&= \frac{1}{|J|^2}\left(y_\eta y_{\xi\eta} - \frac{y_\eta^2}{|J|}|J|_{,\xi} - y_\xi y_{\eta\eta} + \frac{y_\xi y_\eta}{|J|}|J|_{,\eta}\right)
\end{aligned}
\qquad (6.24a)
$$

$$
\begin{aligned}
\frac{\partial^2\xi}{\partial y^2} &= \frac{\partial}{\partial y}(\xi_y) = \frac{1}{|J|}\left(-x_\eta\frac{\partial}{\partial\xi} + x_\xi\frac{\partial}{\partial\eta}\right)\left(-\frac{x_\eta}{|J|}\right)\\
&= \frac{1}{|J|^2}\left(x_\eta x_{\xi\eta} - \frac{x_\eta^2}{|J|}|J|_{,\xi} - x_\xi x_{\eta\eta} + \frac{x_\xi x_\eta}{|J|}|J|_{,\eta}\right)
\end{aligned}
\qquad (6.24b)
$$

In a similar manner, the second-order derivatives of η with respect to x and y can also be obtained

$$\frac{\partial^2\eta}{\partial x^2} = \frac{1}{|J|^2}\left(-y_\eta y_{\xi\xi} + \frac{y_\eta y_\xi}{|J|}|J|_{,\xi} + y_\xi y_{\xi\eta} - \frac{y_\xi^2}{|J|}|J|_{,\eta}\right) \qquad (6.24c)$$

$$\frac{\partial^2\eta}{\partial y^2} = \frac{1}{|J|^2}\left(-x_\eta x_{\xi\xi} + \frac{x_\eta x_\xi}{|J|}|J|_{,\xi} + x_\xi x_{\xi\eta} - \frac{x_\xi^2}{|J|}|J|_{,\eta}\right) \qquad (6.24d)$$

where $|J|_{,\xi}$ and $|J|_{,\eta}$ are the first-order derivatives of the determinant of Jacobian with respect to ξ and η, respectively. Differentiation of $|J|$ in Eq. (6.21) leads to

$$|J|_{,\xi} = x_\xi y_{\xi\eta} - y_\xi x_{\xi\eta} + y_\eta x_{\xi\xi} - x_\eta y_{\xi\xi} \qquad (6.25a)$$

$$|J|_{,\eta} = -x_\eta y_{\xi\eta} + y_\eta x_{\xi\eta} - y_\xi x_{\eta\eta} + x_\xi y_{\eta\eta} \qquad (6.25b)$$

And the mixed derivatives of ξ and η with respect to x and y are given by

$$\frac{\partial^2\xi}{\partial x\partial y} = \frac{1}{|J|^2}\left(-y_\eta x_{\xi\eta} + \frac{y_\eta x_\eta}{|J|}|J|_{,\xi} + y_\xi x_{\eta\eta} - \frac{y_\xi x_\eta}{|J|}|J|_{,\eta}\right) \qquad (6.26a)$$

$$\frac{\partial^2\eta}{\partial x\partial y} = \frac{1}{|J|^2}\left(-y_\xi x_{\xi\eta} - \frac{y_\eta x_\xi}{|J|}|J|_{,\xi} + y_\eta x_{\xi\xi} + \frac{y_\xi x_\xi}{|J|}|J|_{,\eta}\right) \qquad (6.26b)$$

The above formulation of coordinate transformation is general. Thus, various shape functions for transformation can be used. Herein, the quadratic serendipity and cubic serendipity shape functions are illustrated and applied to the numerical examples.

6.3.2 Quadratic sub-domain

The mapping of quadratic serendipity domain in the Cartesian $x - y$ plane to an 8-node square computational domain in the natural coordinates $\xi - \eta$, $-1 \leq \xi, \eta \leq 1$ as shown in Fig. 6.13, can be achieved by using the following relationship (Bath, 1996)

$$x = \sum_{i=1}^{8} N_i(\xi, \eta) \cdot x_i \tag{6.27a}$$

$$y = \sum_{i=1}^{8} N_i(\xi, \eta) \cdot y_i \tag{6.27b}$$

where x_i and y_i are the coordinates of the i-th boundary node in $x - y$ plane, $N_i(\xi, \eta)$ are the quadratic serendipity shape function defined as:

$$N_1 = -\frac{1}{4}(1 - \xi)(1 - \eta)(1 + \xi + \eta), \quad N_2 = \frac{1}{4}(1 + \xi)(1 - \eta)(-1 + \xi - \eta),$$

$$N_3 = \frac{1}{4}(1 + \xi)(1 + \eta)(-1 + \xi + \eta), \quad N_4 = \frac{1}{4}(1 - \xi)(1 + \eta)(-1 - \xi + \eta),$$

$$N_5 = \frac{1}{2}(1 - \xi^2)(1 - \eta), \quad N_6 = \frac{1}{2}(1 + \xi)(1 - \eta^2),$$

$$N_7 = \frac{1}{2}(1 - \xi^2)(1 + \eta), \quad N_8 = \frac{1}{2}(1 - \xi)(1 - \eta^2) \tag{6.28}$$

The first-order derivatives of the physical coordinates with respect to the natural coordinates are calculated as (Chen, 1999)

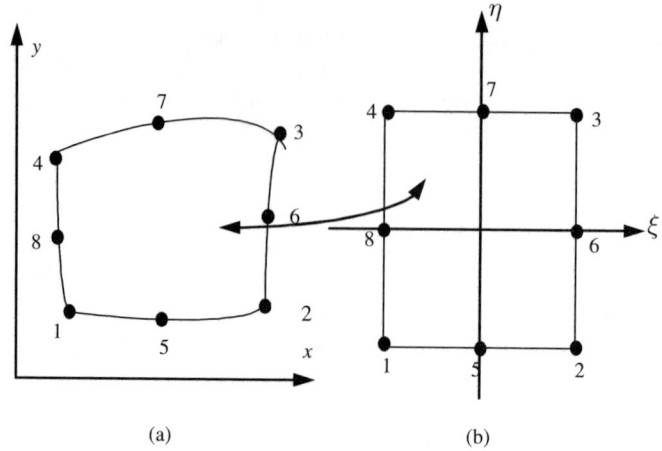

FIGURE 6.13: Mapping of the quadratic serendipity element: (a) the physical domain; (b) the computational domain .

$$x_\xi = \tfrac{1}{4}(1-\eta)(2\xi+\eta)x_1 + \tfrac{1}{4}(1-\eta)(2\xi-\eta)x_2 + \tfrac{1}{4}(1+\eta)(2\xi+\eta)x_3 + \tfrac{1}{4}(1+\eta)(2\xi-\eta)x_4 - \xi(1-\eta)x_5 + \tfrac{1}{2}(1-\eta^2)x_6 - \xi(1+\eta)x_7 - \tfrac{1}{2}(1-\eta^2)x_8$$

$$(6.29a)$$

$$x_\eta = \tfrac{1}{4}(1-\xi)(\xi+2\eta)x_1 + \tfrac{1}{4}(1+\xi)(-\xi+2\eta)x_2 + \tfrac{1}{4}(1+\xi)(\xi+2\eta)x_3 + \tfrac{1}{4}(1-\xi)(-\xi+2\eta)x_4 - \tfrac{1}{2}(1-\xi^2)x_5 - \eta(1+\xi)x_6 + \tfrac{1}{2}(1-\xi^2)x_7 - \eta(1-\xi)x_8$$

$$(6.29b)$$

$$y_\xi = \tfrac{1}{4}(1-\eta)(2\xi+\eta)y_1 + \tfrac{1}{4}(1-\eta)(2\xi-\eta)y_2 + \tfrac{1}{4}(1+\eta)(2\xi+\eta)y_3 + \tfrac{1}{4}(1+\eta)(2\xi-\eta)y_4 - \xi(1-\eta)y_5 + \tfrac{1}{2}(1-\eta^2)y_6 - \xi(1+\eta)y_7 - \tfrac{1}{2}(1-\eta^2)y_8$$

$$(6.29c)$$

$$x_\eta = \tfrac{1}{4}(1-\xi)(\xi+2\eta)y_1 + \tfrac{1}{4}(1+\xi)(-\xi+2\eta)y_2 + \tfrac{1}{4}(1+\xi)(\xi+2\eta)y_3 + \tfrac{1}{4}(1-\xi)(-\xi+2\eta)y_4 - \tfrac{1}{2}(1-\xi^2)y_5 - \eta(1+\xi)y_6 + \tfrac{1}{2}(1-\xi^2)y_7 - \eta(1-\xi)y_8$$

$$(6.29d)$$

And the second-order derivatives are given by (Chen, 1999)

$$x_{\xi\xi} = \frac{1}{2}(x_1+x_2+x_3+x_4-2x_5-2x_7)+\frac{1}{2}\eta(-x_1-x_2+x_3+x_4+2x_5-2x_7)$$

$$x_{\eta\eta} = \frac{1}{2}(x_1+x_2+x_3+x_4-2x_6-2x_8)+\frac{1}{2}\xi(-x_1+x_2+x_3-x_4-2x_6+2x_8)$$

$$y_{\xi\xi} = \frac{1}{2}(y_1+y_2+y_3+y_4-2y_5-2y_7)+\frac{1}{2}\eta(-y_1-y_2+y_3+y_4+2y_5-2y_7)$$

$$y_{\eta\eta} = \frac{1}{2}(y_1+y_2+y_3+y_4-2y_6-2y_8)+\frac{1}{2}\xi(-y_1+y_2+y_3-y_4-2y_6+2y_8)$$

$$x_{\xi\eta} = \tfrac{1}{4}(x_1 - x_2 + x_3 - x_4) + \tfrac{1}{2}\xi(-x_1 - x_2 + x_3 + x_4 + 2x_5 - 2x_7)$$
$$+ \tfrac{1}{2}\eta(-x_1 + x_2 + x_3 - x_4 - 2x_6 + 2x_8)$$

$$y_{\xi\eta} = \tfrac{1}{4}(y_1 - y_2 + y_3 - y_4) + \tfrac{1}{2}\xi(-y_1 - y_2 + y_3 + y_4 + 2y_5 - 2y_7) \qquad (6.30)$$
$$+ \tfrac{1}{2}\eta(-y_1 + y_2 + y_3 - y_4 - 2y_6 + 2y_8)$$

6.3.3 Cubic sub-domain

A curvilinear quadrilateral domain with curve boundaries in the physical coordinate $x - y$ is shown in Fig. 6.14. Each side of the domain can be approximated by a cubic function. The irregular domain can be mapped into a square domain, $-1 \le \xi,\ \eta \le 1$ by use of the following cubic serendipity shape function (Li et al, 1986)

$$x = \sum_{i=1}^{12} N_i(\xi, \eta) \cdot x_i \qquad (6.31a)$$

$$y = \sum_{i=1}^{12} N_i(\xi, \eta) \cdot y_i \qquad (6.31b)$$

where $N_i(\xi, \eta)$ is the cubic serendipity shape function defined by

$$N_i(\xi, \eta) = \frac{1}{32}(1 + \xi_i\xi)(1 + \eta_i\eta)[9(\xi^2 + \eta^2) - 10], \quad i = 1,\ 2,\ 3,\ 4$$

$$N_i(\xi, \eta) = \frac{9}{32}(1 - \xi^2)(1 + \eta_i\eta)(1 + 9\xi\xi_i), \quad i = 5,\ 6,\ 7,\ 8$$

$$N_i(\xi, \eta) = \frac{9}{32}(1 + \xi_i\xi)(1 - \eta^2)(1 + 9\eta_i\eta], \quad i = 9,\ 10,\ 11,\ 12$$

where ξ_i and η_i are the coordinates of the node i in the $\xi - \eta$ plane. All of these shape functions possess the delta function property, i.e., the shape functions are equal to utility at the i-th point and zero at all the other points. The first-order derivative of the physical coordinate with respect to the natural

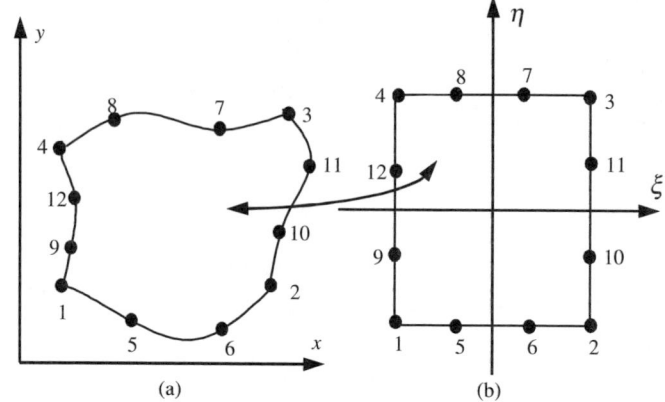

FIGURE 6.14: Mapping of the cubic serendipity element from (a) physical domain to (b) computational domain.

coordinate are derived as

$$x_\xi = \sum_{i=1}^{4} \tfrac{1}{32}(1 + \eta_i\eta)[\xi_i(9\xi^2 + 9\eta^2 - 10) + 18\xi(1 + \xi_i\xi)]x_i +$$
$$\sum_{i=5}^{8} \tfrac{9}{32}(1 + \eta_i\eta)[-2\xi(1 + 9\xi_i\xi) + 9\xi_i(1 - \xi^2)]x_i + \sum_{i=9}^{12} \tfrac{9}{32}(1 - \eta^2)(1 + 9\eta_i\eta)\xi_i x_i$$

$$(6.32a)$$

$$x_\eta = \sum_{i=1}^{4} \tfrac{1}{32}(1 + \xi_i\xi)[\eta_i(9\xi^2 + 9\eta^2 - 10) + 18\eta(1 + \eta_i\eta)]x_i +$$
$$\sum_{i=5}^{8} \tfrac{9}{32}(1 - \xi^2)(1 + 9\xi_i\xi)\eta_i x_i + \sum_{i=9}^{12} \tfrac{9}{32}(1 + \xi_i\xi)[-2\eta(1 + 9\eta_i\eta) + 9\eta_i(1 - \eta^2)]x_i$$

$$(6.32b)$$

$$y_\xi = \sum_{i=1}^{4} \tfrac{1}{32}(1 + \eta_i\eta)[\xi_i(9\xi^2 + 9\eta^2 - 10) + 18\xi(1 + \xi_i\xi)]y_i +$$
$$\sum_{i=5}^{8} \tfrac{9}{32}(1 + \eta_i\eta)[-2\xi(1 + 9\xi_i\xi) + 9\xi_i(1 - \xi^2)]y_i + \sum_{i=9}^{12} \tfrac{9}{32}(1 - \eta^2)(1 + 9\eta_i\eta)\xi_i y_i$$

$$(6.32c)$$

$$y_\eta = \sum_{i=1}^{4} \tfrac{1}{32}(1 + \xi_i\xi)[\eta_i(9\xi^2 + 9\eta^2 - 10) + 18\eta(1 + \eta_i\eta)]y_i +$$
$$\sum_{i=5}^{8} \tfrac{9}{32}(1 - \xi^2)(1 + 9\xi_i\xi)\eta_i y_i + \sum_{i=9}^{12} \tfrac{9}{32}(1 + \xi_i\xi)[-2\eta(1 + 9\eta_i\eta) + 9\eta_i(1 - \eta^2)]y_i$$

$$(6.32d)$$

and the second-order derivatives are obtained as follows:

$$x_{\xi\xi} = \sum_{i=1}^{4} (1 + \eta_i\eta)(18 + 54\xi_i\xi)x_i + \sum_{i=5}^{8} \frac{9}{32}(1 + \eta_i\eta)(-2 - 54\xi_i\xi)x_i \quad (6.32a)$$

$$x_{\eta\eta} = \sum_{i=1}^{4} (1 + \xi_i\xi)(18 + 54\eta_i\eta)x_i + \sum_{i=9}^{12} \frac{9}{32}(1 + \xi_i\xi)(-2 - 54\eta_i\eta)x_i \quad (6.32b)$$

$$y_{\xi\xi} = \sum_{i=1}^{4} (1 + \eta_i\eta)(18 + 54\xi_i\xi)y_i + \sum_{i=5}^{8} \frac{9}{32}(1 + \eta_i\eta)(-2 - 54\xi_i\xi)y_i \quad (6.32c)$$

$$y_{\eta\eta} = \sum_{i=1}^{4} (1 + \xi_i\xi)(18 + 54\eta_i\eta)y_i + \sum_{i=9}^{12} \frac{9}{32}(1 + \xi_i\xi)(-2 - 54\eta_i\eta)y_i \quad (6.32d)$$

$$x_{\xi\eta} = \sum_{i=1}^{4} \frac{1}{32}[18\xi_i\eta + 27\xi_i\eta_i(\xi^2 + \eta^2) - 10\xi_i\eta_i + 18\xi\eta_i]x_i$$
$$+ \sum_{i=5}^{8} \frac{9}{32}\eta_i(-2\xi - 27\xi_i\xi^2 + 9\xi_i)x_i + \sum_{i=9}^{12} \frac{9}{32}\xi_i(-2\eta + 9\eta_i - 27\eta^2\eta_i)x_i \quad (6.32e)$$

$$y_{\xi\eta} = \sum_{i=1}^{4} \frac{1}{32}[18\xi_i\eta + 27\xi_i\eta_i(\xi^2 + \eta^2) - 10\xi_i\eta_i + 18\xi\eta_i]y_i$$
$$+ \sum_{i=5}^{8} \frac{9}{32}\eta_i(-2\xi - 27\xi_i\xi^2 + 9\xi_i)y_i + \sum_{i=9}^{12} \frac{9}{32}\xi_i(-2\eta + 9\eta_i - 27\eta^2\eta_i)y_i \quad (6.32f)$$

6.4 Multi-domain DQ formulation of plane elastic problems

6.4.1 Multi-domain DQ for plane elastic heterogeneous solids

To simulate the discontinuity at the interface of two adjacent sub-domains, the numerical scheme should be first-order accurate at the interface while remain high-order accuracy everywhere. As mentioned before, DQ method is a high-order scheme which cannot capture the first-order accuracy at the discontinuity, thus it fails to furnish correct results for the problem in the presence of discontinuity. In this case, multi-domain DQ approach is adopted. Multi-domain DQ method is based on DQ and domain decomposition as an alternative to simulate complex problems with discontinuities in geometry, boundary conditions and material. Each sub-domain is firmly bonded together through interfaces. Each sub-domain is guaranteed to be homogenous, having the same material property through the sub-domain by the domain decomposition. Then the discontinuity along the interface is transformed by the imposition of compatibility conditions. In general, the heterogeneous solid is composed of irregular sub-domains, which cannot be solved by the present approach. Under the circumstances, the additional mapping technique is utilized to transfer the irregular sub-domain into a square one. Then DQ discretization of the governing equation is applied in the square computational domain. All the corresponding boundary conditions and compatibility conditions along the interfaces are transformed from the physical domain to the computational one. DQ scheme is applied separately in each normalized sub-domain in the natural coordinates. Finally, all the differential quadrature discretized equations are assembled and the solutions are obtained by use of Gauss elimination.

6.4.2 Governing equations and its DQ discretization

In the absence of body force, the governing equation for a two-dimensional problem in linear elasticity on the domain Ω bounded by the boundary Γ is given by

$$\alpha \frac{\partial^2 u}{\partial x^2} + \beta \frac{\partial^2 v}{\partial x \partial y} + \frac{\partial^2 u}{\partial y^2} = 0 \tag{6.33a}$$

$$\alpha \frac{\partial^2 v}{\partial y^2} + \beta \frac{\partial^2 u}{\partial x \partial y} + \frac{\partial^2 v}{\partial x^2} = 0 \tag{6.33b}$$

where x and y are space variables in the domain Ω, u and v represent the components of the displacement in x- and y-direction, respectively. When plane stress case is assumed, the coefficients denoted by α and β are given by

$$\alpha = \frac{2}{1-\mu}, \quad \beta = \frac{1+\mu}{1-\mu} \tag{6.34}$$

where μ is Passion's ratio, which may be varied from one sub-domain to another. For plane strain problem, μ is replaced by $\frac{1+\mu}{1-\mu}$.

In the previous section, the problem domain is assemblies of a certain number of regular sub-domains, and DQ formulation can be carried out on basis of the sub-domains one by one directly without coordinate transformation. Since the coordinates of each sub-domain in the Cartesian plane vary in different sub-domains, therefore, the weighting coefficients for DQ discretization are location-dependent. However, if the sub-domains are of arbitrary shapes, mapping technique is required to transform the irregular domain into a square computational domain, and then DQ discretization can be formulated. Since all the sub-domains are mapped into the identical square domain $-1 \le \xi, \ \eta \le 1$, the weighting coefficients of the first- and second-order derivatives denoted by a_{ij} and b_{ij} are the same.

According to Shu and Richards (1992), the weighting coefficients in the ξ direction are determined by differentiating the Lagrange interpolation formula as

$$a_{ij} = \begin{cases} \frac{M^{(1)}(\xi_i)}{(\xi_i - \xi_j) M^{(1)}(\xi_j)} & i \ne j \\ -\sum\limits_{k=1, k \ne i}^{N} a_{ik} & i = j \end{cases} \tag{6.35}$$

$$b_{ij} = \begin{cases} 2\left[a_{ii} \cdot a_{ij} - \dfrac{a_{ij}}{\xi_i - \xi_j} \right], & i \neq j \\ -\displaystyle\sum_{k=1, k\neq i}^{N} b_{ik} & i = j \end{cases} \qquad (6.36)$$

where

$$M(\xi) = \prod_{j=1}^{N} (\xi - \xi_j)$$

$$M^{(1)}(\xi_k) = \prod_{j=1, j\neq k}^{N} (\xi_k - \xi_j), \ k = 1, 2, \cdots, N$$

where ξ_i denotes the mesh point in the ξ direction in the computational domain. Similarly, the weighting coefficients in the η direction can be obtained by replacing ξ in Eqs. (6.35) and (6.36) by η.

By substituting Eqs. (6.17) and (6.18) into Eq. (6.34), we obtain the following governing equations in Ω_s defined in the natural coordinate

$$F_1(\xi,\eta)\frac{\partial^2 u}{\partial \xi^2} + F_2(\xi,\eta)\frac{\partial^2 u}{\partial \eta^2} + F_3(\xi,\eta)\frac{\partial u}{\partial \xi} + F_4(\xi,\eta)\frac{\partial u}{\partial \eta} + F_5(\xi,\eta)\frac{\partial^2 u}{\partial \xi \partial \eta} +$$
$$\bar{F}_1(\xi,\eta)\frac{\partial^2 v}{\partial \xi^2} + \bar{F}_2(\xi,\eta)\frac{\partial^2 v}{\partial \eta^2} + \bar{F}_3(\xi,\eta)\frac{\partial v}{\partial \xi} + \bar{F}_4(\xi,\eta)\frac{\partial v}{\partial \eta} + \bar{F}_5(\xi,\eta)\frac{\partial^2 v}{\partial \xi \partial \eta} = 0$$
$$H_1(\xi,\eta)\frac{\partial^2 v}{\partial \xi^2} + H_2(\xi,\eta)\frac{\partial^2 v}{\partial \eta^2} + H_3(\xi,\eta)\frac{\partial v}{\partial \xi} + H_4(\xi,\eta)\frac{\partial v}{\partial \eta} + H_5(\xi,\eta)\frac{\partial^2 v}{\partial \xi \partial \eta} +$$
$$\bar{F}_1(\xi,\eta)\frac{\partial^2 u}{\partial \xi^2} + \bar{F}_2(\xi,\eta)\frac{\partial^2 u}{\partial \eta^2} + \bar{F}_3(\xi,\eta)\frac{\partial u}{\partial \xi} + \bar{F}_4(\xi,\eta)\frac{\partial u}{\partial \eta} + \bar{F}_5(\xi,\eta)\frac{\partial^2 u}{\partial \xi \partial \eta} = 0$$
$$(6.37)$$

where

The above transformation functions $F_i(\xi,\eta)$, $H_i(\xi,\eta)$ and $\bar{F}_i(\xi,\eta)$ ($i = 1, 2, \ldots, 5$) can be expressed as functions of ξ and η by using Eqs. (6.22) and (6.24).

The application of DQ discretization to Eq.(6.37) results in a set of simul-

$$F_1(\xi,\eta) = \alpha^s \xi_x^2 + \xi_y^2,$$
$$F_2(\xi,\eta) = \alpha^s \eta_x^2 + \eta_y^2,$$
$$F_3(\xi,\eta) = \alpha^s \xi_{xx} + \xi_{yy},$$
$$F_4(\xi,\eta) = \alpha^s \eta_{xx} + \eta_{yy},$$
$$F_5(\xi,\eta) = 2\alpha^s \xi_x \eta_x + 2\xi_y \eta_y$$

$$H_1(\xi,\eta) = \xi_x^2 + \alpha^s \xi_y^2,$$
$$H_2(\xi,\eta) = \eta_x^2 + \alpha^s \eta_y^2,$$
$$H_3(\xi,\eta) = \xi_{xx} + \alpha^s \xi_{yy},$$
$$H_4(\xi,\eta) = \eta_{xx} + \alpha^s \eta_{yy},$$
$$H_5(\xi,\eta) = 2\xi_x \eta_x + 2\alpha^s \xi_y \eta_y$$

$$\bar{F}_1(\xi,\eta) = \beta^s \xi_x \xi_y$$
$$\bar{F}_2(\xi,\eta) = \beta^s \eta_x \eta_y$$
$$\bar{F}_3(\xi,\eta) = \beta^s \xi_{xy}$$
$$\bar{F}_4(\xi,\eta) = \beta^s \eta_{xy}$$
$$\bar{F}_5(\xi,\eta) = \beta^s (\xi_x \eta_y + \xi_y \eta_x)$$

taneous algebraic equations,

$$F_1(\xi,\eta) \sum_{k_1=1}^{N_\xi} b_{ik_1} u_{k_1 j} + F_2(\xi,\eta) \sum_{k_2=1}^{N_\eta} \bar{b}_{jk_2} u_{ik_2} + F_3(\xi,\eta) \sum_{k_1=1}^{N_\xi} a_{ik_1} u_{k_1 j}$$
$$+F_4(\xi,\eta) \sum_{k_2=1}^{N_\eta} \bar{a}_{jk_2} u_{ik_2} + F_5(\xi,\eta) \sum_{k_1=1}^{N_\xi} a_{ik_1} \sum_{k_2=1}^{N_\eta} \bar{a}_{jk_2} u_{k_1 k_2} + \bar{F}_1(\xi,\eta) \sum_{k_1=1}^{N_\xi} b_{ik_1} v_{k_1 j}$$
$$+\bar{F}_2(\xi,\eta) \sum_{k_2=1}^{N_\eta} \bar{b}_{jk_2} v_{ik_2} + \bar{F}_3(\xi,\eta) \sum_{k_1=1}^{N_\xi} a_{ik_1} v_{k_1 j}$$
$$+\bar{F}_4(\xi,\eta) \sum_{k_2=1}^{N_\eta} \bar{a}_{jk_2} v_{ik_2} + \bar{F}_5(\xi,\eta) \sum_{k_1=1}^{N_\xi} a_{ik_1} \sum_{k_2=1}^{N_\eta} \bar{a}_{jk_2} v_{k_1 k_2} = 0$$

$$(6.38a)$$

$$H_1(\xi,\eta) \sum_{k_1=1}^{N_\xi} b_{ik_1} v_{k_1 j} + H_2(\xi,\eta) \sum_{k_2=1}^{N_\eta} \bar{b}_{jk_2} v_{ik_2} + H_3(\xi,\eta) \sum_{k_1=1}^{N_\xi} a_{ik_1} v_{k_1 j}$$
$$+H_4(\xi,\eta) \sum_{k_2=1}^{N_\eta} \bar{a}_{jk_2} v_{ik_2} + H_5(\xi,\eta) \sum_{k_1=1}^{N_\xi} a_{ik_1} \sum_{k_2=1}^{N_\eta} \bar{a}_{jk_2} v_{k_1 k_2}$$
$$+\bar{F}_1(\xi,\eta) \sum_{k_1=1}^{N_\xi} b_{ik_1} u_{k_1 j} + \bar{F}_2(\xi,\eta) \sum_{k_2=1}^{N_\eta} \bar{b}_{jk_2} u_{ik_2} + \bar{F}_3(\xi,\eta) \sum_{k_1=1}^{N_\xi} a_{ik_1} u_{k_1 j} +$$
$$\bar{F}_4(\xi,\eta) \sum_{k_2=1}^{N_\eta} \bar{a}_{jk_2} u_{ik_2} + \bar{F}_5(\xi,\eta) \sum_{k_1=1}^{N_\xi} a_{ik_1} \sum_{k_2=1}^{N_\eta} \bar{a}_{jk_2} u_{k_1 k_2} = 0$$

$$(6.38b)$$

where N_ξ and N_η are the number of the mesh points in the computational domain in the direction of ξ and η, respectively. Usually, we adopt $N_\xi = N_\eta$ for an easy matching between adjacent sub-domains.

Equation (6.39) can be expressed in the following compact form

$$\sum_{\beta=1}^{N} d_{\alpha\beta}^{11} u_\beta + \sum_{\beta=1}^{N} d_{\alpha\beta}^{12} v_\beta = 0 \tag{6.39a}$$

$$\sum_{\beta=1}^{N} d_{\alpha\beta}^{21} u_\beta + \sum_{\beta=1}^{N} d_{\alpha\beta}^{22} v_\beta = 0 \tag{6.39b}$$

where $\alpha = (i-1)N_\eta + j$, $N = N_\xi \times N_\eta$, $j = 1, \ldots, N_\eta$ and the coefficient matrices are

$$d_{\alpha\beta}^{11} = (F_1 b_{ik_1} + F_3 a_{ik_1})\delta_{jk_2} + (F_2 \bar{b}_{jk_2} + F_4 \bar{a}_{jk_2})\delta_{ik_1} + F_5 a_{ik_1} \bar{a}_{jk_2}$$

$$d_{\alpha\beta}^{21} = d_{\alpha\beta}^{12} = (\bar{F}_1 b_{ik_1} + \bar{F}_3 a_{ik_1})\delta_{jk_2} + (\bar{F}_2 \bar{b}_{jk_2} + \bar{F}_4 \bar{a}_{jk_2})\delta_{ik_1} + \bar{F}_5 a_{ik_1} \bar{a}_{jk_2}$$

$$d_{\alpha\beta}^{22} = (H_1 b_{ik_1} + H_3 a_{ik_1})\delta_{jk_2} + (H_2 \bar{b}_{jk_2} + H_4 \bar{a}_{jk_2})\delta_{ik_1} + H_5 a_{ik_1} \bar{a}_{jk_2}$$

where the delta function $\delta_{ij} = 1$ if $i = j$ and it is zero if $i \neq j$. The matrix form for Eq. (6.40) is

$$\begin{bmatrix} \mathbf{d^{11}} & \mathbf{d^{12}} \\ \mathbf{d^{21}} & \mathbf{d^{22}} \end{bmatrix} \begin{pmatrix} \mathbf{u}^{(s)} \\ \mathbf{v}^{(s)} \end{pmatrix} = \begin{pmatrix} \mathbf{0} \\ \mathbf{0} \end{pmatrix} \tag{6.40}$$

This is a system of linear algebraic equations with $2N$ unknowns. In compact form, the above equation becomes

$$\mathbf{D}^{(s)} \mathbf{w}^{(s)} = \mathbf{0}, \quad s = 1, \ldots M \tag{6.41}$$

where

$$\mathbf{D}^{(s)} = \begin{bmatrix} \mathbf{d^{11}} & \mathbf{d^{12}} \\ \mathbf{d^{21}} & \mathbf{d^{22}} \end{bmatrix}, \quad \mathbf{w}^{(s)} = \begin{pmatrix} \mathbf{u}^{(s)} \\ \mathbf{v}^{(s)} \end{pmatrix} \tag{6.42}$$

In the same manner, the governing equations of matrix form in all the subdomains as Eq. (6.40) can be obtained. In the end, all the sub-matrices, the weighting coefficient matrices as well as the unknown displacement vectors, should be assembled together to arrive at a system of linear algebraic equations. The global set of algebraic equations can be expressed in matrix form as

$$\begin{bmatrix} \mathbf{D^{(1)}} & & & \\ & \mathbf{D^{(2)}} & & \\ & & \ddots & \\ & & & \mathbf{D^{(M)}} \end{bmatrix} \begin{pmatrix} \mathbf{w^{(1)}} \\ \mathbf{w^{(2)}} \\ \vdots \\ \mathbf{w^{(M)}} \end{pmatrix} = \mathbf{0} \tag{6.43}$$

where M is the number of sub-domains.

To make Eq. (6.43) solvable by Gauss elimination method, the enforcement of external boundary conditions and continuity conditions along the interface of two adjacent sub-domains are required. The boundary conditions which are transformed into the computational domain can be implemented at the boundary points directly with ease. The compatibility conditions along the interfaces are illustrated in the following section.

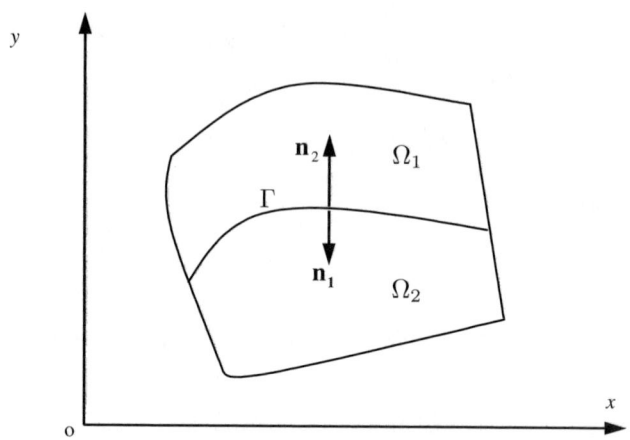

FIGURE 6.15:   Interface of two adjacent sub-domains.

6.4.3   Compatibility conditions

As mentioned before, the compatibility conditions at the interfaces have a dominant influence on the solutions. The successful application of multi-domain DQ method in plane elastic problems with material discontinuity can be achieved if the proper compatibility conditions are satisfied at the interfaces between adjacent sub-domains. For the present problem, the compatibility conditions at each node of the linear/or curvilinear interface between two adjacent sub-domains shown in Fig. 6.15 are given as

$$\mathbf{u}^{(1)} = \mathbf{u}^{(2)}, \ \mathbf{v}^{(1)} = \mathbf{v}^{(2)} \quad on \ \Gamma \tag{6.44}$$

$$\sigma_{\mathbf{n_1}}^{(1)} = \sigma_{\mathbf{n_2}}^{(2)}, \ \tau_{\mathbf{n_1}}^{(1)} = \tau_{\mathbf{n_2}}^{(2)} \quad on \ \Gamma \tag{6.45}$$

where $\mathbf{n_1}$ and $\mathbf{n_2}$ are the outward unit normal vectors at a point in the interface Γ for the two sub-domains, respectively. $\sigma_{\mathbf{n_1}}^{(1)}$ and $\tau_{\mathbf{n_1}}^{(1)}$ are the stress component perpendicular to the interface and shear stress along the interface for the first sub-domain, respectively. Similarly, $\sigma_{\mathbf{n_2}}^{(2)}$ and $\tau_{\mathbf{n_2}}^{(2)}$ are for the second sub-domain.

Let $\mathbf{n} = (l, m)$, where l, m are the cosines of the angles between the normal direction $\mathbf{n}$ and the axes x and y. The expression for the two direction cosines on the four sides of a quadrilateral sub-domain can be given as follows (Zhong and He, 2003)

$$\mathbf{n} = (l, m) = \begin{cases} \frac{\xi}{\sqrt{x_\eta^2 + y_\eta^2}}(y_\eta, -x_\eta) & \xi = \pm 1 \\ \frac{\eta}{\sqrt{x_\xi^2 + y_\xi^2}}(-y_\xi, x_\xi) & \eta = \pm 1 \end{cases} \tag{6.46}$$

Obviously, the displacement compatibility in Eq. (6.45) can be applied into the natural coordinate directly without coordinate transformation. The equilibrium conditions in Eq. (6.46) should be mapped into the computational domain as the functions of ξ and η. The normal and shear stress on the normal $\mathbf{n}$ denoted by $\sigma_{\mathbf{n}}$ and $\tau_{\mathbf{n}}$, respectively, can be expressed as

$$\sigma_{\mathbf{n}} = \frac{E}{1 - \mu^2}(l^2 + m^2\mu)\frac{\partial u}{\partial x} + \frac{E}{1 - \mu^2}(l^2\mu + m^2)\frac{\partial v}{\partial y} + \frac{Elm}{(1 + \mu)}(\frac{\partial u}{\partial y} + \frac{\partial v}{\partial x}) \tag{6.47a}$$

$$\tau_{\mathbf{n}} = \frac{Elm}{1 + \mu}(\frac{\partial v}{\partial y} - \frac{\partial u}{\partial x}) + (l^2 - m^2)\frac{E}{2(1 + \mu)}(\frac{\partial u}{\partial y} + \frac{\partial v}{\partial x}) \tag{6.47b}$$

Equation (6.48) can be rewritten in the form of the natural coordinate $\xi - \eta$ by using Eq. (6.17) as

$$\sigma_{\mathbf{n}} = \frac{E}{1 - \mu^2}(l^2 + m^2\mu)[\xi_x, \eta_x]\begin{Bmatrix} \frac{\partial u}{\partial \xi} \\ \frac{\partial u}{\partial \eta} \end{Bmatrix} + \frac{E}{1 - \mu^2}(l^2\mu + m^2)[\xi_y, \eta_y]\begin{Bmatrix} \frac{\partial v}{\partial \xi} \\ \frac{\partial v}{\partial \eta} \end{Bmatrix}$$
$$+ \frac{Elm}{(1 + \mu)}\left([\xi_y, \eta_y]\begin{Bmatrix} \frac{\partial u}{\partial \xi} \\ \frac{\partial u}{\partial \eta} \end{Bmatrix} + [\xi_x, \eta_x]\begin{Bmatrix} \frac{\partial v}{\partial \xi} \\ \frac{\partial v}{\partial \eta} \end{Bmatrix}\right) \tag{6.48a}$$

$$\tau_{\mathbf{n}} = \frac{Elm}{1 + \mu}\left([\xi_y, \eta_y]\begin{Bmatrix} \frac{\partial v}{\partial \xi} \\ \frac{\partial v}{\partial \eta} \end{Bmatrix} - [\xi_x, \eta_x]\begin{Bmatrix} \frac{\partial u}{\partial \xi} \\ \frac{\partial u}{\partial \eta} \end{Bmatrix}\right)$$
$$+ \frac{E(l^2 - m^2)}{2(1 + \mu)}\left([\xi_y, \eta_y]\begin{Bmatrix} \frac{\partial u}{\partial \xi} \\ \frac{\partial u}{\partial \eta} \end{Bmatrix} + [\xi_x, \eta_x]\begin{Bmatrix} \frac{\partial v}{\partial \xi} \\ \frac{\partial v}{\partial \eta} \end{Bmatrix}\right) \tag{6.48b}$$

By substituting the above equations into Eq. (6.46), the equilibrium conditions at the nodes along the interface are formulated in the natural coordinate $\xi - \eta$.

6.4.4 Numerical examples

In this section, a few numerical examples are presented to demonstrate the validity of multi-domain DQ method for plane elastic heterogeneous solids embedded with various shaped inclusions having different material properties. In all the examples, the non-uniform mesh point distribution used in the work of Shu and Richards (1992) are adopted to generate the mesh points in both ξ and η directions in the computational domain for more accurate solutions and more rapid convergence rate,

$$\begin{cases} \xi_i = -1 + \cos[(i-1)\pi/(N_\xi - 1)], & i = 1, 2, \ldots, N_\xi \\ \eta_j = -1 + \cos[(j-1)\pi/(N_\eta - 1)], & j = 1, 2, \ldots, N_\eta \end{cases} \qquad (6.49)$$

Example 6.5 *A square plate with a square inclusion*

The first example, shown in Fig. 6.16(a)is a square plate with a square inclusion at its center under the uniform distributed forces at its left and right sides. The plate and its inclusion possess different material properties, but the two parts are bonded firmly with each other. Due to its double symmetry, a quarter of the plate in Fig. 6.16(b)is analyzed for simplicity.

According to the material discontinuity and the geometry, the analysis domain can be divided into four rectangular sub-domains. The parameters taken in this problem for the plate are: $E_1 = 3.0 \times 10^7 \, N/m^2$, $\mu_1 = 0.3$ while the parameters of the inclusion are $E_2 = 3.0 \times 10^6 \, N/m^2$, $\mu_2 = 0.25$. The problem is solved for plane stress case with $L = 2m$, $q = 100N$. Mapping technique is employed to transform the four sub-domains in the physical coordinate into the normalized natural coordinate $-1 \le \xi$, $\eta \le 1$ by 8-node quadratic sub-domain. For the sake of comparison, multi-domain DQ method is applied to solve the present problem directly basing on each regular sub-domain in the physical coordinate, namely, avoiding coordinate transformation (CT). The results given by the multi-domain DQ with and without CT are compared to investigate the effects of the application of CT. The grid number 11×11 is adopted for DQ discretization. Figure6.17 and 6.18 show the displacements and stresses components on the line $y = L/4$, respectively. It is shown in Fig.6.17 and 6.18 that both of the computed results are in excellent agreement with those obtained by FEM. In other words, the numerical results of the square plate given by the multi-domain DQ with and without CT are close to each other. It can be concluded from Fig.6.17 and 6.18 that two techniques (with and without CT) are equally suitable to solve the problems which consist of regular sub-domains.

Note that displacement results shown in Fig. 6.17 are continuous across the interface but not smooth because of the different materials of the adjacent sub-domains. There exists a sharp turning point in the interface for the displacements. As illustrated in Fig. 6.18, the normal stress σ_x and shear stress along the line of $y = L/4$ are continuous at the interface as enforced by the compatibility conditions, whereas the normal stress σ_y experiences an abrupt finite jump at the node in the interface. The discontinuity of the normal stress σ_y at the interface can be clearly captured by the multi-domain DQ method which is found to be first-order accurate at the discontinuity.

To further investigate the convergence of the numerical results obtained by multi-domain DQ method coupled with CT, the results at some selected nodes (Points A, B, and C as shown in Fig. 6.16(b)) are summarized in Table 6.2 with different mesh sizes. From Table 6.1, it is clearly seen that the results converge fast as the grid number increases.

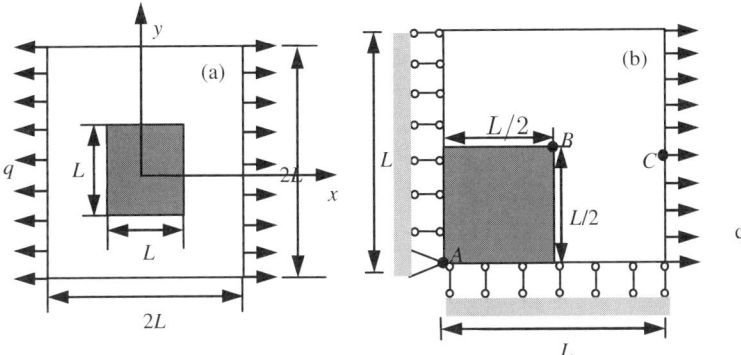

FIGURE 6.16: (a) A square plate with a square inclusion; (b) a quarter of the square plate.

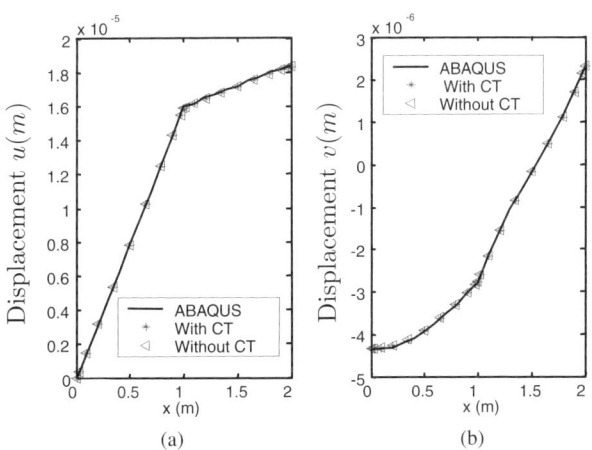

FIGURE 6.17: Comparison of the displacements on the line $y = L/4$.

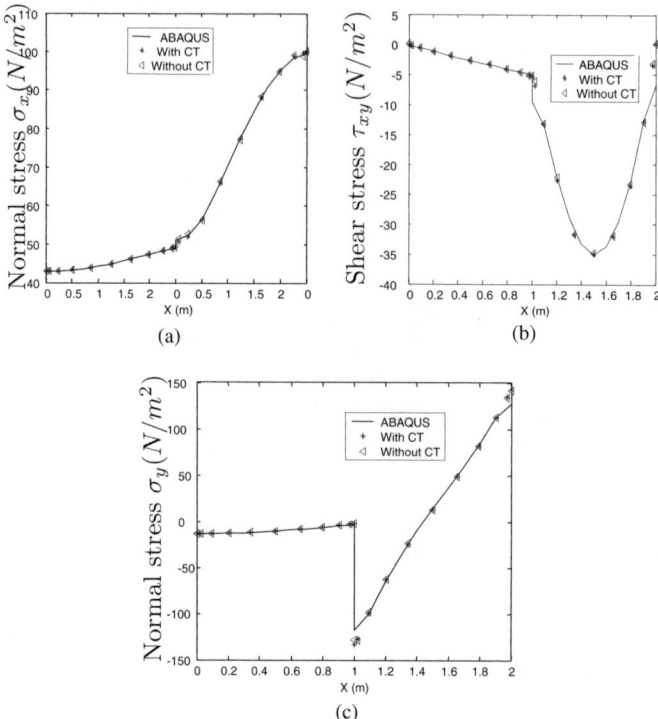

FIGURE 6.18: Comparison of the stresses on the line of $y = L/4$ for the square plate with a square inclusion: (a) Normal stress σ_x; (b) shear stress τ_{xy}; (c) normal stress σ_y. (CT-Coordinate transformation.)

TABLE 6.2: Convergence of the multi-domain DQ results for the square plate with a square inclusion by using CT

Element	σ_{xA}	σ_{yA}	$u_B(\times 10^{-6})$	$v_B(\times 10^{-6})$	σ_{xC}
5×5	51.3238	-12.7276	10.0435	-4.7057	97.7737
7×7	49.4023	-14.2027	9.84989	-4.6889	98.9832
9×9	48.9324	-14.5054	9.76603	-4.6909	99.1440
11×11	48.7672	-14.6197	9.71829	-4.6895	99.3532
13×13	48.6902	-14.6711	9.68950	-4.6888	99.4435
ABAQUS	48.6505	-14.7043	9.60461	-4.7028	100.000

Example 6.6 *Square plate with a circular inclusion*

Now we analyze a square plate with a circular inclusion subjected to uniform distributed loading at its upper and low edges as shown in Fig. 6.19(a). Again, the right upper quadrant of the plate in Fig. 6.19(b) is modeled due to double symmetry. Similarly, the symmetry conditions are implemented at the left and bottom edges. According to the material discontinuity and the geometry, the analysis domain is decomposed into four sub-domains, three irregular sub-domains with curved sides, one rectangular one. All the parameters taken in this problem are the same as the first example except $L = 5m$. The plane stress situation is assumed. To model the circular boundary, quadratic and cubic serendipity sub-domains are employed to solve the problem and the results are compared to examine the effects of different order elements on the accuracy of the solutions. When 9×9 non-uniform grid number in each sub-domain is used, the displacements components u at the $x-$ axes and v at the $y-$axes are illustrated in Figs. 6.20(a) and (b), respectively. Figure 6.21 demonstrates the normal stresses results along the $y-$axes. Figures 6.20 and 6.21 show that the numerical results given by the multi-domain DQ method using quadratic and cubic serendipity sub-domains are close to each other, indicating the relative independence of the results on different order sub-domains. There is negligible difference between the two sets of results. Again, the two sets of results given by the present DQ method agree well with those by FEM. From Fig. 6.20, we note that turning points in the displacement curves occur at the interface owing to the discontinuity of material properties. As for the stress components shown in Fig. 6.21, σ_x experiences an abrupt jump at the interface while σ_y is continuous at the interface according to the imposition of continuity conditions. All of the features in the presence of material discontinuity are captured clearly and exactly by the present method (with CT) but direct DQ method (without CT) is incapable.

The numerical results given by quadratic and cubic sub-domains for points A and B as shown in Fig. 6.19(b) are listed in Table 6.3. The two sets of results by using increasing grid numbers are compared to examine quantitatively the effects of the different order sub-domains on the accuracy of the results. It is readily seen from Table 6.3 that the application of cubic sub-domain produces relatively more accurate results than its quadratic counterpart when compared with those given by ABAQUS. In general, high-order sub-domain can approximate curvilinear boundaries more accurately, as a result, yielding more accurate results than the low-order one. For more information about the different-order sub-domains, the interested reader may refer to the paper of Zhong and He (2003) in which three different-order sub-domains, quadratic, cubic and quintic, are employed to solve the plane elastic problem of a square plate with a circular hole at its center. The same conclusion is reached by Zhong and He (2003), i.e., higher-order sub-domain is superior to its lower-order counterparts in respect of the accuracy of the results. On the other hand, more program work is demanded due to the higher-order coordinate transformation. From Tables 6.3 and 6.4, it is clearly demonstrated that the

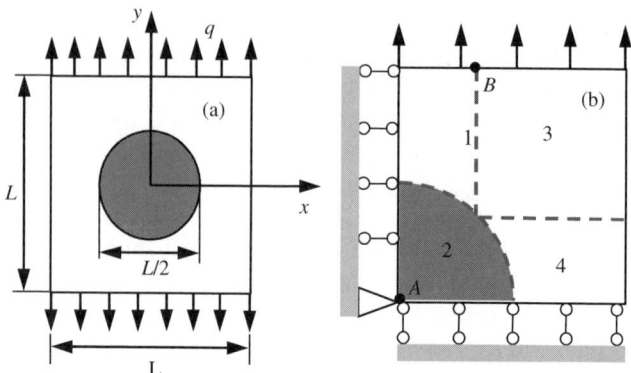

FIGURE 6.19:   (a) A square plate with a circular inclusion; (b) A quarter of the plate.

numerical results converge rapidly with the increase of grid number.

TABLE 6.3:   Results of the square plate with a circular inclusion by (I) Quadratic sub-domain (II) Cubic sub-domain

Element mesh	σ_{xA}		$u_B \times 10^{-6}$		σ_{yB}	
	I	II	I	II	I	II
5×5	-6.89596	-7.7701	1.1279	1.13443	101.0322	100.6259
7×7	-7.79552	-7.9951	1.1313	1.13261	100.1497	100.0216
9×9	-7.93897	-8.0590	1.1447	1.14687	99.97717	100.0187
11×11	-7.97405	-8.0554	1.1461	1.14853	99.98955	99.99313
ABAQUS	-8.0112		1.14905		100.0	

It is worth noting that the stress results are in good agreement with those given by FEM except the results at the interface (see Fig. 6.18(c) and Fig. 6.21 (a)). There exists obvious discrepancy at the interface. This may be due to the implementation of the first-order compatibility conditions at the interface while high-order scheme is used elsewhere. The phenomenon indirectly shows the character of the present method.

Example 6.7 *Rectangular plate with two circular inclusions*

The last example is a rectangular plate with two identical circular inclusions as shown in Fig. 6.22. The plate is subjected to the uniformly distributed loading on its four sides. Plane stress case is considered. Because of symmetry, half of the plate is modeled and the symmetry conditions are imposed on the left edge. In accordance with the material discontinuity and geometry,

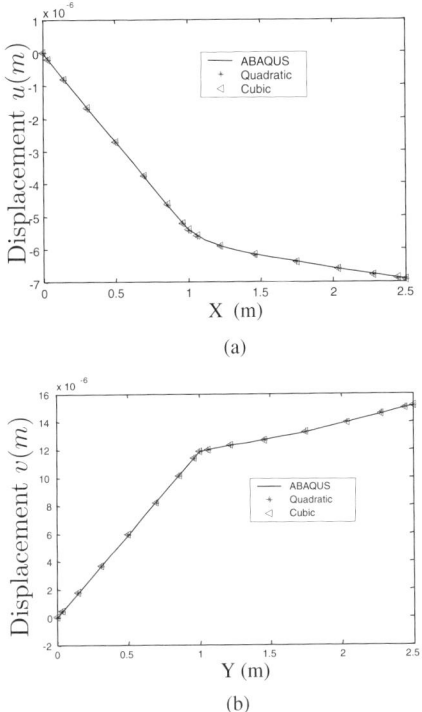

FIGURE 6.20: (a) Displacement component u at x-axis; (b) v at y-axis.

TABLE 6.4: Convergence of the multi-domain DQ results for the square plate with a circular inclusion by cubic sub-domain

Element mesh	σ_{xA}	σ_{yA}	$u_B(\times10^{-6})$	$v_B(\times10^{-5})$	σ_{xB}
5×5	-7.7701	32.6711	1.13443	1.36448	56.0812
7×7	-7.9951	33.4946	1.13261	1.38082	57.2533
9×9	-8.0590	33.6002	1.14687	1.38414	57.2109
11×11	-8.0554	33.6105	1.14853	1.38434	57.1946
ABAQUS	-8.0112	33.5709	1.14905	1.38451	57.2034

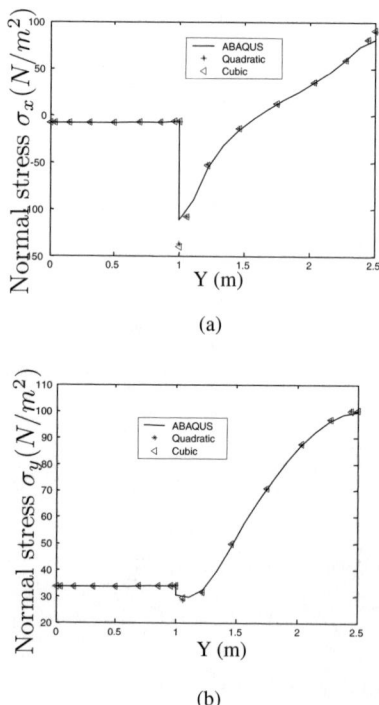

(a)

(b)

FIGURE 6.21:   (a) Normal stress σ_x; (b) σ_y along y-axis.

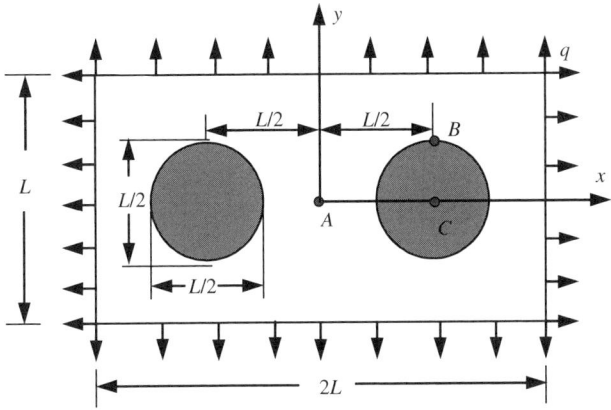

FIGURE 6.22: A rectangular plate with two circular inclusions.

the analysis domain is decomposed into nine irregular sub-domains. The parameters are taken as follows:

Uniform distributed load: $q = 100N$
Length of the plate: $2L = 8m$
Height of the plate: $L = 4m$
Radius of the inclusions: $r = 1m$
Young's modulus for the plate: $E_1 = 3.0 \times 10^{10} \, N/m^2$
Poisson's ratio for the plate: $\mu_1 = 0.17$
Young's modulus for the inclusions: $E_2 = 2.1 \times 10^{11} \, N/m^2$
Poisson's ratio for the inclusions: $\mu_2 = 0.25$

The problem is constructed by use of cubic sub-domain for more accurate results. The numerical results on the three points A, B and C are listed in Table 6.5 with the refinement of the mesh sizes. The convergence of the results can be assured by increasing the grid points. The problem consists of nine sub-domains and the interfaces are up to twelve. As mentioned before, the high-order DQ scheme is applied only at the interior nodes in the sub-domain except the nodes on the interfaces. Instead, the first-order compatibility conditions are enforced. Thus, high-order accuracy is obtained for the nodes inside the domain but only first-order accuracy is achieved at nodes along the interfaces. In view of this, the rate of convergence is influenced significantly. Therefore, it is confirmed that the adverse effect of interfaces on the accuracy of the solutions is unavoidable.

TABLE 6.5: Results of rectangular plate with two circular inclusions

Element mesh	u_B $(\times 10^{-9})$	u_c $(\times 10^{-9})$	σ_{yA}	σ_{xB}	σ_{yB}	σ_{yC}
5×5	5.4905	4.3585	64.67307	112.4342	99.9937	136.9546
7×7	6.6008	4.2807	70.5088	102.9139	99.9873	136.8141
9×9	6.5964	4.3514	72.0298	97.62095	100.0367	147.9048
11×11	5.4463	4.3420	72.9877	98.74339	100.0325	146.6240

6.5 Conclusions

A multi-domain differential quadrature has been exploited in this chapter to solve plane elastic problems with material discontinuity, which cannot be solved by direct DQ. Material discontinuity results in two remarkable phenomena: discontinuous finite jump of a stress component and sharp turning point in the shear stress. The proposed approach effectively captures these two phenomena. By imposing the suitable compatibility condition at the interfaces of two different materials, the numerical scheme used here is first-order accurate at the interface but high-order accurate elsewhere. The accuracy is validated by comparing numerical results obtained by the present procedure and ABAQUS.

Multi-domain differential quadrature approach in conjunction with coordinate transformation has been exploited in this chapter to treat two-dimensional elastic inhomogeneous problems. The multi-domain DQ is an improvement for the original DQ method which is constrained to homogenous problems with simple geometry and simple boundary conditions. For the present problems with discontinuity in material and geometry, multi-domain DQ approach is applicable to simulate the discontinuity where first-order accuracy is required. Based on DQ and domain decomposition idea, the whole physical domain is separated into a certain number of sub-domains according to the various discontinuities. In this regard, multi-domain DQ is similar to FEM in which the whole domain is a combination of small elements. Thus, multi-domain DQ is also termed as differential quadrature element method (DQEM). Mapping technique is introduced in the application of multi-domain DQ to transform the irregular sub-domain into the normalized square one by use of quadratic or cubic serendipity sub-domain. The numerical examples are presented to show the validity and efficiency of the method for plane elastic heterogeneous solid with inhomogeneous inclusions. It is found that the numerical results agree well with those by ABAQUS. From the stress results, the discontinuity of the stress component can be clearly captured by multi-domain DQ method; and the displacements obtained are continuous at the nodes on the interface where sharp turning points appear. Through the analysis of complex structures, it is found that the interfaces have dominant influence on the accuracy of solutions. More research is required to minimize the adverse effect of the interfaces.

Chapter 7

Localized Differential Quadrature (LDQ) Method

Although DQ is a numerical technique of high accuracy, it was realized from the very beginning that DQ is inefficient when the number of grid points is large (Civan and Sliepcevich, 1983). Later it was revealed that it is also sensitive to the grid point distribution. Quan and Chang (1989a,b) numerically compared the performance of different grid point distributions and found that the grid points given by the roots of the Chebyshev polynomials of the first kind are nearly optimum. Bert and Malik (1996) pointed out that the optimum distribution of grid points is also problem-dependent. Moradi and Taheri (1998) investigated the effect of various spacing schemes on the accuracy of DQ results for buckling behaviors of composites. Recently, a systematic error analysis was carried out by Shu et al. (2001) to assess the effect of the grid distribution. From the error analysis it is concluded that the optimal grid distribution may not be given by the roots of orthogonal polynomials. It is clearly shown from the previous works that the grid distribution exerts significant influence on the accuracy of DQ results and the selection of the grid distribution depends on the problems under consideration. To date, great efforts have been devoted to finding the optimum grid distributions for different problems analyzed by direct DQ method (Chen, 2001; Chen et al., 2000; Fung, 2001). Even today, the general rule for the grid distribution is still lacking.

Preferred grid distribution and small number of grid points greatly limit the applications of DQ. An example is wave propagation in space. A traveling wave sweeps all parts of space. Grid points should be uniformly distributed to correctly capture wave profiles at different time steps. However, a preferred grid distribution required by DQ fails to satisfy the accuracy requirement. Moreover, fewer grid points also deteriorates wave profiles as time increases. Motivated by the difficulties encountered by DQ, localized DQ method representing an important research direction is introduced in the present chapter. It is hoped that the localized DQ method can handle more complicated problems such as wave propagation in space.

First in this chapter, an illustrative example is given to reveal some phenomenological observations. A simplified stability analysis is then performed to explain the observations. The stability analysis reveals that more grid points used in DQ method furnish more accurate results but are accompanied by the occurrence of instability. In the engineering analysis, it is crucial to keep the balance between the accuracy and stability of the results. To achieve this balance, localized DQ method is proposed recently. Localized DQ method is characterized by applying DQ approximation to a small neighborhood of the grid point of interest rather than to the whole solution domain. The derivatives at each grid point are then approximated by a weighted sum of function values on its neighboring points, rather than on all grid points. In doing so, we may obtain a very accurate solution without losing stability. The applicability of the method will be validated through the wave propagation examples.

7.1 DQ and its spatial discretization of the wave equation

It is well known that many grid points used in the DQ application may lead to instability of the results, as will be demonstrated below by the following example.

Consider the vibration of a string of unit length clamped at both ends. The governing equation is the following wave equation in one dimension

$$\frac{\partial^2 u}{\partial t^2} = \frac{\partial^2 u}{\partial x^2} + g\left(\frac{\partial u}{\partial x}\right)^2, \quad 0 \le x \le 1 \tag{7.1a}$$

with the boundary and initial conditions

$$u(0,t) = u(1,t) = 0, \quad u(x,0) = u_0 \sin(2\pi x), \quad \frac{\partial u(x,0)}{\partial t} = 0 \tag{7.1b}$$

where $u(x,t)$ is the transverse displacement of the string, u_0 is the amplitude of the initial displacement. First consider the case $g = 0$. In Eq. (7.1), the first two relations are the fixed boundary condition and the last two relations state the initial condition. The solution to Eq. (7.1) is given by the D'Alembert integral as

$$u(x,t) = u_0 \sin(2\pi x) \cos(2\pi t), \quad 0 \le x \le 1 \tag{7.2}$$

Suppose $u(x,t)$ is approximated by its values $u_i(t)$ $(i = 1, \ldots, N)$ on N distinct grid points $0 = x_1 < \ldots < x_N = 1$. $v_i(t)$ denotes velocity on each grid point. Using DQ approximation rule, we rewrite Eq. (7.1a) in the form of

$$\begin{cases} \frac{\partial u_i}{\partial t} = v_i \\ \frac{\partial v_i}{\partial t} = \frac{\partial^2 u_i}{\partial t^2} = \sum_{j=1}^{N} b_{ij} u_i \end{cases}, \quad i = 1, \ldots, N \tag{7.3}$$

The above equation enables us to use the fourth-order Runge–Kutta method to solve. Therefore, we use DQ for spatial discretization while Runge-Kutta method for temporal discretization. The procedure is similar with that in Chapter 3 for dynamic instability problem. Taking time step as $\Delta t = 10^{-4}$s and using three sets of uniform grid distributions, we obtain wave profiles at different time steps as shown in Fig. 7.1. Figure 7.1 shows the string displacements at several time steps for $N = 10$, 15 and 18, respectively. In the figure, the ordinate is the relative displacement $u(x,t)/u_0$, and the abscissa is location x(m). The solid lines denote DQ solutions while dots denote analytical solutions given by Eq. (7.2).

Figure 7.1(a) shows the displacements every 0.15s apart using $N = 10$. The simulations are run for several seconds and went though several cycles without losing stability. The curves are not smooth, however. When N increases from 10 to 15 (Fig. 7.2(b)), the curves become much more smooth, but around $t = 0.75$s, the solution becomes unstable. At time step $t = 0.9$s, the solution is totally invalid. Figure 7.1(c) depicts the displacements for $N = 18$. The results are stable, being in good agreement with the exact solutions before $t = 0.45$s, but instability occurs thereafter.

From the figure we have two phenomenological observations. First, high accuracy requires large N. Good stability, however, requires using fewer grid points. Second, DQ is of high accuracy, but poor numerical stability. Therefore, besides error analysis as given in Eq. (5.10) in Chapter 5, we also need stability analysis. In the following section, a simple stability analysis will be presented.

7.2 Stability analysis

Stability requires that the solution be not far away from the true solution if a small noise is present in the system. Suppose the function value $f(x_i, t)$ is fluctuated by a noise perturbation ε_i. The perturbed derivative is then given by

$$\frac{\partial f(x_i, t)}{\partial x} = \sum_{j=1}^{N} a_{ij}[f(x_i, t) + \varepsilon_j] \qquad (7.4)$$

The change due to the noise perturbation in the first-order derivative is

$$\delta \frac{\partial f(x_i, t)}{\partial x} = \sum_{j=1}^{N} a_{ij}\varepsilon_j \qquad (7.5)$$

For the sake of simplicity, we apply equidistance grid distribution for the analysis. In this case, the coefficient of first-order derivative can be expressed as

$$a_{ij} = \frac{N}{(j-i)L} \prod_{\substack{k=1 \\ k \neq i,j}}^{N} \frac{i-k}{j-k} = Na'_{ij}, \quad a_{ii} = -N \sum_{\substack{j=1 \\ j \neq i}}^{N} a'_{ij} \qquad (7.6)$$

where $L = x_N - x_1$ is the length of the solution domain. Substituting Eq. (7.6) into Eq. (7.5) yields

$$\delta \frac{\partial f(x_i, t)}{\partial x} = N \sum_{j=1}^{N} a'_{ij} \varepsilon_j \qquad (7.7)$$

It is found that $\sum_{j=1}^{N} a'_{ij} \varepsilon_j$ is zero if all ε_j are equal based on the expression of coefficients in Eq.(1.35) in Chapter 1. ε_j are, however, not necessarily mutually equal, so $\sum_{j=1}^{N} a'_{ij} \varepsilon_j$ is generally not zero. If ε_j is small, the deviation of the partial derivative is small, too. For dynamic problems, however, these small errors may accumulate with each time step. A rough estimation for the first-order derivative is then that the accumulated error at each time step is proportional to $N\Delta t$, where Δt is time step. Analogically, the accumulated error at each time step for the second-order derivative is proportional to $(N\Delta t)^2$.

If large N is used, time step must be small so as to keep the errors within controllable range. This is exemplified by Fig.7.1, in which Δt is kept constant. The case for $N=18$ becomes unstable faster than $N=15$, indicating that higher order DQ discretization becomes unstable faster than lower-order DQ discretization.

From the foregoing analysis, it is clearly illustrated that DQ is of high accuracy but poor stability. The high accuracy can be achieved by using more grid points. However, larger grid points unavoidably lead to instability of the results. In other words, grid points exert reverse effects on accuracy and stability.

It should be noted that the above stability analysis is just illustrative rather than a rigorous proof. Stability of Lagrange polynomial interpolation is a complicated problem, and thus is omitted here. In order to keep the balance of accuracy and stability, we may propose that their sum is minimized, that is,

$$\text{Find N such that } \left| \delta \frac{\partial f}{\partial x} \right| + \left| R^1 \right| \to \min \qquad (7.8)$$

Without solving the equation, it is believed that there exists an optimum N since the accuracy is proportional to N or its powers from Eq. (5.10) in Chapter 5, but stability is proportional to the reciprocal of N from Eq. (7.7). This relation is schematically shown in Fig.7.2. In Fig.7.2, the abscissa is the number of grid point N. The ordinate is either the residual $\left| R^1 \right|$ or the

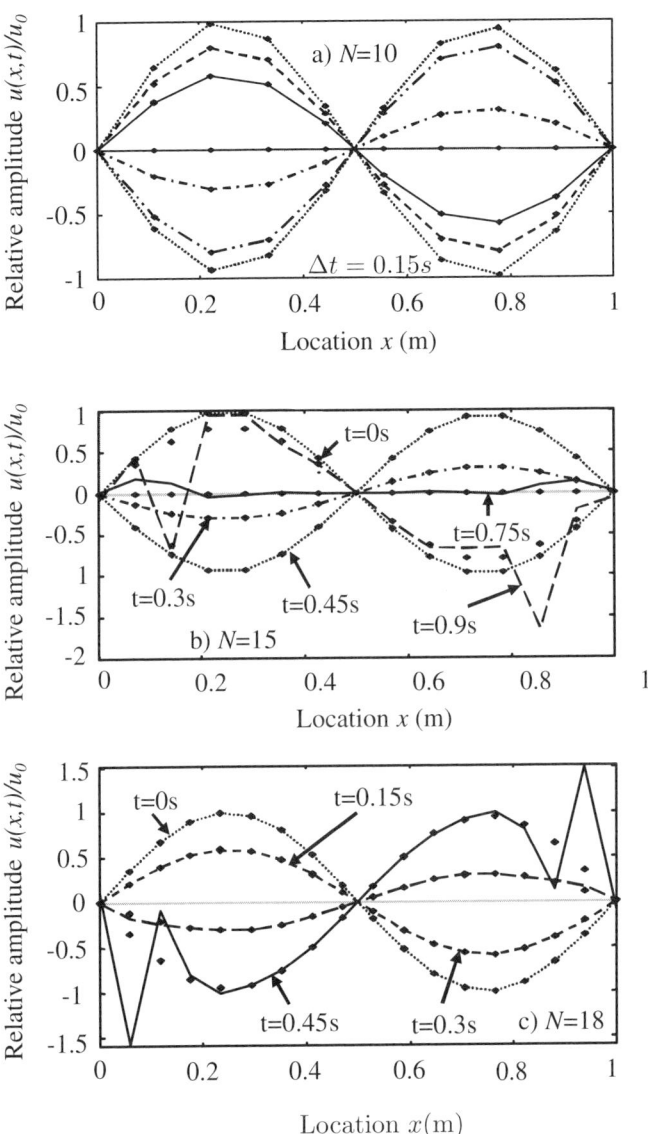

FIGURE 7.1: DQ solution of the string vibration equation using 10, 15 and 18 grid points, respectively.

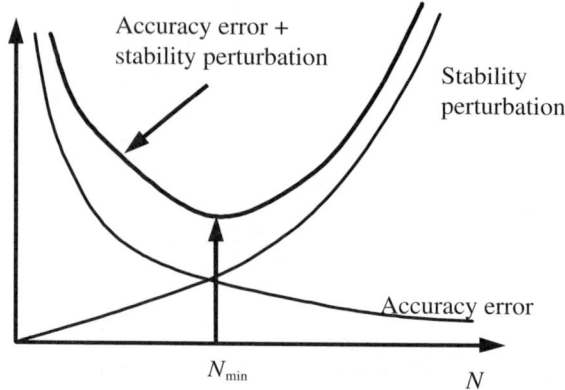

FIGURE 7.2:   Accuracy and stability relationship with the number of grid points, and the existence of a minimum on the curve of their sum.

perturbation of the first-order derivative $\left|\delta\frac{\partial f}{\partial x}\right|$. The residual $\left|R^1\right|$ is a monotonically decreasing function of N while $\left|\delta\frac{\partial f}{\partial x}\right|$ is a monotonically increasing function of N. The $N_{\min}$ which minimizes Eq. (7.8) lies somewhere between the two extremities from advanced calculus.

To keep the balance between these two factors, accuracy and stability, the grid number N cannot be too large. This analysis, which we will not explore further, motivates us to propose the idea of localization of DQ. Instead of using the function values on all grid points to approximate the derivatives as common DQ, we employ only a small portion of nearby grid points to approximate the derivatives as demonstrated in Fig.7.3. For example, if we want to find the derivatives at point A, we then find a neighborhood of point A. The number of points in the neighborhood is assumed to be $m(m << N)$. We use these m points in the neighborhood to approximate the derivatives at point A. Similarly, to find the derivatives at point B, we first specify the neighborhood of point B of size m. The derivatives at point B are then approximated by the weighted sum of the points in this neighborhood as shown in Fig. 7.3.

In summary, for each grid point, there is a neighborhood around it with m points which is greatly smaller than the total number of the grid points N. The derivatives on each point are approximated by the sum of the function values on the grid points inside its neighborhood. Stability is guaranteed if m is not very large, say $5 \sim 8$. Accuracy is also guaranteed by using any large N. This idea leads us to the following two formulations, i.e., coordinate- and spline-based localized DQ.

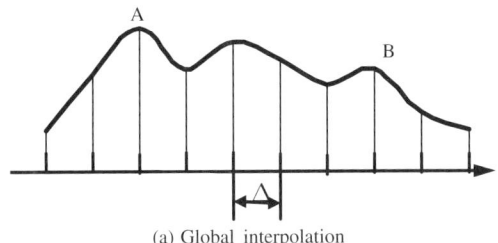

(a) Global interpolation

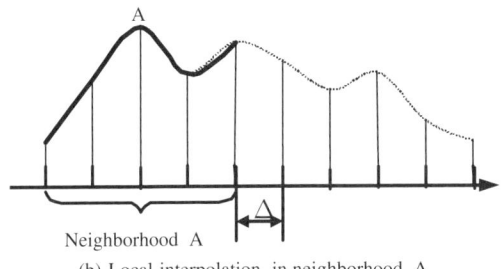

(b) Local interpolation in neighborhood A

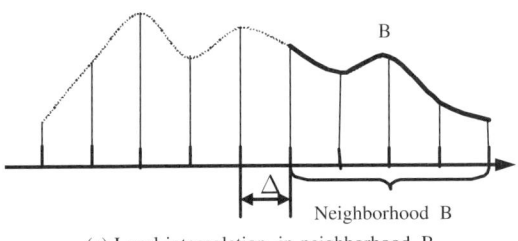

(c) Local interpolation in neighborhood B

FIGURE 7.3: Localization of DQ approximation to the neighborhood of a node of interest.

7.3 Coordinate-based localized DQ

7.3.1 DQ localization in one dimension

The first step to localize DQ method is to find the neighborhood of a grid point of interest. We use

$$r_{il} = |x_i - x_l|, \quad i, l = 1, \ldots, N \tag{7.9}$$

to denote the distance between any two points in the solution domain. By comparison, we may find the permutation $s(1), s(2), \ldots, s(N)$ such that

$$r_{is(1)} \leq r_{is(2)} \leq \cdots \leq r_{is(N)} \tag{7.10}$$

This is a typical permutation problem, and it is easy to find a suitable algorithm to solve the problem.

It is clear that the points falling in the neighborhood of i-th point (x_i) are the first m points which satisfy the above equation. Denote

$$S_i = (s(1), s(2), \ldots, s(m)), \quad i = 1, \ldots, N \tag{7.11}$$

and then S_i defines the neighborhood of the grid point of interest. We may rewrite DQ approximation in this neighborhood in the form of

$$\frac{\partial f(x_i, t)}{\partial x} = \sum_{j \in S_i} a_{ij} f(x_j, t) \tag{7.12a}$$

$$\frac{\partial^2 f(x_i, t)}{\partial x^2} = \sum_{j \in S_i} b_{ij} f(x_j, t) \tag{7.12b}$$

where

$$a_{ij} = \frac{1}{x_j - x_i} \prod_{\substack{k \in S_i \\ k \neq i, j}} \frac{x_j - x_k}{x_j - x_i}, \quad j \in S_i, \quad j \neq i \tag{7.13a}$$

$$a_{ii} = -\sum_{\substack{j \in S_i \\ j \neq i}} a_{ij} \tag{7.13b}$$

For the second-order derivative, the weighting coefficients can be obtained by using Eq. (7.13) as:

$$b_{ij} = 2[a_{ij}a_{ii} - \frac{a_{ij}}{x_j - x_i}], \quad i, j = 1, \ldots, m, \quad i \neq j \tag{7.14a}$$

$$b_{ii} = -\sum_{j \neq i}^{m} b_{ij}, \quad i = 1, \ldots, m \tag{7.14b}$$

These equations are similar to Eq. (1.45) in Chapter 1 for DQ in the neighborhood in one-dimension.

7.3.2 DQ localization in two dimensions

In a straightforward way, the localization can be extended to two dimensions. Suppose the two dimensional domain of interest is discretized by $N \times M$ regular grid points. Consider distance between any two points,

$$r_{il} = |x_i - x_l|, \quad i, l = 1, \ldots, N, \quad \rho_{jl} = |y_j - y_l|, \quad j, l = 1, \ldots, M \quad (7.15)$$

By comparison, it is found that the sequence $s(1), \ldots, s(N)$ and $q(1), \ldots, q(M)$ satisfy

$$r_{is(1)} \le r_{is(2)} \le \cdots \le r_{is(N)}, \rho_{jq(1)} \le \rho_{jq(2)} \le \cdots \le \rho_{jq(M)} \quad (7.16)$$

Denote neighborhood index sets as

$$S_{ij} = (s(1), s(2), \ldots, s(m)), Q_{ij} = (q(1), q(2), \ldots, q(m)),$$
$$i = 1, \ldots, N, j = 1, \ldots, M \quad (7.17)$$

S_{ij} and Q_{ij} determine the neighborhood of any grid point. Then the derivatives are approximated using the points in the neighborhood:

$$\frac{\partial f(x_i, y_j, t)}{\partial x} = \sum_{k \in S_{ij}} a_{ik} f(x_k, y_j, t) \quad (7.18a)$$

$$\frac{\partial^2 f(x_i, y_j, t)}{\partial x^2} = \sum_{k \in S_{ij}} b_{ik} f(x_k, y_j, t) \quad (7.18b)$$

$$\frac{\partial f(x_i, y_j, t)}{\partial y} = \sum_{k \in Q_{ij}} a_{jk} f(x_k, y_j, t) \quad (7.19a)$$

$$\frac{\partial^2 f(x_i, y_j, t)}{\partial y^2} = \sum_{k \in Q_{ij}} b_{jk} f(x_i, y_k, t) \quad (7.19b)$$

where a_{ij} and b_{ij} are of the same forms as in Eqs. (7.13) and (7.14).

Since interpolation takes place only in the neighborhood, the above scheme does not require the regular solution domain.

Consider discretization of the two-dimensional nonlinear wave equation

$$\frac{\partial^2 u}{\partial t^2} = \frac{\partial^2 u}{\partial x^2} + \frac{\partial^2}{\partial y^2} + g \left(\frac{\partial u}{\partial x} \right)^2 \quad (7.20)$$

Suppose the discrete value of $u(x, y, t)$ on the grid point (x_i, y_j) is $u_{ij}(t)$, and its velocity is $v_{ij}(t)$. Then Eq. (7.20) can be rewritten in the following form

$$\frac{\partial u_{ij}}{\partial t} = v_{ij} \quad (7.21a)$$

$$\frac{\partial v_{ij}}{\partial t} = \sum_{k \in S_i} b_{ik} u_{kj} + \sum_{k \in S_i} b_{jk} u_{ik} + g(v_{ij})^2 \quad (7.21b)$$

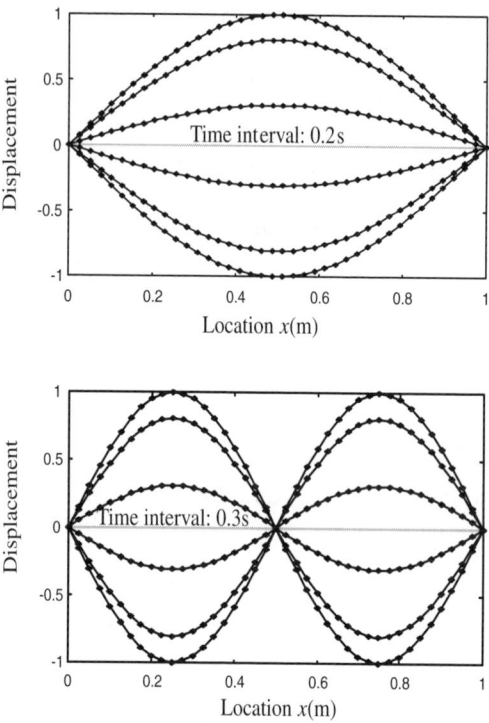

FIGURE 7.4: String vibration.

Runge–Kutta method can then be used to numerically integrate Eqs. (7.21) in time direction.

7.3.3 Numerical examples

In this section, some numerical examples are presented to demonstrate the capability of the localized DQ method.

Example 7.1 *String vibration*

The string vibration problem is governed by the same equation as given by Eq. (7.1). We adopted the following parameters for the computation: the nonlinear effect is not considered with $g=0$, neighborhood size (grid number in the neighborhood) $m = 5$, the total number of grid points $N = 40$, and time step $\Delta t = 10^{-4}$s. The simulation results (solid lines) and analytical solutions (dots) at two time intervals of 0.2s and 0.3s are shown in Fig. 7.4(a) and (b), respectively, for comparison. It is readily seen from Fig. 7.4 that the results given by the new localized DQ method are in excellent agreement with the

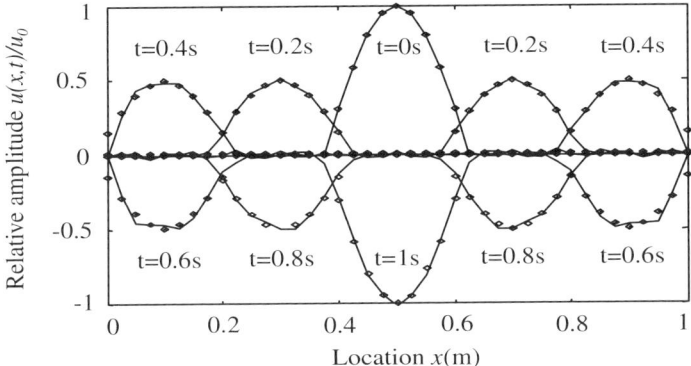

FIGURE 7.5:   Wave propagation on a string.

analytical solutions. Recalling the example of Eq. (7.1), we conclude that the present method is both stable and accurate.

Example 7.2 *Wave propagation on a string*

Wave propagation in a string has the same governing equation of string vibration, but has different initial conditions. The initial wave profile is a localized packet, which will sweep over the whole string at subsequent time steps. The nonlinear effect is not considered with $g=0$ in Eq. (7.20).

Suppose the initial and boundary conditions are

$$u(0,t) = u(1,t) = 0, \quad u(x,0) = u_0 \sin(\frac{x - 0.375}{0.275}\pi) \qquad (7.22)$$

It is a localized wave concentrated on a central small area. With time increasing, it propagates outward toward the right and left ends. After hitting upon both ends, the waves are reflected as a negative wave and travels inwards toward the center. The two reflected waves meet again at center and repeat next cycle. This is a complicated problem, which can serve as a good example to test the present method.

In the numerical computations, the following parameters were used: $m = 5$, $N = 40$ and time step $\Delta t = 10^{-4}$ s. The numerical results (denoted by solid line) and the analytical solutions (denoted by dots) are shown in Fig. 7.5. Again, they are in very good agreement.

For comparison, we conducted another computation with the same parameters as those in Fig.7.5 but with larger total number of grid points $N = 100$. The numerical results are plotted in Fig.7.6. By comparing Fig.7.5 with 7.6(a), it is observed that the curve in Fig.7.6(a) with $N = 100$ is smoother than that in Fig.7.5 with $N = 40$. Furthermore, better agreement between the numerical

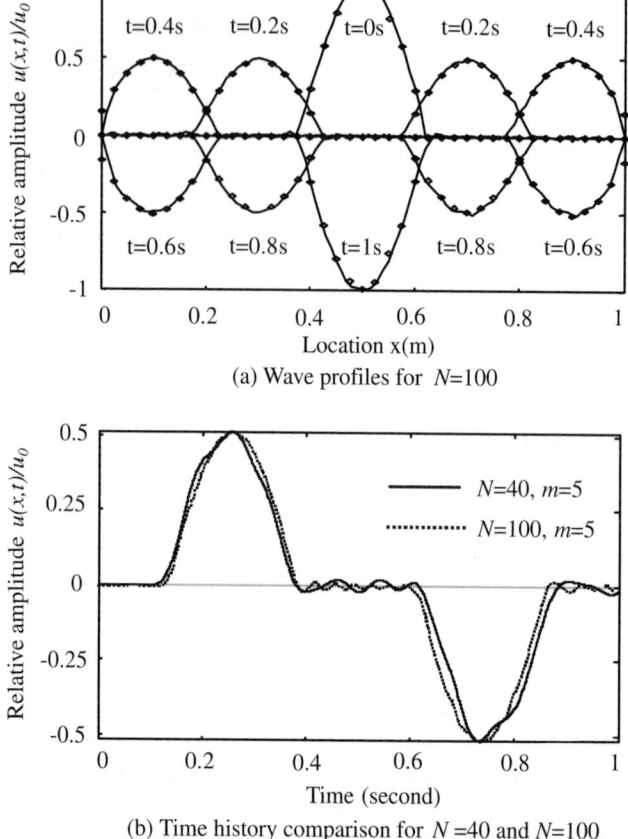

(a) Wave profiles for *N*=100

(b) Time history comparison for *N* =40 and *N*=100

FIGURE 7.6: Comparison between the results obtained by using $N=40$ and $N=100$.

results and the analytical solutions is achieved by using $N = 100$ as shown in Fig.7.6(a). In other words, more grid points used in the localized DQ method lead to more accurate results as demonstrated in Fig. 7.6(b). Even using grid points as large as $N = 100$, stability remains maintained by the present localized method. However, if we applied traditional global DQ approximation for this wave propagation problem with $N = 100$, a very small time step should be used to guarantee the stability.

To further check the effect of the number of grid points on the accuracy of the numerical results, the time history of wave displacements at the point $x = 0.25$ for $N = 40$ and $N = 100$ is compared in Fig. 7.6(b). From Fig. 7.6, it is illustrated that the present method is indeed a local method due to the fact that increasing grid number improves accuracy such as FEM. The more

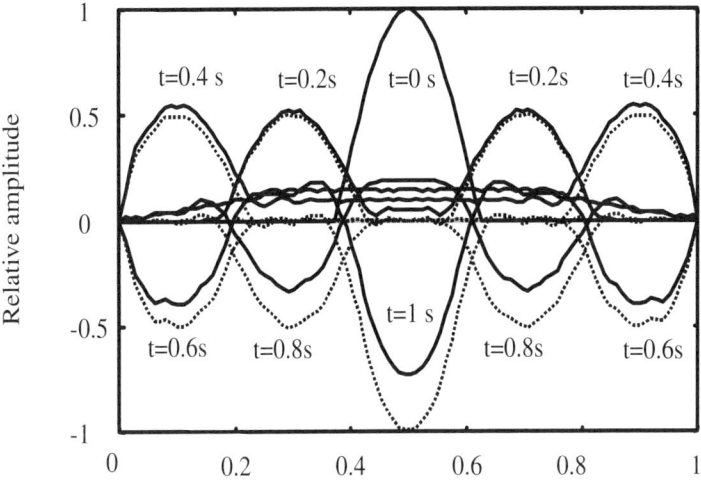

FIGURE 7.7:   Nonlinear wave propagation along a string.

grid points are used, the better the results are, a feature different from global methods.

Example 7.3 *Nonlinear wave propagation on a string*

Everything is the same as the previous example, the only difference being the nonlinear effect is not zero with $g = 0.1$ in Eq. (7.20). The numerical results for this case are given in Fig. 7.7 using solid line. In the figure, the results in Fig. 7.5 are also plotted using dotted line to compare the nonlinear effect.

Figure 7.7 exhibits the significant nonlinear effects on the wave forms at different time steps. In the linear case, the wave form is symmetric about the axis $u= 0$, while in the nonlinear case, the symmetric axis is raised up. Moreover, the nonlinear wave is no longer localized. The humped wave is followed by a long tail. These are typical nonlinear phenomena.

This example also shows the ease to implement nonlinearity numerically.

Example 7.4 *Wave propagation on a membrane*

We now consider a two-dimensional wave propagation problem governed by Eq. (7.20). Suppose $u(x, y, t)$ represents the displacement of a rectangular membrane subject to the following boundary and initial conditions

$$u(0, y, t) = u(1, y, t) = u(x, 0, t) = u(x, 1, t) = 0 \tag{7.23a}$$

$$u(x, y, 0) = u_0 \exp\{-200[(x - 0.5)^2 + (y - 0.5)^2]\} \tag{7.23b}$$

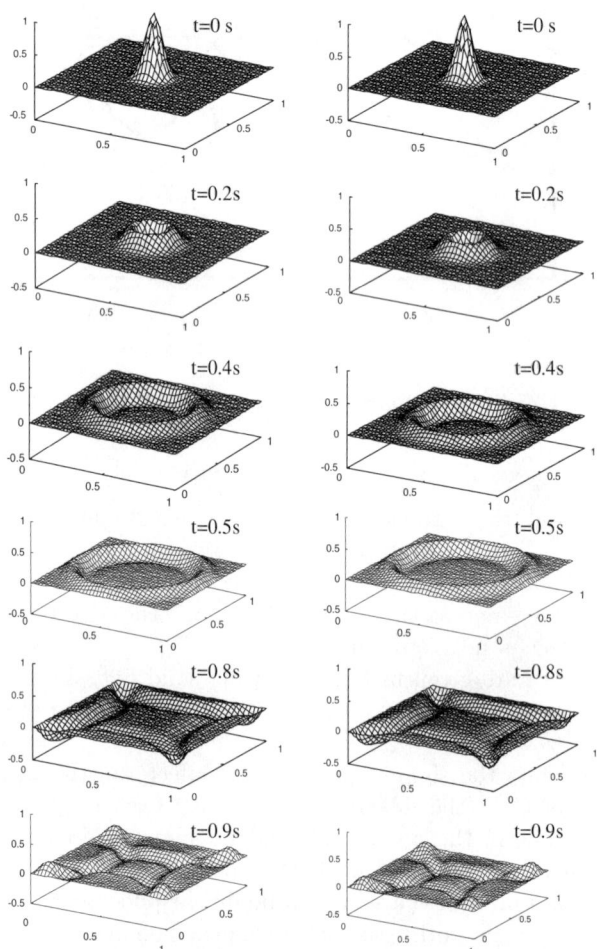

FIGURE 7.8: Wave propagation on a membrane. The left column shows the analytical solutions while the right column shows the numerical results.

The exact solution can be obtained through calculus operations as

$$u(x,y,t) = u_0 \sum_{i=1}^{\infty} \sum_{j=1}^{\infty} C_{ij} \Psi_{ij}(x,y) \cos(2\pi\omega_{ij}t) \tag{7.24}$$

where

$$C_{ij} = \frac{\pi}{50} \sin\left(\frac{i\pi}{2}\right) \sin\left(\frac{j\pi}{2}\right) \exp\left\{-\frac{(i^2+j^2)\pi^2}{800}\right\} \tag{7.25a}$$

$$\Psi_{ij}(x,y) = \sin(i\pi x)\sin(j\pi y) \tag{7.25b}$$

$$\omega_{ij} = \frac{1}{2}\sqrt{i^2+j^2} \tag{7.25c}$$

The parameters used in the calculation were $N = M = 40$ and $\Delta t = 10^{-4}$s. The numerical and analytical results are shown in Fig. 7.8 at six time steps. In the figure, the vertical ordinate shows the relative displacement $u(x,y,t)/u_0$ and the two horizontal axes indicate the x- and y-location. The left and right columns show the analytical and numerical results, respectively.

The results correctly captured several features of wave propagation on a membrane such as wave spreading, reflection at boundaries, wave interaction after reflection. Quantitatively, the numerical results are also in very good agreement with the analytical solutions, indicating the effectiveness of the localized DQ method in the complicated dynamics problems.

Example 7.5 *Wave scattering by a rectangle obstacle*
Consider a rectangular domain with a small obstacle in the domain. Suppose a sinusoidal wave like sound wave originated from left end propagating rightward. The solution domain is assumed as

$$0 \le x \le 2, \quad 0 \le y \le 1.5 \tag{7.26}$$

The obstacle occupies the region

$$0.5 \le x \le 0.75, \quad 0 \le y \le 0.875 \tag{7.27}$$

The governing equation is given by Eq. (7.20). The boundary condition at $x = 0$ is

$$u(0,y,t) = \sin(2\pi t) \tag{7.28}$$

The boundary conditions at other boundaries and on the obstacles are that the normal velocity is zero.

The grid distribution is uniform with $N = 80$ and $M = 60$. Time step is still $\Delta t = 10^{-4}$s. The results are shown in Fig. 7.9 for 12 time steps. In the figure, the axes are locations in x and y directions, while the curves in the figure are contours at different time steps. Fig.7.9 (a) shows a sinusoidal wave propagates from the left side to the right side. At $t = 0.6$ s, the wave front hits the obstacle as shown in Fig.7.9 (b). Because viscosity is neglected in the wave equation in Eq. (7.20), the wave front sweeps the sides of the obstacle without much deformation (see Fig.7.9(c) and(d)). But a reflected wave is generated and bounced back. The incident and reflected waves meet somewhere behind the obstacle, resulting in deformed contour lines behind the obstacle as shown in Fig.7.9(e) and (f). Meanwhile, the wave fronts, after passing the obstacle, meet again in front of the obstacle. These interference effects result in the wave scattering pattern shown in Fig.7.9(g) behind the obstacle, where we can see a closed curve right behind the obstacle. This closed curve grows and re-opens as it moves past the body (see Fig.7.9 (i)~(k)). Until $t = 1.6$ s, a new closed curve is formed again behind the obstacle, showing that a new cycle starts.

This example shows the capability of the present method in dealing with complicated problems. It is almost impossible to use the common DQ to solve this problem due to the complicated solution domain and the problems itself.

Example 7.6 *2D vector wave equation*
With slight modifications, the above approach is applicable to the two-dimensional wave equation in the vector form. Suppose the two-dimensional vector wave equation is given in the following form of

$$A\frac{\partial^2 U}{\partial t^2} = B \nabla^2 U \tag{7.29}$$

where

$$A = \begin{pmatrix} a_{11} & a_{12} \\ a_{21} & a_{22} \end{pmatrix}, \quad B = \begin{pmatrix} b_{11} & b_{12} \\ b_{21} & b_{22} \end{pmatrix}, \quad U = \begin{pmatrix} u_1 \\ u_2 \end{pmatrix} \tag{7.30}$$

are coefficient matrices and displacement vector, respectively. Equation (7.29) has been used to model quite a few wave phenomena in engineering and science, among which wave propagation in soil and bones are of particular interest (Biot, 1956; Cowin, 1999). Both soil and bone are the so-called poroelastic continuum, that is, porous elastic solids with fluid-filled pores. In a poroelastic medium, waves in the solid matrix and in the pore fluid propagate at different phase speeds, interacting with each other according to Eq. (7.29) (Biot, 1956; Cowin, 1999).

Rewriting Eq. (7.29) in the form of

$$\begin{cases} \frac{\partial U}{\partial t} = V \\ \frac{\partial V}{\partial t} = A^{-1}B \nabla^2 U \end{cases} \tag{7.31}$$

we may solve the above equation using the similar procedures as given in Eq. (7.19) or in Eq. (7.21). A particular numerical example is given here. Consider a one-dimensional poroelastic rod of length 1. At both ends, the rod is fixed, that is,

$$U(x = 0, \ t) = \begin{pmatrix} 0 \\ 0 \end{pmatrix}, \quad U(x = 1, \ t) = \begin{pmatrix} 0 \\ 0 \end{pmatrix} \tag{7.32a}$$

Suppose the rod is initially assigned a sinusoidal deformation

$$\mathbf{U}(x, t = 0) = \begin{pmatrix} u_0 \sin(\pi x) \\ u_0 \sin(\pi x) \end{pmatrix}, \quad \frac{\partial \mathbf{U}(x, t = 0)}{\partial t} = \begin{pmatrix} 0 \\ 0 \end{pmatrix} \qquad (7.32b)$$

where u_0 is the amplitude. We solved this problem using the present method with the following coefficient matrices

$$\mathbf{A} = \begin{pmatrix} 2400 & -600 \\ -600 & 700 \end{pmatrix}, \mathbf{B} = \begin{pmatrix} 8 \times 10^5 & 3 \times 10^7 \\ 3 \times 10^5 & 8 \times 10^7 \end{pmatrix} \qquad (7.33)$$

Figure 7.10 shows the time history of the displacements (u_1 and u_2) at the center $x = 0.5$. The solid lines denote the displacement u_1 and u_2. The crosses and dots denote the corresponding analytical solutions (Biot, 1956). It is readily seen from Fig. 7.10 that the numerical results are identical to the analytical solutions.

7.4 Spline-based localized DQ method

The essence of the localized DQ method defined in the previous sections is to approximate the unknown functions by use of multi-segment functions. Theoretically speaking, a function can be approximated to the desired degree of accuracy by using polynomials or multi-segment polynomials. The former is simple, but it becomes unstable as the polynomial order is high. Multi-segment polynomials, on the other hand, are low-order ones which are stable as well as accurate. In view of this, multi-segment functions are widely used for approximation. Among multi-segment polynomials, spline functions are particularly attractive. In practice, cubic B-splines are widely used for approximation, because they provide a suitable balance between flexibility and accuracy (Schultz, 1973; Schumaker, 1981).

The spline-based DQ is a ramification of the DQ family. Taking the advantages of spline functions, the spline-based DQ is free of the limitation of grid point number. So far, DQ based on quintic and sextic B-spline functions has been successfully applied to the solution of various boundary-value problems (Zhong, 2004; Zhong and Guo, 2004). In this section, a spline-based DQ method is introduced.

The present section is organized as follows. In Section 7.4.1, the cardinal cubic spline interpolation functions are constructed and the explicit expressions of the weighting coefficients for derivative approximation are given. The incorporation of initial conditions, the stability of the spline-based DQ method and two ways of application are addressed in Section 7.4.2. Three dynamic systems of Duffing-type nonlinearity are solved in Section 7.4.3. The results are discussed in Section 7.4.4 and some concluding remarks are given in Section 7.4.5.

7.4.1 Cardinal cubic spline interpolation

First of all, a set of uniformly spaced nodes is selected in a normalized interval $[0, 1]$, i.e.

$$X_0 = 0, \ X_N = 1, \ X_{j+1} - X_j = h, \ j = 0, 1, 2, \ldots, N-1, N \qquad (7.34)$$

where $N+1$ is the number of nodes in the interval, h is the length of every segment. The normalized cubic B-spline function is given by (Schultz, 1973; Schumaker, 1981)

$$\varphi_3(x) = \frac{1}{6h^3} \begin{cases} 0, & x \leq -2h; \\ (x+2h)^3, & -2h \leq x \leq -h; \\ (x+2h)^3 - 4(x+h)^3, & -h \leq x \leq 0; \\ (2h-x)^3 - 4(h-x)^3, & 0 \leq x \leq h; \\ (2h-x)^3, & h \leq x \leq 2h; \\ 0, & x \geq 2h. \end{cases} \qquad (7.35)$$

Apparently, it is a piecewise polynomial covering four consecutive segments only. To build a global interpolation function over the interval, usually extra nodes outside the interval $[x_0, x_N]$ are needed to meet the end condition requirements. A typical spline interpolation over the given interval using cubic B-splines is expressed in the form of

$$s_3(x) = \sum_{j=-2}^{N+2} \Phi_j(x) y_j, \ \Phi_j = \Phi_0(x - jh) \qquad (7.36)$$

In order to meet the required interpolation condition

$$s_3(x_i) = y_i, \qquad (7.37)$$

the interpolation functions $\Phi_j(x)$ should satisfy the cardinal condition at every node, i.e.

$$\Phi_j(x_i) = \delta_{ij} = \begin{cases} 1, & i = j; \\ 0, & otherwise, \end{cases} \qquad (7.38)$$
$$i, j = -2, -1, 0, 1, \ldots, N-1, N, N+1, N+2,$$

where $\Phi_j(x)$ are usually given in terms of a combination of translated and scaled spline function φ_3. To acquire the cardinal spline interpolation function $\Phi_j(x)$, the following three auxiliary spline interpolation functions (Li and Qi,

1979) are constructed first:

$$\psi_3(x) = \sum_{j=-2}^{N+2} y_j \varphi_3(x - x_j) \tag{7.39a}$$

$$\psi_3^{(1/2)}(x) = \sum_{j=-2}^{N+2} y_j \varphi_3^{(1/2)}(x - x_j) \tag{7.39b}$$

$$\psi_3^{(1)}(x) = \sum_{j=-2}^{N+2} y_j \varphi_3^{(1)}(x - x_j) \tag{7.39c}$$

where

$$\varphi_3^{(1/2)}(x) \equiv \varphi_3(x + h/2) + \varphi_3(x - h/2) \tag{7.40a}$$

$$\varphi_3^{(1)}(x) \equiv \varphi_3(x + h) + \varphi_3(x - h) \tag{7.40b}$$

With the local non-zero property of the spline function $\phi_3(x)$, all the terms but the one containing y_j on the right sides of Eq. (7.39a), (7.39b) and (7.39c) are eliminated. Thus, the cardinal spline interpolation function is obtained as

$$s_3(x) = \frac{10}{3}\psi_3(x) - \frac{4}{3}\psi_3^{(1/2)}(x) + \frac{1}{6}\psi_3^{(1)}(x) \tag{7.41}$$

Hence

$$\Phi_j(x) = \frac{10}{3}\varphi_3(x - x_j) - \frac{4}{3}\varphi_3^{(1/2)}(x - x_j) + \frac{1}{6}\varphi_3^{(1)}(x - x_j) \tag{7.42}$$

It has been shown that the accuracy of the cardinal spline in Eq. (7.41) is of order $O(h^4)$ (Li and Qi, 1979). It can be readily seen that Eq. (7.37) is satisfied at every node (see Fig.7.11). Since the extra nodes outside the interval are often cumbersome to handle, non-integral nodes within the interval are introduced instead. Namely, $x_{1/2} = h/2$, $x_{3/2} = 3h/2$ and $x_{N-3/2} = (N - 3/2)h$, $x_{N-1/2} = (N - 1/2)h$ are added in the vicinity of the two ends of the interval. As a result, the spline-based DQ meets the self-starting requirement of a good algorithm for initial-value problems (Hughes, 1987). The function values at the non-integral nodes are given by

$$y_{1/2} = \sum_{i=-2}^{N+2} \Phi_i(h/2)y_i = \frac{1}{288}(y_{-2} + y_3) - \frac{7}{96}(y_{-1} + y_2) + \frac{41}{72}(y_0 + y_1) \tag{7.43a}$$

and

$$y_{3/2} = \sum_{i=-2}^{N+2} \Phi_i(3h/2)y_i = \frac{1}{288}(y_{-1} + y_4) - \frac{7}{96}(y_0 + y_3) + \frac{41}{72}(y_1 + y_2) \tag{7.43b}$$

Then, the function values at the extra nodes are given by solving Eqs. (7.43a) and (7.43b)

$$y_{-1} = 21(y_0 + y_3) - y_4 + 288y_{3/2} - 164(y_1 + y_2) \tag{7.44a}$$

$$y_{-2} = 277y_0 + 288y_{1/2} - 3608y_1 + 6048y_{3/2} - 3423y_2 + 440y_3 - 21y_4 \tag{7.44b}$$

Similarly, the function values at the extra nodes outside the right end of the interval are expressed as

$$y_{N+1} = 21(y_{N-3} + y_N) - y_{N-4} + 288y_{N-3/2} - 164(y_{N-2} + y_{N-1}) \tag{7.45a}$$

$$y_{N+2} = 277y_N + 288y_{N-1/2} - 3608y_{N-1} + 6048y_{N-3/2} - 3423y_{N-2} + 440y_{N-3} - 21y_{N-4} \tag{7.45b}$$

Now, the cardinal cubic B-spline interpolation function is rearranged into the following form which is free of extra outside nodes

$$s_3(x) = \sum_{j=0}^{N} \Omega_j(x)y_j, \Omega_j(x_i) = \delta_{ij} = \begin{cases} 1, & i = j, \\ 0, & otherwise, \end{cases} \tag{7.46}$$
$$i, j = 0, 1/2, 1, 3/2, 2, \ldots, N - 2, N - 3/2, N - 1, N - 1/2, N,$$

where

$$\Omega_0(x) = 277\Phi_{-2}(x) + 21\Phi_{-1}(x) + \Phi_0(x),$$
$$\Omega_{1/2}(x) = 288\Phi_{-2}(x),$$
$$\Omega_1(x) = -3608\Phi_{-2}(x) - 164\Phi_{-1}(x) + \Phi_1(x),$$
$$\Omega_{3/2}(x) = 6048\Phi_{-2}(x) + 288\Phi_{-1}(x), \tag{7.47a}$$
$$\Omega_2(x) = -3423\Phi_{-2}(x) - 164\Phi_{-1}(x) + \Phi_2(x),$$
$$\Omega_3 = 440\Phi_{-2}(x) + 21\Phi_{-1}(x) + \Phi_3(x),$$
$$\Omega_4(x) = -21\Phi_{-2}(x) - \Phi_{-1}(x) + \Phi_4(x),$$
$$\Omega_i(x) = \Phi_i(x) \ for \ 5 \leq i \leq N - 5 \tag{7.47b}$$

and

$$\Omega_{N-4} = -21\Phi_{N+2}(x) - \Phi_{N+1}(x) + \Phi_{N-4}(x),$$
$$\Omega_{N-3} = 440\Phi_{N+2}(x) + 21\Phi_{N+1}(x) + \Phi_{N-3}(x),$$
$$\Omega_{N-2}(x) = -3423\Phi_{N+2}(x) - 164\Phi_{N+1}(x) + \Phi_{N-2}(x),$$
$$\Omega_{N-3/2} = 6048\Phi_{N+2}(x) + 288\Phi_{N+1}(x), \tag{7.47c}$$
$$\Omega_{N-1} = -3608\Phi_{N+2}(x) - 164\Phi_{N+1}(x) + \Phi_{N-1}(x),$$
$$\Omega_{N-1/2} = 288\Phi_{N+2}(x),$$
$$\Omega_N(x) = 277\Phi_{N+2}(x) + 21\Phi_{N+1}(x) + \Phi_N(x).$$

It is noted that the selection of non-integral nodes at the vicinity of the two ends is not unique. The selection of different non-integral nodes at the vicinity of the two ends should meet the criterion that the extra nodes outside the domain can be expressed in terms of the introduced nodes as well as the inner nodes at the vicinity of the two ends.

7.4.2 Weighting coefficients for spline-based DQ

The essence of the DQ method is that the derivative of a function with respect to a space variable at a given point is approximated by a weighted linear summation of the function values at all discrete nodes in the domain. Therefore, the approximation of a derivative at a node in spline-based DQ is given by

$$
D_n\{f(x)\}_i = \sum_{j=0}^{N} C_{ij}^{(n)} f(x_j),
$$
$$
i, j = 0, 1/2, 1, 3/2, 2, \ldots, N - 2, N - 3/2, N - 1, N - 1/2, N,
$$

(7.48)

where D_n is a differential operator of order n, the subscript i indicates the value of $D_n\{f(x)\}$ at node x_i, $C_{ij}^{(n)}$ are the weighting coefficients related to the function values $f(x_j)$. In the spline-based DQ method, it is required that Eq. (7.48) be exactly satisfied when function $f(x)$ takes the cardinal spline basis functions $\Omega_j(x)$. Consequently, all weighting coefficients are given in explicit forms

$$
C_{ij}^{(1)} = \Omega_j^{(1)}(x_i), C_{ij}^{(2)} = \Omega_j^{(2)}(x_i),
$$
$$
i, j = 0, 1/2, 1, 3/2, 2, \ldots, N - 2, N - 3/2, N - 1, N - 1/2, N.
$$

(7.49)

It is worth mentioning that the localized nature of splines results in banded weighting coefficient matrices for derivatives. Meanwhile, the following relationships among weighting coefficients hold

$$
C_{ij}^{(1)} = -C_{(N-i)(N-j)}^{(1)}, C_{ij}^{(2)} = C_{(N-i)(N-j)}^{(2)},
$$
$$
i, j = 0, 1/2, 1, 3/2, 2, \ldots, N - 2, N - 3/2, N - 1, N - 1/2, N.
$$

(7.50)

7.4.3 Application of spline-based DQ method to initial-value problems

(a) Incorporation of initial conditions

With the first-order weighting coefficients $C_{ij}^{(1)}$, initial conditions can be incorporated into the spline-based DQ with ease following the same strategy as described by Fung (2001). For simple representation, the superscript of the weighting coefficients for the first-order derivatives is omitted. Then, Eq. (7.48) is written as

$$
s'(x_i) = D_1\{f(x)\}_i = \sum_{j=0}^{N} C_{ij} f(x_j) = C_{i0} f(x_0) + \sum_{j=1/2}^{N} C_{ij} f(x_j).
$$

(7.51)

Likewise, the second-order derivative of the function is written as

$$
\begin{aligned}
s''(x_i) &= C_{i0}f'(x_0) + \sum_{j=1/2}^{N} C_{ij}\left(C_{j0}f(x_0) + \sum_{k=1/2}^{N} C_{jk}f(x_k)\right) \\
&= C_{i0}f'(x_0) + \sum_{j=1/2}^{N} C_{ij}C_{j0}f(x_0) + \sum_{j=1/2}^{N} C_{ij}\sum_{k=1/2}^{N} C_{jk}f(x_k)
\end{aligned}
\tag{7.52}
$$

Obviously, the two initial conditions $f(x_0)$ and $f'(x_0)$ are imposed. Eqs. (7.51) and (7.52) can be rewritten in the matrix forms as follows:

$$
\left\{ \begin{array}{c} \dot{y}_{1/2} \\ \vdots \\ \dot{y}_N \end{array} \right\} = \{C_0\}y_0 + \left\{ \begin{array}{c} y_{1/2} \\ \vdots \\ y_N \end{array} \right\}
\tag{7.53}
$$

$$
\left\{ \begin{array}{c} \ddot{y}_{1/2} \\ \vdots \\ \ddot{y}_N \end{array} \right\} = \{C_0\}\dot{y}_0 + [C]\{C_0\}y_0 + [C][C]\left\{ \begin{array}{c} y_{1/2} \\ \vdots \\ y_N \end{array} \right\}
\tag{7.54}
$$

where $\{C_0\}$ denotes $\{C_{i0}\}$ vector, $[C]$ denotes matrix $[C_{ij}]$ which is of size $(N+4)(N+4)$. In the same manner, high-order derivatives are given by

$$
\left\{ \begin{array}{c} y_{1/2}^{(m)} \\ \vdots \\ y_N^{(m)} \end{array} \right\} = \sum_{p=0}^{m-1} [C]^{m-p-1}\{C_0\}y_0^{(p)} + [C]^m\left\{ \begin{array}{c} y_{1/2} \\ \vdots \\ y_N \end{array} \right\}, \quad m \geq 1
\tag{7.55}
$$

The above way to incorporate the initial conditions is simple and straightforward. In addition, the recursive formula in Eq. (7.55) implies that low-order spline interpolation functions are applicable to the solution of high-order differential equations. This will be demonstrated later in an example.

(b) Two ways of application

Since the spline-based DQ is constructed on the normalized interval $[0, 1]$ and the method is virtually free of limitation on the grid point numbers, there are two practical ways of applying the method to the solution of initial-value problems.

(1) Indirect approach. This approach is conventional since most time integration methods adopt the similar procedures. It is characterized by the division of the entire time domain into several subintervals. The spline-based DQ is applied to each subinterval. The end state conditions for each subinterval are used as the initial conditions for the next subinterval. The process is repeated until the solution at the end of the interval is obtained.

(2) Direct approach. It has been rare to evaluate the state variables at a time directly unless it is close to the initial time. The development of spline-based DQ, however, has made this achievable. This is attributed to the local behavior of the spline functions. Suppose that τ represents the normalized time domain $[0, 1]$, which is required by the spline-based DQ; t represents the actual time variable, which is given as

$$t = L\tau, \tau \in [0, 1], \tag{7.56}$$

where L represents the length of actual time domain. Before the implementation of DQ, all time derivatives are transformed into the normalized time domain

$$\frac{d^{(n)}y}{dt^{(n)}} = \frac{1}{L^{(n)}} \frac{d^{(n)}y}{d\tau^{(n)}}, \; n = 1, 2 \ldots \tag{7.57}$$

7.4.4   Stability of spline-based DQ

The stability of an algorithm is of critical concern in the solution of initial-value problems. Thus, an undamped linear single degree-of-freedom system is considered to investigate the stability of the present spline-based DQ method. The dynamic equation is expressed as

$$\ddot{y}(t) + \omega^2 y(t) = 0 \tag{7.58}$$

with initial conditions

$$y(t = 0) = y_0 \quad \text{and} \quad \dot{y}(t = 0) = v_0 \tag{7.59}$$

The stability is evaluated by calculating the spectral radius of the numerical amplification matrix $[Z]$, which relates the state at the end of the temporal interval to the initial state, i.e.

$$\left\{ \begin{array}{c} y|(t = end) \\ \dot{y}|(t = end) \end{array} \right\} = [Z] \left\{ \begin{array}{c} y_0 \\ v_0 \end{array} \right\} \tag{7.60}$$

The algorithm is said to be stable if the spectral radius $|\lambda|_{max} \leq 1$ or unstable if otherwise.

Figure 7.12 shows the spectral radii for spline-based DQ with $N = 8$, 10 and 20. Clearly, there is no bifurcation or bubble in the curve because the eigenvalues of the amplification matrix are a complex conjugate pair. It is seen that the spline-based DQ is conditionally stable. The unstable computational time interval falls within $[0.2, 1.3]\omega t$. With the increase of number of nodes, however, the unstable interval of the spectral radii remains virtually unaltered and even becomes narrower. As reported by Fung in his comprehensive research about the stability of the conventional DQ method (Fung, 2002), the stable time domain varies with the increase of grid point number. Even for the DQ with Legendre, Radau, Chebyshev grid patterns, A-stability is not guaranteed if the grid number is large. In contrast, the stability criterion for spline-based DQ is virtually invariant against the number of grid points. Furthermore, there is no limitation on the number of grid points in spline-based DQ. This salient advantage makes it possible to achieve long-term integration with satisfactory accuracy regardless of the stability of the method. An example will be given later in which up to 800 nodes are used to furnish solution of the long-term solution of a dynamic system.

7.4.5   Examples

Example 7.7 *Duffing-type nonlinear equation*
    The form of Duffing-type nonlinear differential equation is written as

$$y^{(2)} + y + \varepsilon y^3 = F \sin \omega t \qquad (7.61)$$

where F and ε are given parameters, ω is also a given constant which represents the enforcing frequency. The analytical solution is given approximately by means of a trigonometric series

$$y = a_1 \sin \omega t + a_3 \sin 3\omega t + a_5 \sin 5\omega t + \ldots \qquad (7.62)$$

Three examples of Duffing-type nonlinear equations are calculated to illustrate the capability of the spline-based DQ method.

Case 1

A second-order differential equation which represents the free vibration of pendulum is studied. The frequency of the oscillations depends on the initial displacement of the pendulum. Given parameters $\varepsilon = -1/6$, $\omega = 0.7$ and $F = 0$ in Eq. (7.61) for the unforced Duffing system, the solution (Groves, 1983) is

$$y \cong 2.058 \sin 0.7t + 0.0816 \sin 2.1t + 0.00337 \sin 3.5t \qquad (7.63)$$

for initial conditions

$$y = 0, \ y^{(1)} = 1.62376 \qquad (7.64)$$

Case 2

Given parameters $F = 2$, $\varepsilon = -1/6$, and $\omega = 1$ for the forced Duffing equation, the solution (Groves, 1983) is

$$y \cong -2.5425 \sin t - 0.07139 \sin 3t - 0.00219 \sin 5t \qquad (7.65)$$

for initial conditions

$$y = 0, \ y^{(1)} = -2.7676 \qquad (7.66)$$

Case 3

The following high-order (fourth-order) nonlinear differential equation is studied

$$y^{(4)} + 5y^{(2)} + 4y - \frac{1}{6}y^3 = 0 \qquad (7.67)$$

Its analytical solution (Groves, 1983) for $\omega = 0.9$ is given by

$$y \cong 2.1906 \sin 0.9t - 0.02247 \sin 2.7t + 0.000045 \sin 4.5t \qquad (7.68)$$

for initial conditions

$$y = 0, \ y^{(1)} = 1.91103, \ y^{(2)} = 0, \ y^{(3)} = -1.15874 \qquad (7.69)$$

The computational time domain is chosen as $[0, 100]$ for the three examples unless otherwise stated. The selection of long time domain is to investigate the stability and reliability of the method. As mentioned before, the accuracy of the conventional DQ method decreases with the increase of domain length. For the above three examples, acceptable solutions over only a few cycles were obtained using the conventional DQ method. In the forced Duffing equation case, convergent results were only obtainable for time domain shorter than one half cycle (Groves, 1983).

During the solution of dynamic systems of Duffing-type nonlinearity, the resultant nonlinear algebraic equations are solved using modified Powell hybrid algorithm (Press et al., 1986). To demonstrate the stability of the spline-based DQ method, the indirect approach is adopted in the solution of unforced Duffing system with $\varepsilon = -1/6$, $\omega = 0.7$ and $F = 0$. The solution over 100 s (approximately 11 cycles) is sought and the time domain is divided into 100 subintervals. $N = 8$ is chosen for each subinterval. The results of displacement, velocity and acceleration, which are displayed in Fig. 7.13, are in excellent agreement with the analytical solutions. By using the indirect approach, solution with the same order of accuracy shown in Fig. 7.13 over much longer time domain, say over 30 cycles, is also obtainable. When more subintervals are used in the indirect approach, larger accumulated errors occur eventually. There is a balance between the number of subintervals over the time domain and the number of points N. When fewer intervals are used in the indirect approach and the accumulation error is insignificant, the accuracy of the results is virtually the same as that of the direct approach. In the extreme case, i.e., the direct approach, large N is usually needed to gain satisfactory accuracy. The results with the same accuracy as shown in Fig. 7.13 can be obtained using the direct approach when N is increased to 400. When more intervals are used in the indirect approach and the accumulation error becomes pronounced, the result accuracy differs for the two approaches. For clarity of presentation, only the results over the last few cycles are plotted in Figs.7.13 to 7.15.

The forced dynamic system in Eq. (7.61) with $F = 2$, $\varepsilon = -1/6$, and $\omega = 1$ is dealt with using the indirect approach first. As reported by Liu and Wu (2000), this problem poses challenge to direct DQ method since only one-half-cycle solutions of satisfactory accuracy were obtainable. It is found that results of satisfactory accuracy are obtained only for very short time domain, say, $t \in [0, 10]$. To acquire the solution with good accuracy over long time domain $t \in [0, 100]$, the direct approach and sufficient grid points are needed. When N is increased to 600, the results are in very good agreement with the analytical solution. The solutions of y, $y^{(1)}$, $y^{(2)}$ over $t \in [0, 100]$ (about 16 cycles) for $N = 800$ are shown in Fig. 7.13. It is noteworthy that difficulty in choosing a proper initial solution vector may arise when very large N is used in the direct approach. But this is compensated by the removal of possible accumulation error and stability concern.

To further demonstrate the spline-based DQ method, the high-order system given in Eq. (7.67) is tackled. The solution over 100 s (approximately 14 cycles) is calculated and the time domain is divided into 10 subintervals. The indirect approach is adopted and $N = 70$ is chosen in each subinterval. Very good agreement with the available analytical solution is reached, as shown in Fig.7.14. This confirms the declaration that low-order spline functions can be used to solve high-order differential equations. Actually, solution with reasonable accuracy is obtainable when N takes 40. In addition, further reduction of the number of subintervals results in the need of large N and accordingly the requirement of good initial solution in the iterative process.

7.4.6   Discussions

Based on the construction of cubic cardinal spline functions, a DQ method has been developed and applied to the solution of dynamic systems governed by Duffing-type nonlinear differential equations. Two solution approaches have been presented and their effectiveness has been verified. The stability of the spline-based DQ method has also been studied. Although it is conditionally stable, the virtually invariant stable criterion with respect to the number of grid points makes it a promising numerical tool in the solution of initial-value problems. In addition, low-order spline functions are applicable to the solution of high-order differential equations. In particular, the proposed direct approach is attractive when long-term integration is encountered in the analysis of dynamic systems.

7.5 Conclusions

DQ yields very accurate approximations of derivatives, but it is sensitive to grid distributions and requires a relatively small number of grid points. A simplified stability analysis in this chapter illustrates that stability is proportional to the reciprocal of the number of grid points. The more points are used, the poorer the stability is. It is the stability that deteriorates convergence. In order to overcome the difficulty, a localized DQ method is developed in this chapter. It is characterized by approximating the derivatives at a grid point using the weighted sum of the points in its neighborhood rather than all the grid points. In doing so, stability is enhanced, and accuracy is guaranteed by using any large number of grid points. The effectiveness of the localized DQ method is verified by the complicated dynamics problems. The results obtained by the present method are found to be in good agreement with the analytical results. However, the present method demands more CPU time on permutation and localized DQ computations when compared with traditional DQ approximations.

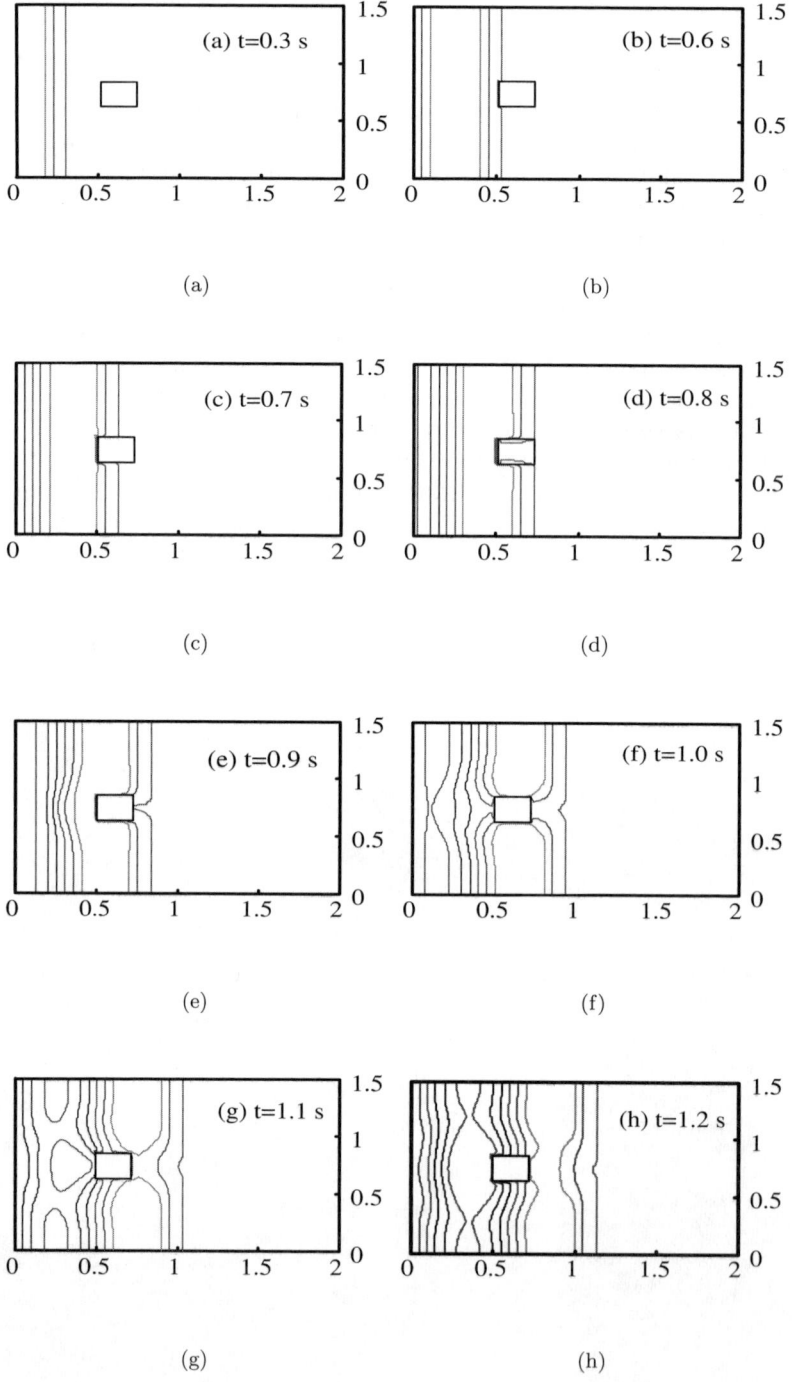

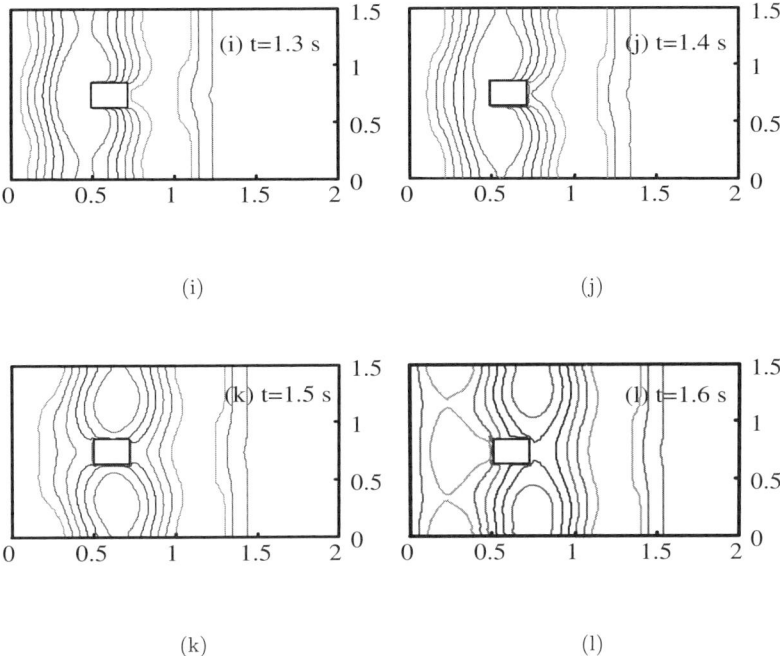

(i)   (j)

(k)   (l)

FIGURE 7.9: Wave scattering by a rectangular obstacle.

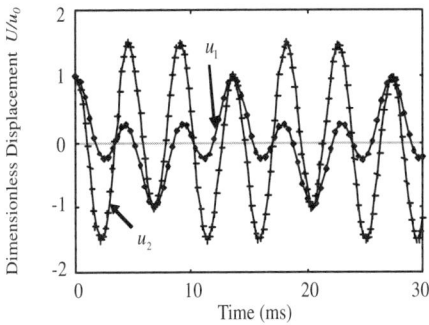

FIGURE 7.10: Time history of displacement at the center point $x= 0.5$ for u_1 and u_2. Crosses and dots denote the corresponding analytical solutions.

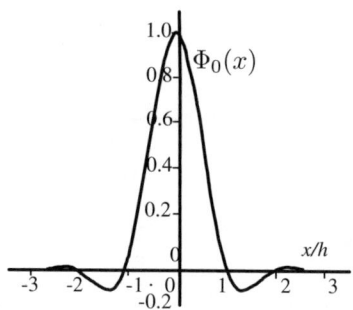

FIGURE 7.11:   A typical cubic cardinal spline function.

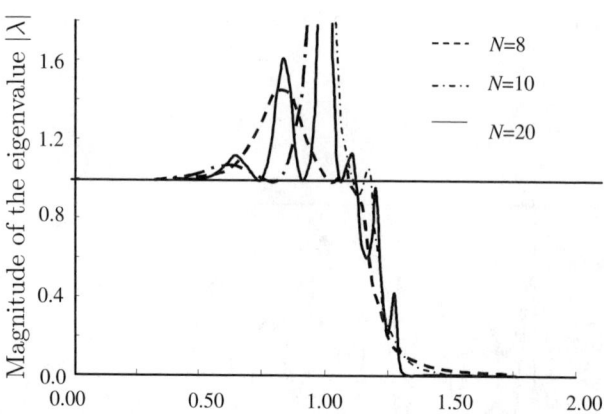

FIGURE 7.12:   Spectral radius of amplification matrix for $N = 8$, 10 and 20.

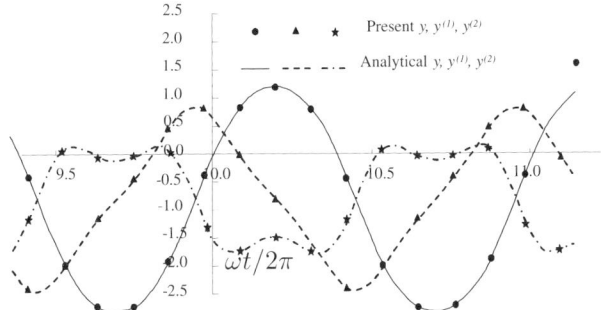

FIGURE 7.13: Solution of Duffing equation of Example 1.

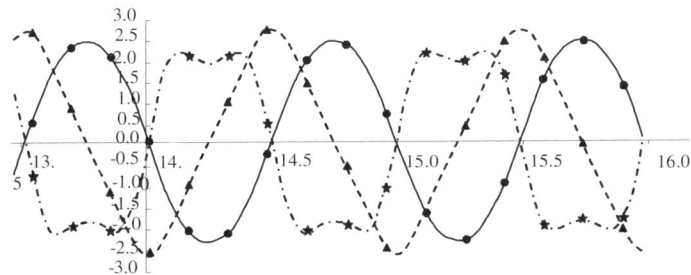

FIGURE 7.14: Solution of Duffing equation of Example 2.

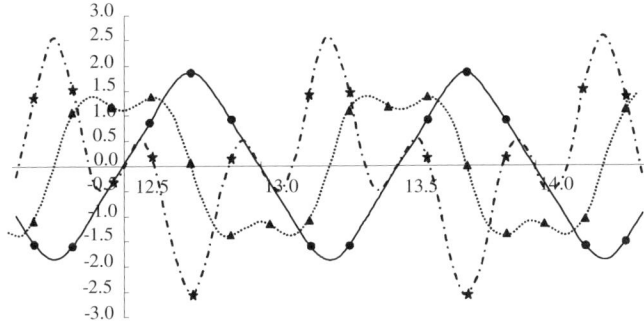

FIGURE 7.15: Solution of Duffing equation of Example 3.

Chapter 8

Mathematical Compendium

The mathematical compendium consists of four independent sections dealing in detail with specific mathematics used in the main text. It is provided as an aid to the readers who may not be familiar with these mathematical applications and thus may find it difficult to collect and study the necessary literature.

8.1 Gauss elimination

In linear algebra, Gauss elimination, named after German mathematician and scientist Carl Friedrich Gauss (1777-1855), is one of the most important numerical methods for solving a system of linear equations. The process of Gauss elimination consists of two steps, i.e., Forward and Backward eliminations. First, Forward elimination is performed to reduce a given system of linear equations to triangular forms, or lead to a degenerate equation with no solution, indicating that the system has no solution. This can be achieved by using elementary row operation. In triangular form, the solutions of the system can than be determined by back-substitution (Backward elimination).

A system of linear equations with n unknowns $x_1, x_2, \cdots, x_n$ is a set of n equations of the form

$$\begin{cases} a_{11}x_1 + a_{12}x_2 + \ldots + a_{1n}x_n = b_1 \\ a_{21}x_1 + a_{22}x_2 + \ldots + a_{2n}x_n = b_2 \\ \vdots \\ a_{n1}x_1 + a_{n2}x_2 + \ldots + a_{nn}x_n = b_n \end{cases} \tag{8.1}$$

In matrix form, we have

$$\mathbf{A}\mathbf{x} = \mathbf{b} \tag{8.2}$$

with

$$\mathbf{A} = \begin{bmatrix} a_{11} & a_{12} & \cdots & a_{1n} \\ a_{21} & a_{22} & \cdots & a_{2n} \\ \vdots & \vdots & \ddots & \vdots \\ a_{n1} & a_{n2} & \cdots & a_{nn} \end{bmatrix}, \quad \mathbf{x} = \begin{bmatrix} x_1 \\ x_2 \\ \vdots \\ x_n \end{bmatrix} \quad \mathbf{b} = \begin{bmatrix} b_1 \\ b_2 \\ \vdots \\ b_n \end{bmatrix} \tag{8.3}$$

The system is homogeneous if all the b_i are zero, otherwise it is non-homogeneous.

Gauss elimination consists of combining the coefficient matrix $\mathbf{A}$ with the right-hand side b to the "augmented" $(n, n + 1)$ matrix

$$\begin{bmatrix} \mathbf{A} \ \mathbf{b} \end{bmatrix} = \begin{bmatrix} a_{11} \ a_{12} \ \cdots \ a_{1n} \ b_1 \\ a_{21} \ a_{22} \ \cdots \ a_{2n} \ b_2 \\ \vdots \ \ \vdots \ \ \ \ \ \ \vdots \ \ \vdots \\ a_{n1} \ a_{n2} \ \cdots \ a_{nn} \ b_n \end{bmatrix} \tag{8.4}$$

A sequence of elementary row operations is then applied to this matrix so as to transform the coefficient part to upper triangular form:

(1) multiply a row by a non-zero real number c,

(2) swap two rows,

(3) add c times one row to another one.

$\begin{bmatrix} \mathbf{A} \ \mathbf{b} \end{bmatrix}$ will then have taken the following form:

$$\begin{bmatrix} a_{11} \ a_{12} \ \cdots \ a_{1n} \ b_1 \\ 0 \ \ a'_{22} \ \cdots \ a'_{2n} \ b'_2 \\ \vdots \ \ \vdots \ \ \ \ \ \ \vdots \ \ \vdots \\ 0 \ \ 0 \ \ \cdots \ a'_{nn} \ b'_n \end{bmatrix} \tag{8.5}$$

and the original equation is transformed to $\mathbf{R}\mathbf{x} = \mathbf{c}$ with an upper triangular matrix R, from which the unknowns x can be found by back substitution.

Assume we have transformed the first column, and we want to continue the elimination with the following matrix

$$\begin{bmatrix} a_{11} \ a_{12} \ \cdots \ a_{1n} \ b_1 \\ 0 \ \ a'_{22} \ \cdots \ a'_{2n} \ b'_2 \\ 0 \ \ a'_{32} \ \ldots \ a'_{3n} \ b'_3 \\ \vdots \ \ \vdots \ \ \ \ \ \ \vdots \ \ \vdots \\ 0 \ \ a'_{n2} \ \cdots \ a'_{nn} \ b'_n \end{bmatrix} \tag{8.6}$$

To zero a'_{32} we want to divide the second row by the "pivot" a'_{22}, multiply it with a'_{32} and subtract it from the third row. If the pivot is zero we have to swap two rows. This procedure frequently breaks down, not only for ill-conditioned matrices. Therefore, most programs perform "partial pivoting", i.e. they swap with the row that has the maximum absolute value of that column.

"Complete pivoting", always putting the absolute biggest element of the whole matrix into the right position, implying reordering of rows and columns, is normally not necessary.

Another variant is Gauss-Jordan elimination, which is closely related to Gaussian elimination. With the same elementary operations it does not only zero the elements below the diagonal but also above. The resulting augmented matrix will then look like:

$$\begin{bmatrix} a_{11} & 0 & \cdots 0 & b_1 \\ 0 & a'_{22} & \cdots 0 & b'_2 \\ & & & \\ 0 & 0 & \cdots a'_{nn} & b'_n \end{bmatrix} \tag{8.7}$$

Therefore, back substitution is not necessary and the values of the unknowns can be computed directly.

Example 8.1 *A system of linear equations with three unknowns*

In the following, a simple system of linear equations with three unknowns is considered to demonstrate the procedure of Gauss elimination. The given system is written as

$$\begin{cases} x_1 & + 2x_2 & + 2x_3 & = & 2 & L_1 \\ x_1 & + 3x_2 & - 2x_3 & = & -1 & L_2 \\ 3x_1 & + 5x_2 & + 8x_3 & = & 8 & L_3 \end{cases} \tag{8.8}$$

with

$$\mathbf{A} = \begin{bmatrix} 1 & 2 & 2 \\ 1 & 3 & -2 \\ 3 & 5 & 8 \end{bmatrix}, \quad \mathbf{x} = \begin{bmatrix} x_1 \\ x_2 \\ x_3 \end{bmatrix}, \quad \mathbf{b} = \begin{bmatrix} 2 \\ -1 \\ 8 \end{bmatrix} \tag{8.9}$$

First, we eliminate x_2 in the lines L_2 and L_3:

$$\begin{cases} x_1 & + & 2x_2 & + & 2x_3 & = & 2 & L_1 \\ & & x_2 & - & 4x_3 & = & -3 & L_2 \leftarrow L_2 - L_1 \\ & & -x_2 & + & 2x_3 & = & 2 & L_3 \leftarrow L_3 - 3L_1 \end{cases} \tag{8.10}$$

Next, x_1 is eliminated in the line L_3 as

$$\begin{cases} x_1 & + & 2x_2 & + & 2x_3 & = & 2 & L_1 \\ & & x_2 & - & 4x_3 & = & -3 & L_2 \\ & & & - & 2x_3 & = & -1 & L_3 \leftarrow L_3 + L_2 \end{cases} \tag{8.11}$$

Now it is obvious that the coefficient matrix A is reduced in triangular form as

$$\mathbf{A} = \begin{bmatrix} 1 & 2 & 2 \\ 0 & 1 & -4 \\ 0 & 0 & -2 \end{bmatrix} \tag{8.12}$$

The first step of Gauss elimination−Forward elimination, is completed. The second step, backward elimination, aims to solve the triangular form of linear equations in reverse order. It can be readily seen from the line L_3 that

$$x_3 = \frac{1}{2} \quad (L_3) \tag{8.13}$$

Then back-substituting $x_3 = \frac{1}{2}$ into the line L_2, we have

$$x_2 = -1 \quad (L_2) \tag{8.14}$$

Finally, $x_3 = 1/2$ and $x_2 = -1$ can be substituted into line L_1 for the solution of x_1 as

$$x_1 = 3 \quad (L_1) \tag{8.15}$$

Thus the solutions of the system are determined.

For a system of linear equations with n unknowns solved by Gauss elimination, approximately $2n^3/3$ operations are required. Therefore, Gauss elimination requires $O(n^3)$ order of computation. The algorithm can be coded by any computer language and applied for systems with thousands of equations and unknowns. However, for millions of equations, the algorithm is computationally expensive. In this case, these large systems are generally solved by interactive methods instead (Kreyszig, 1999; Press et al., 1986).

8.2 Successive over-relaxation (SOR) method

There are two approaches that can be used to determine the solutions of a linear system, i.e., direct methods and iterative methods. Direct methods, such as Gauss elimination method, aim to solve a system of linear equations by performing a finite number of operations to obtain the exact solutions. Direct methods are effective for a relatively small system of equations. However, the intensive computation prohibits the direct methods from solving large systems. In this case, we should resort to iterative methods, a very effective one of which is Successive Over-Relaxation (SOR) iterative method.

Suppose a system of linear equations is of the form

$$\mathbf{Ax} = \mathbf{b} \tag{8.16}$$

The coefficient matrix A can be expressed as the sum of the diagonal, lower triangular, and upper triangular matrices denoted by D, L and U, respectively. The three matrices can be expressed as

$$\mathbf{L} = \begin{pmatrix} 0 & 0 & \cdots & 0 \\ l_{21} & 0 & \cdots & 0 \\ \vdots & \vdots & \ddots & \vdots \\ l_{n1} & l_{n2} & \cdots & 0 \end{pmatrix}, \quad \mathbf{D} = \begin{pmatrix} d_{11} & 0 & \cdots & 0 \\ 0 & d_{22} & \cdots & 0 \\ \vdots & \vdots & \ddots & \vdots \\ 0 & 0 & \cdots & d_{nn} \end{pmatrix}, \quad \mathbf{U} = \begin{pmatrix} 0 & u_{12} & \cdots & u_{1n} \\ 0 & 0 & \cdots & u_{2n} \\ \vdots & \vdots & \ddots & \vdots \\ 0 & 0 & \cdots & 0 \end{pmatrix}$$

Equation (8.16) can then be rewritten as

$$(\mathbf{L} + \mathbf{D} + \mathbf{U})\mathbf{x} = \mathbf{b} \tag{8.17}$$

Multiplying Eq. (8.17) across by the SOR relaxation factor ω

$$\omega(\mathbf{L} + \mathbf{D} + \mathbf{U})\mathbf{x} = \omega\mathbf{b} \tag{8.18}$$

Adding $\mathbf{Dx}$ to both sides of Eq. (8.18), we have

$$(\mathbf{D} + \omega\mathbf{L})\mathbf{x} = [(1 - \omega)\mathbf{D} - \omega\mathbf{U}]\mathbf{x} + \omega\mathbf{b} \tag{8.19}$$

Multiplying Eq. (8.19) across by the inverse of $(\mathbf{D} + \omega\mathbf{L})$ leads to the recurrence relation for the SOR iteration

$$\mathbf{x}^{(k+1)} = (\mathbf{D} + \omega\mathbf{L})^{-1}[(1 - \omega)\mathbf{D} - \omega\mathbf{U}]\mathbf{x}^{(k)} + (\mathbf{D} + \omega\mathbf{L})^{-1}\omega\mathbf{b} \tag{8.20}$$

where $\mathbf{x}^{(k)}$ denotes the k-th iteration results.

If we introduce matrics

$$\mathbf{M}_{SOR} = (\mathbf{D} + \omega\mathbf{L})^{-1}[(1 - \omega)\mathbf{D} - \omega\mathbf{U}]$$

$$\mathbf{C}_{SOR} = (\mathbf{D} + \omega\mathbf{L})^{-1}\omega\mathbf{b} \tag{8.21}$$

Eq. (8.20) can be written as

$$\mathbf{x}^{(k+1)} = \mathbf{M}_{SOR}\mathbf{x}^{(k)} + \mathbf{C}_{SOR} \tag{8.22}$$

Note that for $\omega = 1$, the iteration reduces to the Gauss–Seidel iteration. The iteration is repeated until the specified tolerance (the changes made by iteration) is reached.

The selection of relaxation factor ω depends on the properties of the coefficients matrix $\mathbf{A}$. For symmetric, positive definite matrices it can be proved that ω in the range of 0 and 2 will lead to convergence.

Example 8.2 *A system of linear equations with two unknowns*

Below, we shall illustrate the solution procedure of SOR method via a simple system of linear equations defined by:

$$\begin{pmatrix} 4 & -1 \\ -1 & 4 \end{pmatrix} \begin{pmatrix} x_1 \\ x_2 \end{pmatrix} = \begin{pmatrix} 100 \\ 100 \end{pmatrix} \tag{8.23}$$

The diagonal $\mathbf{D}$, lower triangular $\mathbf{L}$, and upper triangular matrices $\mathbf{U}$ are given by

$$\mathbf{D} = \begin{pmatrix} 4 & 0 \\ 0 & 4 \end{pmatrix} \quad \mathbf{L} = \begin{pmatrix} 0 & 0 \\ -1 & 0 \end{pmatrix} \quad \mathbf{U} = \begin{pmatrix} 0 & -1 \\ 0 & 0 \end{pmatrix}$$

The expressions of $\mathbf{M}_{SOR}$ and $\mathbf{C}_{SOR}$ can then be written as

$$\mathbf{M}_{SOR} = (\mathbf{D} + \omega\mathbf{L})^{-1}[(1 - \omega)\mathbf{D} - \omega\mathbf{U}]$$

$$\mathbf{M}_{SOR} = \left[\begin{pmatrix} 4 & 0 \\ 0 & 4 \end{pmatrix} + \omega\begin{pmatrix} 0 & 0 \\ -1 & 0 \end{pmatrix}\right]^{-1}\left[(1 - \omega)\begin{pmatrix} 4 & 0 \\ 0 & 4 \end{pmatrix} - \omega\begin{pmatrix} 0 & -1 \\ 0 & 0 \end{pmatrix}\right]$$

$$\Rightarrow \mathbf{M}_{SOR} = \begin{pmatrix} 1-\omega & \frac{\omega}{4} \\ \frac{\omega(1-\omega)}{4} & \frac{\omega^2}{16}+1-\omega \end{pmatrix} \tag{8.24}$$

$$\mathbf{C}_{SOR} = (\mathbf{D}+\omega\mathbf{L})^{-1}\omega\mathbf{b}$$

$$\mathbf{C}_{SOR} = \left[\begin{pmatrix} 4 & 0 \\ 0 & 4 \end{pmatrix} + \omega \begin{pmatrix} 0 & 0 \\ -1 & 0 \end{pmatrix}\right]^{-1}\left[\omega\begin{pmatrix} 100 \\ 100 \end{pmatrix}\right]$$

$$\Rightarrow \mathbf{C}_{SOR} = \begin{pmatrix} 25\omega \\ 25\omega + \frac{25\omega^2}{4} \end{pmatrix} \tag{8.25}$$

By substituting Eqs. (8.24) and (8.25) into Eq. (8.23), one obtains

$$\begin{pmatrix} x_1^{(k+1)} \\ x_2^{(k+1)} \end{pmatrix} = \begin{pmatrix} 1-\omega & \frac{\omega}{4} \\ \frac{\omega(1-\omega)}{4} & \frac{\omega^2}{16}+1-\omega \end{pmatrix} \begin{pmatrix} x_1^{(k)} \\ x_2^{(k)} \end{pmatrix} + \begin{pmatrix} 25\omega \\ 25\omega + \frac{25\omega^2}{4} \end{pmatrix} \tag{8.26}$$

Assuming $\omega = 1.1$, four iterations of the SOR method are performed on this linear system.

0th **iteration** Take initial approximations as

$$\begin{pmatrix} x_1^{(0)} \\ x_2^{(0)} \end{pmatrix} = \begin{pmatrix} 0 \\ 0 \end{pmatrix}$$

1st **iteration** Based on the initial values, the solutions are given by

$$\begin{pmatrix} x_1^{(1)} \\ x_2^{(1)} \end{pmatrix} = \begin{pmatrix} -0.1 & \frac{1.1}{4} \\ -\frac{0.11}{4} & 0.39 \\ & 16 \end{pmatrix} \begin{pmatrix} x_1^{(0)} \\ x_2^{(0)} \end{pmatrix} + \begin{pmatrix} 27.5 \\ 35.0625 \end{pmatrix} = \begin{pmatrix} 27.5 \\ 35.0625 \end{pmatrix}$$

2nd **iteration** Substituting the results in the previous iteration step gives

$$\begin{pmatrix} x_1^{(2)} \\ x_2^{(2)} \end{pmatrix} = \begin{pmatrix} -0.1 & \frac{1.1}{4} \\ -\frac{0.11}{4} & \frac{0.39}{16} \end{pmatrix} \begin{pmatrix} x_1^{(1)} \\ x_2^{(1)} \end{pmatrix} + \begin{pmatrix} 27.5 \\ 35.0625 \end{pmatrix} = \begin{pmatrix} 34.39218750 \\ 33.45160156 \end{pmatrix}$$

3rd **iteration** In a similar manner, one obtains the solutions after 3rd iteration:

$$\begin{pmatrix} x_1^{(3)} \\ x_2^{(3)} \end{pmatrix} = \begin{pmatrix} -0.1 & \frac{1.1}{4} \\ -\frac{0.11}{4} & \frac{0.39}{16} \end{pmatrix} \begin{pmatrix} x_1^{(2)} \\ x_2^{(2)} \end{pmatrix} + \begin{pmatrix} 27.5 \\ 35.0625 \end{pmatrix} = \begin{pmatrix} 33.25997168 \\ 33.30133206 \end{pmatrix}$$

4th **iteration** Further iteration leads to

$$\begin{pmatrix} x_1^{(4)} \\ x_2^{(4)} \end{pmatrix} = \begin{pmatrix} -0.1 & \frac{1.1}{4} \\ -\frac{0.11}{4} & \frac{0.39}{16} \end{pmatrix} \begin{pmatrix} x_1^{(3)} \\ x_2^{(3)} \end{pmatrix} + \begin{pmatrix} 27.5 \\ 35.0625 \end{pmatrix} = \begin{pmatrix} 33.33186915 \\ 33.33613081 \end{pmatrix}$$

It is clearly shown that the results converge to the true solutions $x_1 = x_2 = \frac{100}{3}$ after fourth iteration (Kreyszig, 1999; Press et al., 1986).

8.3   One-dimensional band storage

One-dimensional band storage, as its name implies, stores only the non-zero elements while it neglects the zero ones in a sparse matrix by using a one-dimensional matrix. When compared with the direct data storage scheme by storing all elements, one-dimensional band storage can save computer memory considerably and thus enhance computational efficiency. When operating sparse matrices on a computer, it is of great significance to employ one-dimensional band storage. This technique is widely applied in finite element methods (FEM) for the storage of the sparse stiffness matrix. For example, the following sparse symmetric matrix **K** can be stored in a one-dimensional matrix **A** as

$$[K] = \begin{pmatrix} K_{11} & K_{12} & & K_{14} & & & & \\ & K_{22} & K_{23} & 0 & & & & \\ & & K_{33} & K_{34} & & K_{36} & & \\ & & & K_{44} & K_{45} & K_{46} & & \\ & & & & K_{55} & K_{56} & & K_{58} \\ & Sym & & & & K_{66} & K_{67} & K_{68} \\ & & & & & & K_{77} & 0 \\ & & & & & & & K_{88} \end{pmatrix}$$

$$= \begin{pmatrix} A(1) & A(3) & & A(9) & & & & \\ & A(2) & A(5) & A(8) & & & & \\ & & A(4) & A(7) & & A(15) & & \\ & & & A(6) & A(11) & A(14) & & \\ & & & & A(10) & A(13) & & A(21) \\ & & & & & A(12) & A(17) & A(20) \\ & & & & & & A(16) & A(19) \\ & & & & & & & A(18) \end{pmatrix}$$

By doing so, the 8×8 matrix is represented by a one-dimensional matrix with 21 entries only. Computer memory is reduced greatly. This reduction is significant for larger matrices (Kreyszig, 1999; Press et al., 1986).

8.4   Runge–Kutta method (constant time step)

Runge–Kutta method is a method of numerically integrating ordinary differential equations

$$\frac{dy}{dt} = f(y, t), \quad y(0) = y_0 \tag{8.27}$$

by using a trial step at the midpoint of an interval to cancel out lower-order error terms. Suppose the value of $y(t)$ at time t^n, denoted by y^n, is known. The integration time step is h. The second-order formula is

$$
\begin{aligned}
k_1 &= hf(t^n, y^n) \\
k_2 &= hf(t^n + \tfrac{h}{2}, y^n + \tfrac{k_1}{2}) \\
y^{n+1} &= y^n + k_2 + O(h^3)
\end{aligned}
\tag{8.28}
$$

sometimes known as RK2, and the fourth-order formula is,

$$
\begin{aligned}
k_1 &= hf(t^n, y^n) \\
k_2 &= hf(t^n + \tfrac{h}{2}, y^n + \tfrac{k_1}{2}) \\
k_3 &= hf(t^n + \tfrac{h}{2}, y^n + \tfrac{k_2}{2}) \\
k_4 &= hf(t^n + h, y^n + k_3) \\
y^{n+1} &= y^n + \tfrac{1}{6}k_1 + \tfrac{1}{3}k_2 + \tfrac{1}{3}k_3 + \tfrac{1}{6}k_4 + O(h^5)
\end{aligned}
\tag{8.29}
$$

sometimes known as RK4. This method is reasonably simple and robust and is a good general candidate for numerical solution of differential equations when combined with an intelligent adaptive step-size routine.

Differential equations of order higher than one are generally handled by converting them into systems of first order equation by the use of dummy variables.

The initial value problem

$$
\begin{aligned}
\frac{dy_i}{dt} &= f_i(y_1, \cdots, y_m, t) \\
y_i(t_0) &= y_{i0}, i = 1, 2, \cdots, m
\end{aligned}
\tag{8.30}
$$

which can be expressed in vector (matrix) form as

$$
\frac{d\mathbf{y}}{dt} = \mathbf{f}(t, \mathbf{y}); \ \ \mathbf{y}(t_0) = \mathbf{y_0}
\tag{8.31}
$$

can be handled by methods very similar to those given above for single equations.

We include the Runge–Kutta method as a typical example.

$$
\begin{aligned}
\mathbf{k}_1 &= h\mathbf{f}(t^n, \mathbf{y}^n) \\
\mathbf{k}_2 &= h\mathbf{f}(t^n + \frac{h}{2}, \mathbf{y}^n + \frac{\mathbf{k}_1}{2}) \\
\mathbf{k}_3 &= h\mathbf{f}(t^n + \frac{h}{2}, \mathbf{y}^n + \frac{\mathbf{k}_2}{2}) \\
\mathbf{k}_4 &= h\mathbf{f}(t^n + h, \mathbf{y}^n + \mathbf{k}_3) \\
\mathbf{y}^{n+1} &= \mathbf{y}^n + \frac{1}{6}\mathbf{k}_1 + \frac{1}{3}\mathbf{k}_2 + \frac{1}{3}\mathbf{k}_3 + \frac{1}{6}\mathbf{k}_4 + O(h^5) \\
t^{n+1} &= t^n + h
\end{aligned}
\tag{8.32}
$$

The equations above (except the last equation) are vector equations. Denoting the i-th component of $\mathbf{k}$ by k_{ij} ($j = 1, 2, 3, 4$) and denoting the i-th component of $\mathbf{y}_k$ by y_{ik} ($k = 1, 2, 3,...$), we obtain by equating components

$$k_{i1} = hf_i(t^n, y_1^n, \cdots, y_m^n)$$

$$k_{i2} = hf_i(t^n + \frac{h}{2}, y_1^n + \frac{a_{11}}{2}, \cdots, y_m^n + \frac{a_{m1}}{2})$$

$$k_{i3} = hf_i(t^n + \frac{h}{2}, y_1^n + \frac{a_{12}}{2}, \cdots, y_m^n + \frac{a_{m2}}{2}) \quad i = 1, 2, \cdots, m \qquad (8.33)$$

$$k_{i4} = hf_i(t^n + h, y_1^n + a_{13}, \cdots, y_m^n + a_{m3})$$

$$y_i^{n+1} = y_i^n + \frac{1}{6}(k_{i1} + 2k_{i2} + 2k_{i3} + k_{i4}) + O(h^5)$$

$$t^{n+1} = t^n + h$$

8.5 Complex analysis

8.5.1 Complex variable

Complex analysis, traditionally known as the theory of functions of a complex variable, is the branch of mathematics investigating functions of complex numbers. It is useful in many branches of mathematics, including number theory and applied mathematics, and in physics.

Complex analysis is particularly concerned with the analytic functions of complex variables, which are commonly divided into two main classes: the holomorphic functions and the meromorphic functions. Because the separable real and imaginary parts of any analytic function must satisfy Laplace's equation, complex analysis is widely applicable to two-dimensional problems in physics.

A complex function is a function in which the independent variable and the dependent variable are both complex numbers. More precisely, a complex function is a function whose domain Ω is a subset of the complex plane and whose range is also a subset of the complex plane.

For any complex function, both the independent variable and the dependent variable may be separated into real and imaginary parts:

$$z = x + iy \qquad (8.34a)$$
$$w = f(z) = u(z) + iv(z) \qquad (8.34b)$$

where $x, y \in \mathbf{R}$ and $u(z)$ and $v(z)$ are real-valued functions. In other words, the components of the function $f(z)$

$$u = u(x, y) \qquad (8.35a)$$
$$v = v(x, y) \qquad (8.35b)$$

can be interpreted as real valued functions of the two real variables, x and y. The basic concepts of complex analysis are often introduced by extending the elementary real functions (e.g., exponentials, logarithms, and trigonometric functions) into the complex domain.

Just as in real analysis, a "smooth" complex function $w = f(z)$ may have a derivative at a particular point in its domain Ω. In fact, the definition of the derivative

$$f'(z) = \lim_{h \to 0} \frac{f(z+h) - f(z)}{h} \tag{8.36}$$

is analogous to the real case but with one very important difference. In real analysis, the limit can only be approached by moving along the one-dimensional number line. In complex analysis, the limit can be approached from any direction in the two-dimensional complex plane. (The claim that "in real analysis, the limit can only be approached by moving along the one dimensional number line" should not be confused with directional derivatives. It may be explained that in directional derivatives, one still moves along the one dimensional x line but it can be in "discrete" units; that is, if one follows $y = x^2$ curve, that does not mean that one is moving on the plane (instead of the one dimensional x line) but means that one is approaching in steps of discrete units.)

If this limit, the derivative, exists for every point z in Ω, then $f(z)$ is said to be differentiable on Ω. It can be shown that any differentiable $f(z)$ is analytic. This is a much more powerful result than the analogous theorem that can be proved for real-valued functions of real numbers. In the calculus of real numbers, we can construct a function $f(x)$ that has a first derivative everywhere, but for which the second derivative does not exist at one or more points in the function's domain. But in the complex plane, if a function $f(z)$ is differentiable in a neighborhood it must also be infinitely differentiable in that neighborhood.

By applying the methods of vector calculus to compute the partial derivatives of the two real functions $u(x, y)$ and $v(x, y)$ into which $f(z)$ can be decomposed, and by considering two paths leading to a point z in Ω, it can be shown that the existence of derivative implies

$$f'(z) = \frac{\partial u}{\partial x} + i\frac{\partial v}{\partial x} = \frac{\partial v}{\partial y} - i\frac{\partial u}{\partial y} \tag{8.37}$$

Equating the real and imaginary parts of these two expressions we obtain the traditional formulation of the Cauchy-Riemann Equations:

$$\frac{\partial u}{\partial x} = \frac{\partial v}{\partial y}, \text{and} \quad \frac{\partial u}{\partial y} = -\frac{\partial v}{\partial x} \tag{8.38}$$

or, in another common notation

$$u_x = v_y, \text{and} \quad u_y = -v_x \tag{8.39}$$

By differentiating this system of two partial differential equations first with respect to x, and then with respect to y, we can easily show that

$$\frac{\partial^2 u}{\partial x^2} + \frac{\partial^2 u}{\partial y^2} = 0, \text{ and } \quad \frac{\partial^2 v}{\partial x^2} + \frac{\partial^2 v}{\partial y^2} = 0 \qquad (8.40)$$

In other words, the real and imaginary parts of a differentiable function of a complex variable are harmonic functions because they satisfy Laplace's equation.

8.5.2 Holomorphic functions

Holomorphic functions are complex functions defined on an open subset of the complex plane which are differentiable. Complex differentiability has much stronger consequences than usual (real) differentiability. For instance, holomorphic functions are infinitely differentiable, a fact that is far from true for real differentiable functions. Most elementary functions, including the exponential function, the trigonometric functions, and all polynomial functions, are holomorphic.

8.5.3 Meromorphic functions

A meromorphic function on an open subset D of the complex plane is a function that is holomorphic on all D except a set of isolated points, which are poles for the function.

Every meromorphic function on D can be expressed as the ratio between two holomorphic functions (with the denominator not constant 0) defined on D: the poles then occur at the zeroes of the denominator. The Gamma function is meromorphic in the whole complex plane.

Intuitively then, a meromorphic function is a ratio of two nice (holomorphic) functions. Such a function will still be "nice", except at the points where the denominator of the fraction is zero, when the value of the function will be infinite.

From an algebraic point of view, if D is connected, then the set of meromorphic functions is the field of fractions of the integral domain of the set of holomorphic functions. This is analogous to the relationship between $\mathbf{Q}$, the rational numbers, and $\mathbf{Z}$, the integers.

8.6   QR algorithm

8.6.1   QR algorithm

The QR algorithm can be seen as a more sophisticated variation of the basic "power" eigenvalue algorithm. Recall that the power algorithm repeatedly multiplies A times a single vector, normalizing after each iteration. The vector converges to the eigenvector of the largest eigenvalue. Instead, the QR algorithm works with a complete basis of vectors, using QR decomposition to renormalize (and orthogonalize). For a symmetric matrix A, upon convergence, $\mathbf{AQ}=\mathbf{Q}\Lambda$, where Λ is the diagonal matrix of eigenvalues to which $\mathbf{A}$ converged, and where $\mathbf{Q}$ is a composite of all the orthogonal similarity transforms required to get there. Thus the columns of $\mathbf{Q}$ are the eigenvectors.

In numerical analysis of matrices, a QR algorithm is an eigenvalue algorithm; that is, a procedure to calculate the eigenvalues and eigenvectors of a matrix. The basic idea is to perform a QR decomposition, writing the matrix as a product of an orthogonal matrix and an upper triangular matrix, multiply the factors in the other order, and iterate.

Formally, let $\mathbf{A}$ be the matrix of which we want to compute the eigenvalues, and let $\mathbf{A}_0:=\mathbf{A}$. At the k-th step (starting with $k = 0$), we write $\mathbf{A}_k$ as the product of an orthogonal matrix $\mathbf{Q}_k$ and an upper triangular matrix $\mathbf{R}_k$ and we form $\mathbf{A}_{k+1} = \mathbf{R}_k\mathbf{Q}_k$. Note that

$$\mathbf{A}_{k+1} = \mathbf{R}_k\mathbf{Q}_k = \mathbf{Q}_k^T\mathbf{Q}_k\mathbf{R}_k\mathbf{Q}_k = \mathbf{Q}_k^T\mathbf{A}_k\mathbf{Q}_k = \mathbf{Q}_k^{-1}\mathbf{A}_k\mathbf{Q}_k \qquad (8.41)$$

so all the $\mathbf{A}_k$ are similar and hence they have the same eigenvalues. The algorithm is numerically stable because it proceeds by orthogonal similarity transforms. Under certain conditions, the matrices $\mathbf{A}_k$ converge to a triangular matrix. The eigenvalues of a triangular matrix are listed on the diagonal, and the eigenvalue problem is solved. In testing for convergence it is impractical to require exact zeros, but the Gershgorin circle theorem provides a bound on the error.

In this crude form the iterations are relatively expensive. This can be mitigated by first bringing the matrix $\mathbf{A}$ to upper Hessenberg form (which costs $\frac{5}{3}n^3 + O(n^2)$ using Householder reduction) with a finite sequence of orthogonal similarity transforms, much like a QR decomposition. Determining the QR decomposition of an upper Hessenberg matrix costs $6n^2 + O(n)$.

If the original matrix is symmetric, then the upper Hessenberg matrix is also symmetric and thus tridiagonal, and so are all the $\mathbf{A}_k$. This procedure costs $\frac{2}{3}n^3 + O(n^2)$ using Householder reduction. Determining the QR decomposition of a tridiagonal matrix costs $O(n)$.

The rate of convergence depends on the separation between eigenvalues, so a practical algorithm will use shifts, either explicit or implicit, to increase separation and accelerate convergence. A typical symmetric QR algorithm isolates each eigenvalue (then reduces the size of the matrix) with only one or two iterations, making it efficient as well as robust.

8.6.2 QR decomposition

A QR decomposition of a real square matrix $\mathbf{A}$ is a decomposition of $\mathbf{A}$ as

$$\mathbf{A} = \mathbf{QR} \tag{8.42}$$

where $\mathbf{Q}$ is an orthogonal matrix (meaning that $\mathbf{Q}^T\mathbf{Q} = \mathbf{I}$) and $\mathbf{R}$ is an upper triangular matrix (also called right triangular matrix). Analogously, we can define the QL, RQ, and LQ decompositions of $\mathbf{A}$ (with L being a left triangular matrix in this case).

More generally, we can factor a complex $m \times n$ matrix (with $m \geq n$) as the product of an $m \times m$ unitary matrix and an $m \times n$ upper triangular matrix. An alternative definition is decomposing a complex $m \times n$ matrix (with $m \geq n$) as the product of an $m \times n$ matrix with orthogonal columns and an $n \times n$ upper triangular matrix. This is also called the *thin QR factorization.*

If $\mathbf{A}$ is nonsingular, then this factorization is unique if we require that the diagonal elements of $\mathbf{R}$ are positive.

There are several methods for actually computing the QR decomposition, such as by means of the Gram–Schmidt process, Householder transformations, or Givens rotations. Each has a number of advantages and disadvantages. They are separately introduced in the following.

(a) Computing QR by means of Gram-Schmidt
Consider the Gram–Schmidt process, with the vectors to be considered in the process as the columns of the matrix

$$\mathbf{A} = (\mathbf{a_1} | \cdots | \mathbf{a_n}) \tag{8.43}$$

we define

$$\text{proj}_\mathbf{e}\mathbf{a} = \frac{\langle \mathbf{e}, \mathbf{a} \rangle}{\langle \mathbf{e}, \mathbf{e} \rangle}\mathbf{e} \tag{8.44}$$

where $\langle \mathbf{v}, \mathbf{w} \rangle = \mathbf{v}^T\mathbf{w}$. Then

$$\mathbf{u}_1 = \mathbf{a}_1, \mathbf{e}_1 = \frac{\mathbf{u}_1}{\|\mathbf{u}_1\|}$$

$$\mathbf{u}_2 = \mathbf{a}_2 - \text{proj}_{\mathbf{e}_1}\mathbf{a}_2, \mathbf{e}_2 = \frac{\mathbf{u}_2}{\|\mathbf{u}_2\|}$$

$$\mathbf{u}_k = \mathbf{a}_k - \sum_{j=1}^{k-1} \text{proj}_{\mathbf{e}_k}\mathbf{a}_k, \mathbf{e}_k = \frac{\mathbf{u}_k}{\|\mathbf{u}_k\|} \tag{8.45}$$

We then rearrange the equations above so that the $\mathbf{a}_i$ are on the left, producing the following equations

$$\mathbf{a}_1 = \mathbf{e}_1 \|\mathbf{u}_1\|$$

$$\mathbf{a}_2 = \text{proj}_{\mathbf{e}_1} \mathbf{a}_2 + \mathbf{e}_2 \, \|\mathbf{u}_2\|$$

$$\mathbf{a}_3 = \text{proj}_{\mathbf{e}_1} \mathbf{a}_3 + \text{proj}_{\mathbf{e}_2} \mathbf{a}_3 + \mathbf{e}_3 \, \|\mathbf{u}_3\|$$

$$\mathbf{a}_k = \sum_{j=1}^{k-1} \text{proj}_{\mathbf{e}_j} \mathbf{a}_j + \mathbf{e}_k \, \|\mathbf{u}_k\| \tag{8.46}$$

Note that since the $\mathbf{e}_i$ are unit vectors, we have the following

$$\mathbf{a}_1 = \mathbf{e}_1 \, \|\mathbf{u}_1\|$$

$$\mathbf{a}_2 = \langle \mathbf{e}_1, \mathbf{a}_2 \rangle \, \mathbf{e}_1 + \mathbf{e}_2 \, \|\mathbf{u}_2\|$$

$$\mathbf{a}_3 = \langle \mathbf{e}_1, \mathbf{a}_3 \rangle + \langle \mathbf{e}_2, \mathbf{a}_3 \rangle + \mathbf{e}_3 \, \|\mathbf{u}_3\|$$

$$\mathbf{a}_k = \sum_{j=1}^{k-1} \langle \mathbf{e}_j, \mathbf{a}_k \rangle + \mathbf{e}_k \, \|\mathbf{u}_k\| \tag{8.47}$$

Now the right sides of these equations can be written in matrix form as follows:

$$(\mathbf{e}_1 \,|\dots|\, \mathbf{e}_n\,) \begin{pmatrix} \|\mathbf{u}_1\| & \langle \mathbf{e}_1, \mathbf{a}_2 \rangle & \langle \mathbf{e}_1, \mathbf{a}_3 \rangle & \cdots \\ 0 & \|\mathbf{u}_2\| & \langle \mathbf{e}_2, \mathbf{a}_3 \rangle & \cdots \\ 0 & 0 & \|\mathbf{u}_3\| & \cdots \\ \vdots & \vdots & \vdots & \ddots \end{pmatrix} \tag{8.48}$$

But the product of each row and column of the matrices above give us a respective column of $\mathbf{A}$ that we started with, and together, they give us the matrix $\mathbf{A}$, so we have factorized $\mathbf{A}$ into an orthogonal matrix $\mathbf{Q}$ (the matrix of $\mathbf{e}_k$s), via Gram−Schmidt, and the obvious upper triangular matrix as a remainder $\mathbf{R}$.

Alternatively, $\mathbf{R}$ can be calculated as follows. Recall that $\mathbf{Q} = (\mathbf{e_1} \,|\cdots|\, \mathbf{e_n}\,)$ Then, we have

$$\mathbf{R} = \mathbf{Q}^T \mathbf{A} = \begin{pmatrix} \langle \mathbf{e}_1, \mathbf{a}_1 \rangle & \langle \mathbf{e}_1, \mathbf{a}_2 \rangle & \langle \mathbf{e}_1, \mathbf{a}_3 \rangle & \cdots \\ 0 & \langle \mathbf{e}_2, \mathbf{a}_2 \rangle & \langle \mathbf{e}_2, \mathbf{a}_3 \rangle & \cdots \\ 0 & 0 & \langle \mathbf{e}_3, \mathbf{a}_3 \rangle & \cdots \\ \vdots & \vdots & \vdots & \ddots \end{pmatrix} \tag{8.49}$$

Note that $\langle \mathbf{e}_j, \mathbf{a}_j \rangle = \|\mathbf{u}_j\|$ and $\langle \mathbf{e}_j, \mathbf{a}_k \rangle = 0$ for $j > k$ and $\mathbf{Q}\mathbf{Q}^T = \mathbf{I}$, so $\mathbf{Q}^T = \mathbf{Q}^{-1}$.

Example 8.3 *QR decomposition of a third order matrix*
Consider the decomposition of

$$A = \begin{pmatrix} 12 & -51 & 4 \\ 6 & 167 & -68 \\ -4 & 24 & -41 \end{pmatrix}$$

Recall that an orthogonal matrix $\mathbf{Q}$ has the property

$$\mathbf{Q}\mathbf{Q}^T = \mathbf{I}$$

Then, we can calculate $\mathbf{Q}$ by means of Gram–Schmidt as follows:

$$\mathbf{U} = \begin{pmatrix} \mathbf{u}_1 & \mathbf{u}_2 & \mathbf{u}_3 \end{pmatrix} = \begin{pmatrix} 12 & -69 & -58/5 \\ 6 & 158 & 6/5 \\ -4 & 30 & -33 \end{pmatrix}$$

$$\mathbf{Q} = \begin{pmatrix} \frac{\mathbf{u}_1}{\|\mathbf{u}_1\|} & \frac{\mathbf{u}_2}{\|\mathbf{u}_2\|} & \frac{\mathbf{u}_3}{\|\mathbf{u}_3\|} \end{pmatrix} = \begin{pmatrix} 6/7 & -69/175 & -58/175 \\ 3/7 & 158/175 & 6/175 \\ -2/7 & 6/35 & -33/35 \end{pmatrix}$$

Thus, we have

$$\mathbf{A} = \mathbf{Q}\mathbf{Q}^T\mathbf{A} = \mathbf{Q}\mathbf{R}$$

$$\mathbf{R} = \mathbf{Q}^T\mathbf{A} = \begin{pmatrix} 14 & 21 & -14 \\ 0 & 175 & -70 \\ 0 & 0 & 35 \end{pmatrix}$$

(b) Computing QR by means of Householder reflections
A Householder reflection (or Householder transformation) is a transformation that takes a vector and reflects it about some plane. We can use this property to calculate the QR factorization of a matrix.

$\mathbf{Q}$ can be used to reflect a vector in such a way that all coordinates but one disappear.Let $\mathbf{X}$ be an arbitrary real m-dimensional column vector such that $\|\mathbf{X}\| = |\alpha|$ for a scalarα. If the algorithm is implemented using floating-point arithmetic, then α should get the opposite sign as the first coordinate of $\mathbf{X}$ to avoid loss of significance. If $\mathbf{X}$ is a complex vector, then the definition

$$\alpha = -e^{j \arg x_1}\|\mathbf{X}\| \tag{8.50}$$

should be used.

Then, where $\mathbf{e}_1$ is the vector $(1, 0, \ldots, 0)^T$, and $\|\cdot\|$ the Euclidean norm, set

$$\begin{aligned} \mathbf{u} &= \mathbf{x} - \alpha\mathbf{e}_1, \\ \mathbf{v} &= \frac{\mathbf{u}}{\|\mathbf{u}\|}, \\ \mathbf{Q} &= \mathbf{I} - 2\mathbf{v}\mathbf{v}^T. \end{aligned} \tag{8.51}$$

$\mathbf{Q}$ is a Householder matrix and

$$\mathbf{Q}\mathbf{x} = (\alpha, 0, \cdots, 0)^T \tag{8.52}$$

This can be used to gradually transform an m-by-n matrix $\mathbf{A}$ to upper triangular form. First, we multiply $\mathbf{A}$ with the Householder matrix $\mathbf{Q}_1$ we obtain when we choose the first matrix column for $\mathbf{x}$. This results in a matrix $\mathbf{Q}_1\mathbf{A}$ with zeros in the left column (except for the first row).

$$\mathbf{Q}_1\mathbf{A} = \begin{bmatrix} \alpha_1 & * \cdots * \\ 0 & \\ \vdots & A' \\ 0 & \end{bmatrix} \tag{8.53}$$

This can be repeated for $\mathbf{A}'$, (obtained from $\mathbf{Q}_1\mathbf{A}$ by deleting the first row and first column), resulting in a Householder matrix $\mathbf{Q}'_2$. Note that $\mathbf{Q}'_2$ is smaller than $\mathbf{Q}_1$. Since we want it really to operate on $\mathbf{Q}_1\mathbf{A}$ instead of $\mathbf{A}$ we need to expand it to the upper left, filling in a 1, or in general:

$$\mathbf{Q}_k = \begin{pmatrix} \mathbf{I}_{k-1} & 0 \\ 0 & \mathbf{Q}'_k \end{pmatrix} \tag{8.54}$$

After t iterations of this process, $t = \min(m-1, n)$

$$\mathbf{R} = \mathbf{Q}_t \cdots \mathbf{Q}_2\mathbf{Q}_1\mathbf{A}$$

is an upper triangular matrix. So, with

$$\mathbf{Q} = \mathbf{Q}_1^T \mathbf{Q}_2^T \cdots \mathbf{Q}_t^T \tag{8.55}$$

$\mathbf{A} = \mathbf{QR}$ is a QR decomposition of $\mathbf{A}$. This method has greater numerical stability than the Gram–Schmidt method above.

Table 8.1 gives the number of operations in the k-th step of the QR-Decomposition by the Householder transformation, assuming a square matrix with size n.

Summing these numbers over the $(n-1)$ steps (for a square matrix of size n), the complexity of the algorithm is given by

$$\frac{2}{3}n^3 + n^2 + \frac{1}{3}n - 2 = O\left(n^3\right) \tag{8.56}$$

Example 8.4 *QR decomposition of a third order matrix*

TABLE 8.1: Operation numbers of QR-decomposition

Operation	Number of operations in the k-th step
multiplications	$2(n-k+1)^2$
additions	$(n-k+1)^2 + (n-k+1)(n-k) + 2$
division	1
square root	1

Let us calculate the decomposition of

$$A = \begin{pmatrix} 12 & -51 & 4 \\ 6 & 167 & -68 \\ -4 & 24 & -41 \end{pmatrix}$$

First, we need to find a reflection that transforms the first column of matrix A, vector $\mathbf{a}_1 = (12, 6, -4)^{\mathrm{T}}$, to $\|\mathbf{a}_1\|\,\mathbf{e}_1 = (14, 0, 0)^{\mathrm{T}}$. Now,

$$\mathbf{u} = \mathbf{x} - \alpha \mathbf{e}_1 \tag{8.57}$$

and

$$\mathbf{v} = \frac{\mathbf{u}}{\|\mathbf{u}\|} \tag{8.58}$$

Here, $\alpha = 14$ and $\mathbf{x} = \mathbf{a}_1 = (12, 6, -4)^{\mathrm{T}}$ and then

$$\begin{aligned}
\mathbf{Q}_1 &= \mathbf{I} - \frac{2}{\sqrt{14}\sqrt{14}} \begin{pmatrix} -1 \\ 3 \\ -2 \end{pmatrix} \begin{pmatrix} -1 & 3 & -2 \end{pmatrix} = \mathbf{I} - \frac{1}{7} \begin{pmatrix} 1 & & -32 \\ -39 & -6 \\ 2 & & -64 \end{pmatrix} \\
&= \begin{pmatrix} 6/7 & 3/7 & -2/7 \\ 3/7 & -2/7 & 6/7 \\ -2/7 & 6/7 & 3/7 \end{pmatrix}
\end{aligned} \tag{8.59}$$

Now observe:

$$\mathbf{Q}_1 \mathbf{A} = \begin{pmatrix} 14 & 21 & -14 \\ 0 & -49 & -14 \\ 0 & 168 & -77 \end{pmatrix}$$

so we already have almost a triangular matrix. We only need to zero the $(3, 2)$ entry. Take the $(1, 1)$ minor, and then apply the process again to

$$\mathbf{A}' = \mathbf{M}_{11} = \begin{pmatrix} -49 & -14 \\ 168 & -77 \end{pmatrix}$$

By the same method as above, we obtain the matrix of the Householder transformation

$$\mathbf{Q}_2 = \begin{pmatrix} 1 & 0 & 0 \\ 0 & -7/25 & 24/25 \\ 0 & 24/25 & 7/25 \end{pmatrix}$$

after performing a direct sum with 1 to make sure the next step in the process works properly. Now, we find

$$\mathbf{Q} = \mathbf{Q}_1^T \mathbf{Q}_2^T = \begin{pmatrix} 6/7 & -69/175 & 58/175 \\ 3/7 & 158/175 & -6/175 \\ -2/7 & 6/35 & 33/35 \end{pmatrix}$$

$$\mathbf{R} = \mathbf{Q}_2 \mathbf{Q}_1 \mathbf{A} = \mathbf{Q}^T \mathbf{A} = \begin{pmatrix} 14 & 21 & -14 \\ 0 & 175 & -70 \\ 0 & 0 & -35 \end{pmatrix}$$

The matrix $\mathbf{Q}$ is orthogonal and $\mathbf{R}$ is upper triangular, so $\mathbf{A} = \mathbf{QR}$ is the required QR-decomposition.

(c)Computing QR by means of Givens rotations

QR decompositions can also be computed with a series of Givens rotations. Each rotation zeros an element in the subdiagonal of the matrix, forming the R matrix. The concatenation of all the Givens rotations forms the orthogonal Q matrix.

In practice, Givens rotations are not actually performed by building a whole matrix and doing a matrix multiplication. A Givens rotation procedure is used instead which does the equivalent of the sparse Givens matrix multiplication, without the extra work of handling the sparse elements. The Givens rotation procedure is useful in situations where only a relatively few of diagonal elements need to be zeroed, and is more easily parallelized than Householder transformations.

Example 8.5 *QR by means of Givens rotation*

Let us calculate the decomposition of

$$A = \begin{pmatrix} 12 & -51 & 4 \\ 6 & 167 & -68 \\ -4 & 24 & -41 \end{pmatrix}$$

First, we need to form a rotation matrix that will zero the lowermost left element, $a_{31} = -4$. We form this matrix using the Givens rotation method, and call the matrix G_1. We will first rotate the vector $(6, -4)$, to point along the X axis. This vector has an angle $\theta = \arctan\left(\frac{-4}{6}\right)$. We create the orthogonal Givens rotation matrix, $\mathbf{G}_1$:

$$\mathbf{G}_1 = \begin{pmatrix} 1 & 0 & 0 \\ 0 & \cos(\theta) & \sin(\theta) \\ 0 & -\sin(\theta) & \cos(\theta) \end{pmatrix} \approx \begin{pmatrix} 1 & 0 & 0 \\ 0 & 0.83205 & -0.55470 \\ 0 & 0.55470 & 0.83205 \end{pmatrix}$$

And the result of $\mathbf{G}_1\mathbf{A}$ now has a zero in the a_{31} element.

$$\mathbf{G}_1\mathbf{A} \approx \begin{pmatrix} 12 & -51 & 4 \\ 7.21110 & 125.6396 & -33.83671 \\ 0 & 112.6041 & -71.83368 \end{pmatrix}$$

We can similarly form Givens matrices $\mathbf{G}_2$ and $\mathbf{G}_3$, which will zero the sub-diagonal elements a_{21} and a_{32}, forming a triangular matrix $\mathbf{R}$. The orthogonal matrix $\mathbf{Q}^T$ is formed from the concatenation of all the Givens matrices $\mathbf{Q}^T = \mathbf{G}_3\mathbf{G}_2\mathbf{G}_1$. Thus, we have $\mathbf{G}_3\mathbf{G}_2\mathbf{G}_1\mathbf{A} = \mathbf{Q}^T\mathbf{A} = \mathbf{R}$, and the QR decomposition is $\mathbf{A} = \mathbf{QR}$.

8.6.3 Connection to a determinant or a product of eigenvalues

We can use QR decomposition to find the absolute value of the determinant of a square matrix. Suppose a matrix is decomposed as $\mathbf{A} = \mathbf{Q}\mathbf{R}$. Then we have

$$\det(\mathbf{A}) = \det(\mathbf{Q}) \cdot \det(\mathbf{R}) \qquad (8.60)$$

Since $\mathbf{Q}$ is unitary $|\det(\mathbf{Q})| = 1$, thus

$$|\det(\mathbf{A})| = |\det(\mathbf{R})| = \left| \prod_i r_{ii} \right| \qquad (8.61)$$

where r_{ii} are the entries on the diagonal of $\mathbf{R}$.

Furthermore, because the determinant equals the product of the eigenvalues, we have

$$\left| \prod_i r_{ii} \right| = \left| \prod_i \lambda_i \right| \qquad (8.62)$$

where λ_i are eigenvalues of $\mathbf{A}$.

We can extend the above properties to non-square complex matrix $\mathbf{A}$ by introducing the definition of QR-decomposition for non-square complex matrix and replacing eigenvalues with singular values.

Suppose a QR decomposition for a non-square matrix $\mathbf{A}$:

$$\mathbf{A} = \mathbf{Q} \begin{pmatrix} \mathbf{R} \\ \mathbf{O} \end{pmatrix}, \quad \mathbf{Q}^*\mathbf{Q} = \mathbf{I} \qquad (8.63)$$

where $\mathbf{O}$ is a zero matrix and $\mathbf{Q}$ is a unitary matrix.

From the properties of SVD (Singular Value Decomposition) and determinant of matrix, we have

$$\left| \prod_i r_{ii} \right| = \prod_i \sigma_i \qquad (8.64)$$

where σ_i are singular values of A.

Note that the singular values of A and R are identical, although the complex eigenvalues of them may be different. However, if A is square, it holds that

$$\prod_i \sigma_i = \left| \prod_i \lambda_i \right| \qquad (8.65)$$

In conclusion, QR decomposition can be used efficiently to calculate a product of eigenvalues or singular values of matrix.

Chapter 9

Codes

Three sets of FORTRAN codes for DQ, complex DQ method and Coordinate-based Localized DQ method are respectively given in this chapter for reference. These methods are explained in Section 1.3, Section 2.2 and Section 7.3.1 and corresponding examples can also be found in these sections.

If these codes do help you in your research, publication or else, your proper acknowledgement is deeply appreciated.

9.1 DQ for numerical evaluation of function cos(x)

Example 1.4 is studied using the following code.

```
c
c    Main programme
c
c    n      node number
c    xi(i)     nodes
c    yi(i)    function value at node x(i)
c
      parameter    (nmax=100)
      dimension    xi(nmax), yi(nmax), a(nmax, nmax), b(nmax, nmax)
      dimension    zmi(nmax)
         open(2, file='dqcos.dat')
      n    = 15
      pi   = 3.1415926
      dx   = 2.*pi/float(n-1)
      x    =-dx
      do i= 1, n
      x = x + dx
         xi(i) = x
         yi(i) = fun(x)
      end do
```

```
c
      do i=1, n
           term=1.
         do j= 1, n
             if(j .ne. i) then
                term=term*(xi(i)-xi(j))
             end if
           end do
         zmi(i)=term
      end do
c
      do i=1, n
        do j=1, n
          if(j .ne. i) then
           a(i,j)=zmi(i)/zmi(j)/(xi(i)-xi(j))
          end if
        end do
      end do
      do i=1, n
        sum=0.
         do j=1, n
           if(j .ne. i) then
            sum=sum+a(i,j)
           end if
         end do
       a(i,i)=-sum
      end do
c
      do i=1, n
        do j=1, n
          if(j .ne. i) then
             b(i,j)=2.*(a(i,j)*a(i,i)-a(i,j)/(xi(i)-xi(j)))
          end if
        end do
      end do
c
      do i=1, n
        sum=0.
         do j=1, n
           if(j .ne. i) then
            sum=sum+b(i,j)
           end if
         end do

       b(i,i)=-sum
```

```
      end do
c
      do i=1, n
        sum1=0.
        sum2=0.
        do j=1, n
          sum1=sum1+a(i,j)*fun(xi(j))
          sum2=sum2+b(i,j)*fun(xi(j))
        end do
        write(2,2) xi(i), sum1,-sin(xi(i)), sum2,-cos(xi(i))
   2    format(2x, 6e15.6)
      end do
      stop
      end
c
c     subroutine calculating function value at x
c
      function    fun(x)
        fun    = cos(x)
      return
      end
```

9.2 CDQ for harmonic problem

Example 2.2 is studied using the following code.

```
c
c     Main programme: CDQ for harmonic equations
c
c     n node number along the contour C
c
      implicit real*8 (a–h,o–z)
      parameter    (nmax=2000,pi=3.1415926)
      complex      zi(nmax), xni(nmax), xi(nmax), phi(nmax)
      dimension    zmax(nmax,nmax), f1(nmax),  alpha(nmax),
      beta(nmax)
c
c     parameters
```

```
c
      scale =1.e0
      n    = 30
      n1   = 5*2
      n2   = 5*2
      n3   = 5*2
      n4   = 5*2
      m    = 0*2
      n    = n+m
c     write(*,*)  'ca'
c
      call  circular(n,1.d0,1.d0,xi ,zi ,xni ,f1 ,alpha ,beta)
c     call  canti(n,n1 ,n2 ,n3 ,n4 , xi , zi , f1 , alpha , beta)
c     call  crack(n-m,n1 ,n2 ,n3 ,n4 ,m, xi , zi , f1 , alpha , beta)
c     call  hole(n-m,n1 ,n2 ,n3 ,n4 ,m, xi , zi , f1 , alpha , beta)
c     call  canti2(n,n1 ,n2 ,n3 ,n4 , xi , zi , f1 , alpha , beta)
c     do  l=1,2*n
c       write(4,*)  real(xi(l)),imag(xi(l)),f1(l+2*n)
c       write(*,*)  real(xi(l)),imag(xi(l))
c   4        format(2x,5e15.6)
c     enddo
c
      call  coef(n,alpha , beta , zi , xi , xni ,zmax)
      do  i=1,2*n
       do  j=1, 2*n
        zmax(i , j)=scale*zmax(i ,j)
       enddo
      enddo
c
      call  gs1(2*n, zmax,  f1)
c
      do  k=1,n
       phi(k)=cmplx(f1(k) , f1(k+n))*scale
      enddo
c
c     do  l=1,n
c      fq= phi(l)+zi(l)*conjg(derphi(l))+conjg(chi(l))
c     enddo
c
      call  plotcir(n,20 ,40 ,1.d0, zi , phi)
c     call  plotcant(n,20 ,20 ,2.d0 ,1.d0, zi , chi , phi)
c     call  plotcrack(n,30 ,20 ,2.d0 ,1.d0, zi , chi , phi)
c     call  plotcant2(n,20 ,20 ,2.d0 ,1.d0, zi , chi , phi)
      stop
      end
```

```fortran
c
c     subroutine for calculating parameters for circle
c      subroutine   circular (n, rr , rq , xi , zi , xni , f1 , alpha , beta )
      implicit real*8 (a-h,o-z)
      parameter    (nmax=2000,pi=3.1415926)
      complex       zi (nmax) , fi (nmax) , xi (nmax) , xni (nmax)
      dimension     f1 (nmax) , alpha (nmax) , beta (nmax)
      m    = 0
      do  i=1,m
       alpha(i)    = 1.
c      alpha(i+n)  = 0.
       beta(i)     = 0.
c      beta(i+n)   = 1.
      enddo
      do  i=m+1,2*n
       alpha(i)    = 0.
       beta(i)     = 1.
      enddo
c      alpha(1)    = 1.
c      beta(1)     = 0.
c      alpha(n)    = 1.
c      beta(n)     = 0.
c      dth = 2.*pi/float (2*n)
      th   = -1.*dth
      do  l=1,2*n+1
        th     = th+dth
c       th     = float (l-1)*dth
c       if (l .le.  n) then
c        th    =0.5*pi*(1.-dcos ( float (l-1)/float (n-1)*pi ))
c       else
c        th    =pi+0.5*pi*(1.-dcos ( float (l-n)/float (n)*pi ))
c       endif
        x       = rr*cos (th)
        y       = rr*sin (th)
        xn      = cos (th)
        yn      = sin (th)
c       b(l)    = 2.*rq*cmplx (x,y)
c       if (th .ge.  0.  .and.  th .lt.  pi)then
c        fi (l) = cmplx (0., -10.)
c       else
c        fi (l)=cmplx (0.0,0)
c       endif
c       fi (l)=cmplx ( cos (th) ,0)
```

```
c        zi(1)  =  cmplx(x,y)
         xni(1)=  cmplx(xn,yn)
         xi(1)  =  cmplx(x,y)
         fi(1)  =  alpha(1)*xi(1)**2+3.*beta(1)*conjg(xni(1))
     *          *xi(1)**2
         enddo
c
         dth = 2.*pi/float(n)
         th  = -1.*dth
         do 1=1,n+1
          th = th+dth
          zi(1)   = rq*cmplx(cos(th),sin(th))
         enddo
c        write(2,2)  real(zi(1)),imag(zi(1))
c
         do i=1,2*n
          f1(i) = real(fi(i))
c         f1(i) = real(xni(i))*real(xi(i))+imag(xni(i))
     *           *Imag(xi(i))
         enddo
         return
         end
c
c       plotting  for  circule
c
         subroutine   plotcir(n,mm,nn,rr,zi,phi)
         implicit real*8 (a-h,o-z)
         parameter    (nmax=2000,pi=3.1415926)
         complex      zi(nmax),phi(nmax)
         complex      z,fq,der1,der2
c
           open(2,file='laplace_1.dat')
c          open(3,file='laplace_2.dat')
c
         znu = 0.3
         ye  = 2.11e8
c
         dr   = rr/float(mm)
         dth  = 2.*pi/float(nn)
         r    = 0
         do i=1,mm+1
           r  = r + dr
          th  = -dth
          do 1=1,nn+1
            th=th+dth
```

```
      x = r*cos(th)
      y = r*sin(th)
      z = cmplx(x,y)
      call fun(n,zi,phi,z,fq)
      call deriv(n,zi,z,phi,der1,der2)
c     call stress(n,zi,z,phi,phi,sx,sy,sxy,u,v)
      write(2,2) x,y,real(fq),1.*real(z**3),imag(fq),
     1*imag(z**3)
c     write(2,2) x+u,y+v,u,v
  2   format(2x,6e15.6)
      enddo
      write(2,2)
c     write(3,2)
      enddo
      close(2)
c     close(3)
      return
      end
c
c     subroutine for data preparation for cantilever
c
      subroutine  canti(n,n1,n2,n3,n4,xi,zi,f1,alpha,beta)
      implicit real*8 (a-h,o-z)
      parameter   (nmax=2000,pi=3.1415926)
c
      complex     zi(nmax),fi(nmax),xi(nmax)
      dimension   f1(nmax),alpha(nmax),beta(nmax)
c
      znu = 0.3
      a   = 2.
      b   = 1.0
      x   = 0.
c
      dy  = b/float(n1-1)
      y   = -dy
      do i=1,n1
       y  = y+dy
c      y  = 0.5*b*(1.-dcos(float(i-1)/float(n1-1)*pi))
       zi(i)=cmplx(x,y)
      enddo
c
      y   = b
      dx  = a/float(n2)
      do i=1, n2
       x  = x-dx
```

```
c       x   = -0.5*a*(1.-dcos(float(i)/float(n2)*pi))
        zi(i+n1)=cmplx(x,y)
        enddo
c
        dy  = b/float(n3)
        x   = -a
        do i=1,n3
         y  = y-dy
         y  = 0.5*b*(1+dcos(float(i)/float(n3)*pi))
         zi(i+n1+n2)=cmplx(x,y)
        enddo
c
        dx  = a/float(n4+1)
        y   = 0.
        do i=1,n4
         x  = x+dx
c        x  = -0.5*a*(1+dcos(float(i)/float(n4+1)*pi))
         zi(i+n1+n2+n3)=cmplx(x,y)
        enddo
c
        do l=1,n
         xi(2*l-1)=zi(l)
c        write(*,*) l,zi(l)
        enddo
c
        xi(2*n+1) = xi(1)
        do l=1,n
         xi(2*l)=0.5*(xi(2*l-1)+xi(2*l+1))
c        write(*,*) l,xi(2*l)
        enddo
c
        do l=1,2*n1-1
         alpha(l)=(3.-znu)/(1.+znu)
         beta(l) = -1.
        enddo
c
        do l=2*n1,2*n
         alpha(l)=1.
         beta(l) =1.
        enddo
c
        do l=1,2*n1-1
         fi(l)    = cmplx(0.,0.)
        enddo
c
```

```
      do  l=2*n1,2*n1+2*n2
       fi(l)   = cmplx(-real(xi(l)),0.0)
      enddo
c
      do  l=2*(n1+n2)+1,2*n
       fi(l)   = cmplx(a,0.)
      enddo
c
      do  i=1,2*n
       f1(i)  = real(fi(i))
       f1(i+2*n)=imag(fi(i))
      enddo
c
      return
      end
c
c    plotting  for  cantilever
c
      subroutine   plotcant(n,mm,nn,aa,bb,zi,chi,phi)
      implicit  real*8  (a-h,o-z)
      parameter    (nmax=2000,pi=3.1415926)
      complex       zi(nmax),phi(nmax),chi(nmax)
      complex       z
c
        open(2,file='ecant_1.dat')
        open(3,file='ecant_2.dat')
c
      dx   = aa/float(mm-1)
      dy   = bb/float(nn-1)
      x     = -aa-dx
      do  i=1,mm
       x   = x + dx
       y   = -dy
       do  l=1,nn
        y = y+dy
        z = cmplx(x,y)
        call  stress(n,zi,z,chi,phi,sx,sy,sxy,u,v)
        sx0  = -6.*(0.5*bb-y)*(x+aa)**2/bb**3+(0.5*bb-y)
     /bb**3.*(4.*(0.5*bb-y)**2-0.6*bb**2)
        sy0  = -0.5*(1.+(0.5*bb-y)/bb)
               *(1.-2*(0.5*bb-y)/bb)**2
        sxy0=-6.*(x+aa)*(0.25*bb**2-(0.5*bb-y)**2)/bb**3
c       if(dabs(x+1) .le. 1.e-1) then
        write(3,2)  x,y,sx,sx0,sy,sy0,sxy,-sxy0
        write(2,2)  x+u,y+v,sx,sy,sxy
```

```
  2        format (2x,8e15.6)
c        endif
         enddo
         write (2 ,2)
         write (3 ,2)
         enddo
         close (2)
         close (3)
         return
         end
c
         subroutine    canti2 (n ,n1 ,n2 ,n3 ,n4 , xi , zi , f1 , alpha , beta )
         implicit   real*8  (a–h ,o–z)
         parameter    (nmax=2000 , pi =3.1415926)
c
         complex        zi (nmax) , fi (nmax) , xi (nmax)
         dimension      f1 (nmax) , alpha (nmax) , beta (nmax)
c
         znu  =  0.3
         a    =  2.
         b    =  1.
         x    =  0.
c
         dy   =  b/ float (n1−1)
         y    =  −dy
         do  i =1 ,n1
          y   =  y+dy
c         y   =  0.5*b*(1.− dcos ( float (i −1)/ float (n1−1)*pi ))
          zi ( i )=cmplx (x ,y)
         enddo
c
         y    =  b
         dx   =  a/ float (n2)
         do  i =1,  n2
          x   =  x−dx
c         x   =  −0.5*a*(1.− dcos ( float (i )/ float (n2)*pi ))
          zi ( i+n1)=cmplx (x ,y)
         enddo
c
         dy   =  b/ float (n3)
         x    =  −a
         do  i =1 ,n3
          y   =  y−dy
c         y   =  0.5*b*(1+dcos ( float (i )/ float (n3)*pi ))
          zi ( i+n1+n2)=cmplx (x ,y)
```

```
      enddo
c
      dx   = a/float(n4+1)
      y    = 0.
      do  i=1,n4
       x  = x+dx
c      x  = -0.5*a*(1+dcos(float(i)/float(n4+1)*pi))
       zi(i+n1+n2+n3)=cmplx(x,y)
      enddo
c
      do  l=1,n
       xi(2*l-1)=zi(l)
c      write(*,*)  l,zi(l)
      enddo
c
      xi(2*n+1)  =  xi(1)
      do  l=1,n
       xi(2*l)=0.5*(xi(2*l-1)+xi(2*l+1))
c      write(*,*)  l,xi(2*l)
      enddo
c
      do  i=1,n
       zi(i)=(zi(i)-cmplx(0.,0.))*1.
      enddo
c
      do  l=1,2*n1-1
       alpha(l)=(3.-znu)/(1.+znu)
       beta(l)  = -1.
      enddo
c
      do  l=2*n1,2*n
       alpha(l)=1.
       beta(l)  =1.
      enddo
c
      do  l=1,2*(n1+n2+n3)-1
       fi(l)    = cmplx(0.,0.)
      enddo
c
      do  l=2*(n1+n2+n3),2*n
       fi(l)    = cmplx(1.,0.0)
      enddo
c
      do  i=1,2*n
       f1(i)  = real(fi(i))
```

```
      f1 (i+2*n)=imag( fi (i))
      enddo
c

      return
      end
c
c     plotting
c
      subroutine   plotcant2 (n,mm, nn , aa , bb , zi , chi , phi )
      implicit  real*8  (a-h , o-z )
      parameter   (nmax=2000, pi =3.1415926)
      complex     zi (nmax) , phi (nmax) , chi (nmax)
      complex     z
c
        open (2 , file ='elastic .dat ')
        open (3 ,  file ='rand . dat ')
c
      dx   = aa/ float (mm-1)
      dy   = bb/ float (nn-1)
      x    = -aa-dx
      do  i=1 ,mm
       x   = x + dx
       y   = -dy
       do  l=1 ,nn
        y = y+dy
        z  = cmplx (x , y )
        call  stress (n , zi  , z , chi , phi , sx , sy , sxy , u , v )
        sx0 =  -12.*(0.5*bb+y ) * ( x+aa ) **2/ bb**3
          sy0 = 0.
        sxy0 =1.5*(1. -4.*(0.5*bb+y ) **2/ bb**2 )/ bb
        write (3 ,2)  x , y , sx , sx0 , sxy , - sxy0
        write (2 ,2)  x+u , y+v , sx , sy , sxy
    2      format (2x ,6 e15 .6 )
       enddo
       write (2 ,2)
       write (3 ,2)
      enddo
      close (2)
      close (3)
      return
      end
c
c     data preparation for  crack
c
      subroutine   crack (n , n1 , n2 , n3 , n4 , m , xi , zi , f1 , alpha , beta )
```

```fortran
      implicit real*8 (a-h,o-z)
      parameter   (nmax=2000,pi=3.1415926)
c
      complex     zi(nmax),fi(nmax),xi(nmax)
      dimension   f1(nmax),alpha(nmax),beta(nmax)
c
      znu = 0.3
      a   = 2.
      b   = 2.0
      x   = 0.
c
      dy  = b/float(n1-1)
      y   = -dy
      do i=1,n1
       y   = y+dy
       y   = 0.5*b*(1.-dcos(float(i-1)/float(n1-1)*pi))
       zi(i)=cmplx(x,y)
      enddo
c
      y   = b
      dx  = a/float(n2)
      do i=1, n2
       x   = x-dx
       x   = -0.5*a*(1.-dcos(float(i)/float(n2)*pi))
       zi(i+n1)=cmplx(x,y)
      enddo
c
      dy  = b/float(n3)
      x   = -a
      do i=1,n3
       y   = y-dy
       y   = 0.5*b*(1+dcos(float(i)/float(n3)*pi))
       zi(i+n1+n2)=cmplx(x,y)
      enddo
c
      dx  = a/float(n4+1)
      y   = 0.
      do i=1,n4
       x   = x+dx
       x   = -0.5*a*(1+dcos(float(i)/float(n4+1)*pi))
       zi(i+n1+n2+n3)=cmplx(x,y)
      enddo
c
      do l=1,n
       xi(2*l-1)=zi(l)
```

```
c     write (*,*)  l , zi ( l )
      enddo
c
      xi (2*n+1) =  xi (1)
      do  l=1,n
       xi (2*l)=0.5*( xi (2*l-1)+xi (2*l+1))
c     write (*,*)  l , xi (2*l)
      enddo
c
c     hole
c
      dth  =  0.
      dy   =  0.
      if (m .gt.  0)  dth  =  2.*pi/float (m)
      if (m .gt.  0)  dy   =  b/float (m)
      th   =  2.*pi+dth
      y    =  -0.5*dy
      do  i=1,m
       x   =  0.1
       y   =  y+dy
c      th  =  th-dth
c      x   =  -0.5*a+.2*cos (th )
c      y   =  0.5*b+.2*sin (th )
       zi ( i+n)=cmplx (x,y)
      enddo
      dth  =  0.
      if (m .gt.  0)  dth  =  2.*pi/float (2*m)
      th   =  2.*pi+dth
      do  i=1,2*m
       th  =  th-dth
       x   =  -0.5*a+0.2*cos (th )
       y   =  0.5*b+0.2*sin (th )
       xi ( i+2*n)=cmplx (x,y)
      enddo
c
c     do  l=1,2*n1-1
c        write (*,*)  l , xi ( l )
c  alpha ( l )=(3.-znu )/(1.+znu )
c        beta ( l )  =  -1.
c     enddo
c
      do  l=1,2*(n+m)
       alpha ( l )=1.
       beta ( l )  =1.
      enddo
```

```
c
c     do  l=1,2*n1-1
c       fi(l)=0.
c     enddo
c
      do  l=1,2*n
        fi(l)   = cmplx(0.,imag(xi(l)))
c       fi(l)   = cmplx(-real(xi(l)),0.)
      enddo
c
      do  l=2*n+1,2*(n+m)
         fi(l)  = cmplx(0.,0.)
      enddo
c
       do  i=1,2*(n+m)
       f1(i)  = real(fi(i))
       f1(i+2*(n+m))=imag(fi(i))
       enddo
c
      return
      end
c
c     data preparation for a hole in a plate
c
      subroutine   hole(n,n1,n2,n3,n4,m,xi,zi,f1,alpha,beta)
      implicit real*8 (a-h,o-z)
      parameter   (nmax=2000,pi=3.1415926)
c
      complex      zi(nmax),fi(nmax),xi(nmax)
      dimension    f1(nmax),alpha(nmax),beta(nmax)
c
      znu = 0.3
      a   = 2.
      b   = 1.0
      x   = 0.
c
      dy  = b/float(n1+1)
      y   = 0.
      do  i=1,n1
       y  = y+dy
c      y  = 0.5*b*(1.-dcos(float(i-1)/float(n1-1)*pi))
       zi(i)=cmplx(x,y)
      enddo
c
      y   = b
```

```
      dx  = a/float(n2+1)
      x   = 0.
      do i=1, n2
       x  = x-dx
c      x  = -0.5*a*(1.-dcos(float(i)/float(n2)*pi))
       zi(i+n1)=cmplx(x,y)
      enddo
c
      dy  = b/float(n3+1)
      x   = -a
      y   = b
      do i=1,n3
       y  = y-dy
c      y  = 0.5*b*(1+dcos(float(i)/float(n3)*pi))
       zi(i+n1+n2)=cmplx(x,y)
      enddo
c
      dx  = a/float(n4+1)
      y   = 0.
      x   = -a
      do i=1,n4
       x  = x+dx
c      x  = -0.5*a*(1+dcos(float(i)/float(n4+1)*pi))
       zi(i+n1+n2+n3)=cmplx(x,y)
      enddo
c
      do l=1,n
       xi(2*l-1)=zi(l)
c      write(*,*) l,zi(l)
      enddo
c
      xi(2*n+1) = xi(1)
      do l=1,n
       xi(2*l)=0.5*(xi(2*l-1)+xi(2*l+1))
c      write(*,*) l,xi(2*l)
      enddo
c
c     hole
c
      goto 3
      dth = 0.
      dy  = 0.
      if(m .gt. 0) dth = 2.*pi/float(m)
      if(m .gt. 0) dy  = b/float(m)
      th  = 2.*pi+dth
```

```
c     y    = -0.5*dy
      do  i=1,m
c     x    =  0.1  c      y    = y+dy
      th  = th-dth
      x    = -0.5*a+.2*cos(th)
      y    =  0.5*b+.2*sin(th)
       zi(i+n)=cmplx(x,y)
      enddo
      dth  =  0.
      if(m .gt.  0)  dth  = 2.*pi/float(2*m)
      th   = 2.*pi+dth
      do  i=1,2*m
       th  = th-dth
       x    = -0.5*a+0.2*cos(th)
       y    =  0.5*b+0.2*sin(th)
        xi(i+2*n)=cmplx(x,y)
      enddo
      c
x     dth  =  0.
x     if(m .gt.  0)  dth  = 2.*pi/float(2*m)
x     th   = 2.*pi+dth
x     do  i=1,2*m
x      th  = th-dth
x      x    = -0.5*a+0.2*cos(th)
x      y    =  0.5*b+0.2*sin(th)
x       xi(i+2*n)=cmplx(x,y)
x     enddo
c
c     do  l=1,2*n1-1
c      alpha(l)=(3.-znu)/(1.+znu)
c      beta(l)  = -1.
c     enddo
c
  3 continue
      do  l=1,2*n
       alpha(l)=1.
       beta(l)  =1.
      enddo
c
x     do  l=1,2*n1-1
x      fi(l)   = cmplx(0.,imag(xi(l)))
x     enddo
x     do  l=2*n1,2*(n1+n2)
x      fi(l)   = cmplx(0.,b)
x     enddo
```

```
x      do  l=2*(n1+n2)+1,2*(n1+n2+n3)
x        fi(l)   = cmplx(0.,imag(xi(l)))
x      enddo
x      do  l=2*(n1+n2+n3)+1,2*n
x        fi(l)   = cmplx(0.,0.)
x      enddo
c


c
       do  l=1,2*n
         fi(l)   = cmplx(0.,imag(xi(l)))
       enddo
       fi(2*n1)=0.
       fi(2*(n1+n2))=0.
       fi(2*(n1+n2+n3))=0.
       fi(2*n)=0.
c
       do  l=2*n+1,2*(n+m)
         fi(l)   = cmplx(0.,0.)
       enddo
c
       do  i=1,2*(n+m)
         f1(i) = real(fi(i))
         f1(i+2*(n+m))=imag(fi(i))
       enddo
c
       return
       end
c
c      plotting
c
       subroutine   plotcrack(n,mm,nn,aa,bb,zi,chi,phi)
       implicit  real*8  (a-h,o-z)
       parameter   (nmax=2000,pi=3.1415926)
       complex      zi(nmax),phi(nmax),chi(nmax)
       complex      z
c
         open(2,file='crack_1.dat')
         open(3,file='crack_2.dat')
c
       dx   = aa/float(mm-1)
       dy   = bb/float(nn-1)
       x    = -aa-dx
       do  i=1,mm
         x  = x + dx
```

```
      y   = -dy
      do  l=1,nn
       y = y+dy
       z = cmplx(x,y)
       call  stress(n,zi,z,chi,phi,sx,sy,sxy,u,v)
       sx0 = -6.*(0.5*bb-y)*(x+aa)**2/bb**3
      +(0.5*bb-y)/bb**3.*(4.*(0.5*bb-y)**2-0.6*bb**2)
         sy0 = -0.5*(1.+(0.5*bb-y)/bb)
         *(1.-2*(0.5*bb-y)/bb)**2
       sxy0=-6.*(x+aa)*(0.25*bb**2-(0.5*bb-y)**2)/bb**3
       if(cabs(z-cmplx(-1,0.5)) .le. 0.) then
         sx=0.
         sy=0.
         sxy=0.
         u=0.
         v=0.
       endif
       write(3,2) x,y,sx,sx0,sy,sy0,sxy,-sxy0
       write(2,2) x+u,y+v,sx,sy,sxy
    2    format(2x,8e15.6)
      enddo
      write(2,2)
      write(3,2)
     enddo
     close(2)
     close(3)
     return
     end
c
c     subroutine  for  stress  calculation
c
      subroutine   stress(n,zi,z,chi,phi,sx,sy,sxy,u,v)
      implicit  real*8 (a-h,o-z)
      parameter    (nmax=2000,pi=3.1415926)
      complex      zi(nmax)
      complex      phi(nmax),chi(nmax)
      complex      fq,z,der1,der2
c
      zmu = 1.e3
      znu = 0.3
      zkappa=(3.-znu)/(1.+znu)
      ye  = 2.11e8
      c
c     do k=1,n
c        call deriv(n,zi,zi(k),phi,fq)
```

```
c          derphi(k)=fq
c              call deriv(n,zi,zi(k),chi,fq)
c          derchi(k)=fq
c       enddo
c
c          call fun(n,zi,chi,z,fq)
c          call deriv(n,zi,z,derchi,fq)
           call deriv(n,zi,z,chi,fq,der2)
           call deriv(n,zi,z,phi,der1,der2)
c          call deriv(n,zi,z,derphi,der2)
           sy =real(2.*der1+conjg(z)*der2+fq)
           sxy=imag(conjg(z)*der2+fq)
           sx = 4*real(der1)-sy
           call fun(n,zi,phi,z,der2)
           call fun(n,zi,chi,z,fq)
           u=real(zkappa*der2-z*conjg(der1)-conjg(fq))
           v=imag(zkappa*der2-z*conjg(der1)-conjg(fq))
           u=0.5*u/zmu
           v=0.5*v/zmu
        return
        end
c
c       subroutine for forming L
c
        subroutine zmi(n,l,zi,z,zl)
        implicit real*8 (a-h,o-z)
        parameter  (nmax=2000)
        complex    zi(nmax)
        complex    prod,zl,z
         prod=(1.,0.)
        do k=1,n
         if(k .ne. l) then
           prod=prod*(z-zi(k))
          endif
        enddo
        zl=prod
        return
        end
c
c       complex DQ
c
        subroutine dq(n,zi,xi,a)
        implicit real*8 (a-h,o-z)
        parameter  (nmax=2000)
        complex    zi(nmax), a(nmax,nmax), xi(nmax)
```

```
      complex        zll ,sum, prod
      do  i=1,2*n
       do  k=1,n
         call  zmi(n,k, zi , zi (k) , zll )
         sum=cmplx (0. ,0.)
         do  l=1,n
           if( l  .eq.  k)  goto  2
           prod=1.
            do  j=1,n
             if(j  .ne.  l  .and.  j  .ne.  k)
                 prod=prod*(xi ( i)− zi (j ))
            enddo
           sum=sum+prod
  2         continue
         enddo
        a( i , k)=sum/ zll
      enddo
      enddo
c
      return
      end
c
c     Multiplication
c
      subroutine   zint (n, zi , xi , zlbd )
      implicit  real*8  (a−h, o−z)
      parameter    (nmax=2000,mmax=50)
      complex       zi (nmax) , zlbd (nmax, nmax) , xi (nmax)
      complex       zl , zll
       do  i =1,2*n
        do  k=1,n
          call  zmi(n,k, zi , xi ( i ) , zl )
          call  zmi(n,k, zi , zi (k) , zll )
          zlbd ( i , k)=zl / zll
        enddo
       enddo
c
      return
      end
c
c     Function
c
      subroutine   fun (n, zi , fi , z , fq )
      implicit  real*8  (a−h, o−z)
      parameter    (nmax=2000,mmax=50)
```

```
      complex      zi(nmax), fi(nmax),zlambda(nmax)
      complex      zl, zll ,sum,fq , z
        do  j=1,n
          call  zmi(n,j,zi ,z ,zl )
          call  zmi(n,j,zi , zi(j ) , zll )
          zlambda(j)=zl/zll
      enddo
c
      sum=cmplx(0. ,0.)
      do  i=1,n
        sum=sum+zlambda(i)*fi(i)
      enddo
c
      fq=sum
      return
      end
c
c     coef  matrix
c
      subroutine   coef(n,alpha ,beta , zi , xi , xni ,zmax)
      implicit  real*8  (a-h ,o-z)
      parameter    (nmax=2000,mmax=50)
      complex      c1(nmax,nmax) ,c2(nmax,nmax) , xni(nmax)
      complex      zi(nmax), xi(nmax), zlbd(nmax,nmax) ,
                    a(nmax,nmax)
      dimension    zmax(nmax,nmax) , alpha(nmax) , beta(nmax)
        call  dq(n,zi , xi ,a)
        call  zint(n,zi , xi ,zlbd)
        do  i=1,2*n
          do  k=1,n
            c1(i ,k)=  alpha(i)*zlbd(i ,k)
            c2(i ,k)=  beta(i)*conjg(xni(i))*a(i ,k)
          enddo
        enddo
c
        do  i=1,2*n
          do  k=1,n
            zmax(i ,k)         = real(c1(i ,k)+c2(i ,k))
            zmax(i ,k+n)       = -imag(c2(i ,k)+c1(i ,k))
          enddo
        enddo
c
      prod      = 1.
      do  i=1,2*n
        prod=prod*beta(i)
```

```
      enddo
c
      if(dabs(prod) .le. 1.d−10) zmax(1,n+1)=1.d10
c
      return
      end
c
c     differentiation
c
      subroutine   deriv(n,zi,z,phi,der1,der2)
      implicit real∗8 (a−h,o−z)
      parameter   (nmax=2000)
      complex     zi(nmax),phi(nmax),a(nmax),b(nmax),zl(nmax)
      complex     zll,z,sum,der1,der2,prod
c
      goto 4
      der1=cmplx(0.,0.)
       do k=1,n
         call zmi(n,k,zi,zi(k),zll)
         zl(k)=zll
         sum=cmplx(0.,0.)
         do l=1,n
           if(l .eq. k) goto 2
           prod=1.
            do j=1,n
              if(j .ne. l .and. j .ne. k) prod=prod∗(z−zi(j))
            enddo
           sum=sum+prod
  2        continue
         enddo
         a(k)=sum/zll
         der1=der1+sum/zll∗phi(k)
       enddo
c
      goto 5
  4   do k=1,n
         call zmi(n,k,zi,zi(k),zll)
         zl(k)=zll
       enddo
c
      do k=1,n
         if(cabs(z−zi(k)) .ge. 1.d−5) then
         sum=cmplx(0.,0.)
         do l=1,n
           if(l .ne. k) then
```

```
         call  zmi(n,l,zi,z,zll)
             sum=sum+zll
           endif
         enddo
         a(k)=sum/(z-zi(k))/zl(k)
         endif
       enddo
c
       do  k=1,n
       if(cabs(z-zi(k))  .lt.  1.d-5) then
       sum=cmplx(0.,0.)
       do  l=1,n
         if(l  .ne.  k)  then
          sum=sum+a(l)*zl(l)/zl(k)
c         sum=sum-a(l)
         endif
       enddo
       a(k)=sum
       endif
       enddo
c
    5  do  k=1,n
       if(cabs(z-zi(k))  .ge.  1.d-5) then
       sum=cmplx(0.,0.)
       do  l=1,n
         if(l  .ne.  k)  then
          sum=sum+a(l)*zl(l)/zl(k)
         endif
       enddo
       b(k)=(sum-a(k))/(z-zi(k))
       endif
       enddo
c
       do  k=1,n
       if(cabs(z-zi(k))  .lt.  1.d-5) then
       sum=cmplx(0.,0.)
       do  l=1,n
         if(l  .ne.  k)  then
          sum=sum+b(l)*zl(l)/zl(k)
         endif
       enddo
       b(k)=cmplx(.5,0.)*sum
       endif
       enddo
c
```

```
        der1=cmplx(0.,0.)
        der2=cmplx(0.,0.)
         do k=1,n
           der1=der1+a(k)*phi(k)
           der2=der2+b(k)*phi(k)
         enddo
c
      return
      end
c     enddo
c
c     Gauss elimination
c
      subroutine gs1(n, a, b)
      implicit real*8 (a-h,o-z)
      parameter    (nmax=2000)
      dimension a(nmax,nmax),b(nmax)
      dimension m(nmax)
      ep = 1.e-12
      do 10 i = 1, n
 10      m(i) = i
      do 20 k = 1, n
         p = 0.
         do 30 i = k, n
         do 30 j = k, n
           if(abs(a(i,j)) .le. abs(p)) goto 30
           p = a(i,j)
           io = i
           jo = j
 30        continue
           if(abs(p) - ep) 200, 200, 300
 200          write(*,*) "Solution failure"
           return
 300      if(jo .eq. k) goto 400
      do 40 i = 1, n
         t = a(i, jo)
         a(i,jo) = a(i,k)
 40          a(i,k) = t
         j = m(k)
         m(k) = m(jo)
         m(jo) = j
 400        if(io .eq. k) goto 501
        do 50 j = k, n
          t = a(io, j)
          a(io,j) = a(k,j)
```

```
50          a(k,j) = t
        t = b(io)
        b(io) = b(k)
        b(k) = t
501         p = 1/p
        in = n - 1
     if(k .eq. n) goto 600
        do 60 j = k, in
60         a(k, j+1) = a(k, j+1) * p
600        b(k) = b(k) * p
       if(k .eq. n) goto 20

       do 70 i = k, in
       do 80 j = k, in
80          a(i+1, j+1) = a(i+1, j+1) - a(i+1,k)*a(k,j+1)
70          b(i+1) = b(i+1) - a(i+1,k) * b(k)
20       continue
       do 90 i1 = 2, n
         i = n + 1 - i1
         do 90 j = i, in
90            b(i) = b(i) - a(i, j+1) * b(j+1)
       do 1 k = 1, n
         i = m(k)
1      a(1, i) = b(k)
       do 2 k = 1, n
2      b(k) = a(1, k)
     return
     end
```

9.3   Localized DQ method

Example 7.2 is studied using the following code.

```
c
c    Main programme
c
implicit real*8 (a-h,o-z)
     parameter    (nmax=1000, mmax=50)
     dimension    dy(nmax), y0(nmax), tout(10000)
     common  /coord/xi(nmax)
     common  /corder/mset, jcount(nmax, mmax)
```

```
      open(2,  file='ldq1.inp')
      open(3,  file='ldq1.dat')
      open(4,  file='ldqtime.dat')
c
c     Initilization
c
      pi    = 3.1415926
      mset = 3
      nnode= 100
      nout = 200
      do k=1, nout
        tout(k)=(k-1)*2e-1
      end do
c
      xi(1)=0.
      dx   = 1./float(nnode-1)
      do j=2, nnode
        xi(j)=xi(j-1)+dx
      end do
c
      do i=1, 2*nnode+1
        y0(i)=0.
      end do
c
c     Ordering
c
      call smthlng(nnode, mset, xi, jcount)
      do i=1, nnode
      write(*,*) i,(jcount(i,j),j=1,5)
      end do
c
c     do i=37, 62
      do i=1,nnode
c       y0(i)=sin((xi(i)-xi(15))/(xi(25)-xi(15))*pi)
        y0(i+1)=zinitial(xi(i))
      end do
c       y0(19)=1.
c       y0(20)=1.
c       y0(21)=1.
c       y0(18)=1.
c     do i=1, nnode
c       y0(i+1)=sin(1*xi(i)*pi)
c     end do
      t0 = 0.
      tf = 0.001
```

```
          nt = 100
          dt = (tf-t0) /float(nt)
          dt = 1.e-4
          nt = 10000
          t  = - dt
c
      do i = 1, nt+1
          t  = t+dt
          tms=t*1000
c         y0(nnode+2)=exp(-t*5)
c         y0(2)=sin(2.*pi*t)
c         y0(2)=(1.-exp(-t/1.e-3))
c
          y0(2)=0.
          y0(nnode+1)=0.
c         y0(nnode+2)=0.
c         y0(2*nnode+1)=0.
          call rk(2*nnode+1, dt, y0, dy)
c
c         write(2,2) tms,
c         write(3,2) y0(1)*1000, y0(2), y0(3), y0(4)
c         write(4,2) t, y0(26)
c         write(4,2) t, y0(12)
          do k=1, nout
          if(abs(t-tout(k)) .lt. 0.5*dt) then
          write(*,*) i,t
          do j=1, nnode
c            write(3, 2) xi(j), y0(j+1),
c    .                (zinitial(xi(j)-t)+zinitial(xi(j)+t))*0.5
             write(3, 2) xi(j), y0(j+1), sin(1.*pi*xi(j))
     *cos(1.*pi*t)
          end do
          write(3,2)
    2     format(4x, 7e15.6)
    4     format(4x, 1e15.6)
          end if
          end do
c         write(3,4)
      end do
      stop
      end
c
      subroutine rk(n, h, y, dy)
      implicit real*8 (a-h,o-z)
      parameter   (nmax=1000)
```

```
      dimension  y(nmax),  dy(nmax),  yc(nmax),  y1(nmax),  a(4)
        a(1)  =  0.5  *  h
        a(2)  =  a(1)
        a(3)  =  h
        a(4)  =  h
        do  i  =  1,  n
          y1(i)  =  y(i)
        end  do
        do  k  =  1,  3
          do  i  =  1,  n
            yc(i)  =  y1(i)  +  a(k)  *  dy(i)
            y(i)   =  y(i)  +  a(k+1)  *  dy(i)  /  3.
          end  do
          call  diffun((n−1)/2,  yc,  dy)
        end  do
        do  i  =  1,  n
          y(i)  =  y(i)  +  a(1)  *  dy(i)  /  3.
        end  do
          call  diffun((n−1)/2,  yc,  dy)
        return
        end
c
      subroutine  diffun(nnode,  y0,  dy)
      implicit  real*8  (a–h,o–z)
      parameter   (nmax=1000,  mmax=50)
      dimension   dy(nmax),  disp(nmax),  y0(nmax),  vel(nmax)
      dimension   y2(nmax),  xint(nmax),yint(nmax),
         yint2(nmax)
      common    /coord/xi(nmax)
      common    /corder/mset,jcount(nmax,  mmax)
c
      dy(1)  =  1.
        do  i=1,  nnode
          disp(i)=y0(i+1)
          vel(i)=y0(i+1+nnode)
        end  do
c
      do  i=1,  nnode
        do  j=1,  mset
          yint(j)=disp(jcount(i,j))
          xint(j)=xi(jcount(i,j))
        end  do
        call  dq(mset,  xint,  yint,  yint2)
          y2(i)=yint2(1)
      enddo
```

```
      do  j=1, nnode
        dy(j+1)=vel(j)
        dy(j+1+nnode)=1.*y2(j)
      end  do
c
      return
      end
c
      subroutine  dq(n, xi, yi, y2)
      implicit  real*8  (a-h,o-z)
      parameter    (nmax=1000)
        dimension  yi(nmax),  a(nmax, nmax),  b(nmax, nmax)
      dimension    zmi(nmax),  y2(nmax),  xi(nmax)
c
      do  i=1, n
      term=1.
      do  j= 1, n
        if(j .ne. i) then
         term=term*(xi(i)-xi(j))
        end  if
      end  do
      zmi(i)=term
      end  do
c
      do  i=1, n
       do  j=1, n
        if(j .ne. i) then
         a(i,j)=zmi(i)/zmi(j)/(xi(i)-xi(j))
        end  if
       end  do
      end  do
      do  i=1, n
        sum=0.
        do  j=1, n
         if(j .ne. i) then
          sum=sum+a(i,j)
         end  if
        end  do
      a(i,i)=-sum
      end  do
c
      do  i=1, n
       do  j=1, n
         if(j .ne. i) then
          b(i,j)=2.*(a(i,j)*a(i,i)-a(i,j)/(xi(i)-xi(j)))
```

```fortran
        end if
       end do
      end do
c
      do i=1, n
        sum=0.
        do j=1, n
          if(j .ne. i) then
            sum=sum+b(i,j)
          end if
        end do
       b(i,i)=-sum
      end do
c
      do i=1, n
        sum1=0.
        sum2=0.
        do j=1, n
         sum1=sum1+a(i,j)*yi(j)
         sum2=sum2+b(i,j)*yi(j)
        end do
         y2(i)=sum2
c
c         write(*,*) i, sum2
c         write(2,2) xi(i), sum1,-sin(xi(i)),sum2,-cos(xi(i))
    2    format(2x, 6e15.6)
      end do
      return
      end
c
c     Smoothing length
c
      subroutine  smthlng(nnode, mset, x, jcount)
      implicit  real*8 (a-h,o-z)
      parameter   (nmax=1000, mmax=50)
      dimension   x(nmax), hi(nmax), alfr(nmax)
      integer     jcount(nmax,mmax), icount(nmax)
c
      np=nnode
      m =mset
      do i = 1, np
        do j=1, np
          alfr(j) = dist(i,j,x)
        end do
        call order(alfr, np, icount)
```

```
      do j = 1, mset
        jcount(i,j) = icount(j)
      end do
      hi(i) = dist(i,jcount(i,m), x)
    end do
    do i = 1, np
      hi(i)=0.5*hi(i)
    end do
    return
    end
c
c   Distance

    function    dist(i,j,x)
    implicit real*8 (a-h,o-z)
    parameter   (nmax=1000)
    dimension   x(nmax)
      dist=abs(x(i)-x(j))
    return
    end
c
    SUBROUTINE ORDER(ALFR,N, NO)
    implicit real*8 (a-h,o-z)
    PARAMETER        (nmax=1000)
    dimension        ALFR(Nmax), SR(Nmax)
      INTEGER        N
    INTEGER          NO(nmax)
    DO I = 1, N
      SR(I) = ALFR(I)
    END DO
c
    DO I = 1, N
      SMIN = 1.e20
      DO J = 1, N
        IF(SR(J) .LE. SMIN) THEN
          SMIN = SR(J)
          NO(I)= J
        END IF
      END DO
      SR(NO(I)) = 2.e20
    END DO
c
    DO I = 1, N
      SR(I) = ALFR(NO(I))
    END DO
```

```
      DO I = 1, N
        ALFR(I) = SR(I)
      END DO
c
      RETURN
      END
c
      function     zinitial(x)
      implicit real*8 (a-h,o-z)
      parameter    (nmax=1000, mmax=50)
c
      common  /coord/xi(nmax)
      pi        = 3.1415926
      zinitial = 0.
      if(x .gt. 0.375 .and. x .lt. 0.625) then
        zinitial =sin((x -0.375)/0.25*pi)
      end if
      return
      end
```

References

ABAQUS User Guide, ver 6.3, Hibbitt, Karlsson and Sorensen Inc, 2002.

Achieser, N.I. (1992). *Theory of Approximation*, Dover, New York.

Adjerid, S. and Flaherty, J.E. (1986). "A moving finite element method with error estimation and refinement for one-dimensional time dependent partial differential equations." SIAM Journal on Numerical Analysis 23,778-796.

Atkinson, K.E. (1989). *An Introduction to Numerical Analysis.* 2nd ed., Wiley, New York.

Bath, K.J. (1996). *Finite Element Procedures*, Prentice-Hall, Englewood Cliffs, New Jersey.

Bellman, R. and Casti, J. (1971). "Differential quadrature and long-term integration." *Journal of Mathematical Analysis and Applications* 34, 235-238.

Bellman, R., Kashef, B.G. and Casti, J. (1972). "Differential quadrature: A technique for the rapid solution of nonlinear partial differential equations." *Journal of Computational Physics* 10, 40-52.

Bert, C.W., Jang, S.K. and Striz, A.G. (1988). "Two new approximate methods for analyzing free vibration of structural components." *AIAA Journal* 26, 612-618.

Bert, C.W., Wang, X. and Striz, A.G. (1993). "Differential quadrature for static and free vibration analysis of anisotropic plates." *International Journal of Solids and Structures* 30, 1737-1744.

Bert, C.W., Wang, X. and Striz, A.G. (1994a). "Static and free vibration analysis of beams and plates by differential quadrature method." *Acta Mechanica* 102, 11-24.

Bert, C.W., Wang, X. and Striz, A.G. (1994b). "Convergence of the DQ method in the analysis of anisotropic plates." *Journal of Sound and Vibration* 170, 140-144.

Bert, C.W. and Malik, M. (1996a). "Differential quadrature method in computational mechanics: A review." *Applied Mechanics Reviews* 49, 1-28.

Bert, C.W. and Malik, M. (1996b). "Free vibration analysis of thin cylindrical shells by the differential quadrature method." *ASME Journal of Pressure Vessel Technology* 118, 1-12.

Bert, C.W. and Malik, M. (1996c). "Free vibration analysis of tapered rectangular plates by the differential quadrature method: A semi-analytical approach." *Journal of Sound and Vibration* 190, 41-63.

Bert, C.W. and Malik, M. (1996d). "The differential quadrature method for irregular domains and application to plate vibrations." *International Journal of Mechanical Sciences* 38, 589-606.

Bhimaraddi, A. (1993). "Three-dimensional elasticity solution for static response of orthotropic doubly curved shallow shells on rectangular planform." *Composite Structures* 24, 67-77.

Biot, M.A. (1941). "The theory of propagation of elastic waves in a fluid-saturated porous solid." *Journal of Applied Mechanics* 23, 91-95.

Biot, M.A. (1956). "The theory of propagation of elastic waves in a fluid-saturated porous solid." *Journal of Applied Mechanics* 23, 91-95.

Boyce, W.E. and Diprima, R.C. (1986). *Elementary Differential Equations and Boundary Value Problems*, 4th ed., Wiley, New York.

Buchanan, J.L. and Turner, P.R. (1997). *Numerical Methods and Analysis*, McGraw-Hill, New York.

Bugl, P. (1995). Differential Equations- Matrices and Models, Prentice Hall, New Jersey.

Caffrey, P.J. and Lee, J.M. (1994). *MSC/NASTRAN User's Guide: Linear Static Analysis*, ver 68, The Macneal-Schwendler Corporation.

Carpenter, M.H., Gottlieb, D. and Abarbanel, S. (1994). "Time-stable boundary-conditions for finite-difference schemes solving hyperbolic systems-methodology and application to high-order compact schemes." Journal of Computational Physics 111, 220-236.

Chen, C.N. (1999). "The development of irregular elements for differential quadrature element method steady-state heat conduction analysis." *Computer Methods in Applied Mechanics and Engineering* 170, 1-14.

Chen, C.N. (2001). "Differential quadrature finite difference method for structural mechanics problem." *Communication in Numerical Methods in Engineering* 170, 423-441.

Chen, W.L., Striz, A.G. and Bert, C.W. (2000). "High-accuracy plane stress and plate elements in the quadrature element method." *International Journal of Solids and Structures* 37, 627-647.

Churchill, R.V and Brown, J.W. (1990). *Complex Variables and Applications*. 5th ed., McGraw-Hill, New York.

Civan, F. and Sliepcevich, C.M. (1983). "Application of differential quadrature to transport processes." *Journal of Mathematical Analysis and Applications* 93, 206-221.

Civan, F. and Sliepcevich, C.M. (1984). "Differential quadrature for multidimensional problems." *Journal of Mathematical Analysis and Applications* 101, 423-443.

Cowin, S.C. (1999). "Bone poroelasticity." *Journal of Biomechanics* 32, 217-238.

Deif, A.S. (1991). *Advanced Matrix Theory for Scientists and Engineers*, 2nd ed., Abacus Press, New York.

Dong, C.Y., Lo, S.H. and Cheung, Y.K. (2002). "Application of the boundary-domain integral equation in elastic inclusion problems." *Engineering Analysis with Boundary Elements* 26, 471-477.

Dong, C.Y., Lo, S.H. and Cheung, Y.K. (2003). "Stress analysis of inclusion problems of various shapes in an infinite anisotropic elastic medium." *Computer Methods in Applied Mechanics and Engineering* 192, 683-696.

Donnell, L.H. (1976). *Beams, Plates, and Shells.* McGraw-Hill, New York.

Eshelby, J.D. (1957). "The determination of elastic field of an ellipsoidal inclusion, and related problems." *Proceedings of the Royal Society (London) A* 241, 376-396.

Fan, J. and Zhang, J. (1992). "Analytical solutions for thick doubly curved laminated shells." *Journal of Engineering Mechanics* 118, 1338-1356.

Fung, T.C. (2001). "Solving initial value problems by differential quadrature method - Part 2: second-and higher-order equations." *International Journal of Numerical Methods in Engineering* 50, 1429-1454.

Fung, T.C. (2002). "Stability and accuracy of differential quadrature method in solving dynamics problems." *Computer Methods in Applied Mechanics and Engineering* 191, 1311–1331.

Gong, S.X. and Meguid, S.A. (1992). "A general treatment of the elastic field of an elliptical inhomogeneity under antiplane shear." *Journal of Applied Mechanics* 59, S131-S135.

Gong, S.X. and Meguid, S.A. (1993). "On the elastic fields of an elliptical inhomogeneity under plane deformation." *Proceedings of the Royal Society (London) A* 443, 457-471.

Groves, Jr. F.R. (1983). "Numerical solution of nonlinear differential equations using computer algebra." *International Journal of Computer Mathematics* 13, 301–309.

Henrici, P. (1986). *Applied And Computational Complex Analysis, Vol. III*, Wiley, New York.

Honein, T. and Hermann, G. (1990). "On the bounded inclusions with circular or straight boundaries in plane elastostatics." *Journal of Applied Mechanics* 57, 850-856.

Huang, N.N. and Tauchert, T.R. (1992). "Thermal stresses in doubly-curved cross-ply laminates." *International Journal of Solids and Structures* 29, 991-1000.

Hughes, T.J.R. (1987). *The Finite Element Method: Linear Static and Dynamic Finite Element Analysis*, Prentice-Hall, Englewood Cliffs, New Jersey.

http://en.wikipedia.org/wiki/Complex_analysis

Jaswon, M.A. and Bhargava, R.D. (1961). "Two-dimensional elastic inclusion problems." *Proceedings of the Cambridge Philosophical Society* 57, 669-680.

Kapania, R.K. (1989). "A review on the analysis of laminated shells." *Journal of Pressure Vessel Technology* 111, 88-96.

Karami, G. and Malekzadeh, P. (2002). "A new differential quadrature methodology for beam analysis and the associated differential quadrature element method." *Computational Methods in Applied Mechanics and Engineering* 191, 3509-3526.

Kobayashi, H. and Sonoda, K. (1989). "Rectangular plates on elastic foundations." *International Journal of Mechanical Sciences* 31, 679-692.

Kreyszig, E. (1999) *Advanced Engineering Mathematics*, John Wiley & Son Ltd, New York.

Lam, S.S.E. (1993). "Application of the differential quadrature method to two-dimensional problems with arbitrary geometry." *Computers and Structures* 47, 459-464.

Lam, K.Y., Zhang, J. and Zong, Z. (2004). "A numerical study of wave propagation in poroelastic medium by use of localized differential quadrature method." *Applied Mathematical Modelling* 28, 487-511.

Lang, J., Erdmann, B. and Seebass, M. (1999). "Impact of nonlinear heat transfer on temperature control in regional hyperthermia." *IEEE Transactions on Biomedical Engineering* 46, 1129-1138.

Li, W.Y., Cheung, Y.K. and Tham, L.G. (1986). "Spline finite strip analysis of general plates." ASCE Journal of Engineering Mechanics 111, 43-54.

Li, Y.S. and Qi, D.X. (1979). *Spline Function Method (in Chinese)*, Science Press, Beijing.

Liew, K.M., Teo, T.M. and Han, J.B. (2001). "Three-dimensional static solutions of rectangular plates by variant differential quadrature method." *International Journal of Mechanical Science* 43, 1611-1628.

Lin, J.H., Shen, W.P. and Williams, F.W. (1995). "A high precision direct integration scheme for structures subjected to transient dynamic loading." *Computers and Structures* 56, 113-12.

Lin, J.H., Shen, W.P. and Williams, F.W. (1997). "Accurate high-speed computation of non-stationary random structural response." *Engineering Structures* 19, 586-593.

Liu, G.R. and Wu, T.Y. (2000). "Numerical solution for differential equations of Duffing-type non-linearity using the generalized differential quadrature." *Journal of Sound and Vibration* 237, 805-817.

Liu, G.R. and Wu, T.Y. (2001a). "Vibration analysis of beams using the generalized differential quadrature rule and domain decomposition." *Journal of Sound and Vibration* 246, 461-481.

Liu, G.R. and Wu, T.Y. (2001b). "In-plane vibration analyses of circular arches by the generalized differential quadrature rule." *International Journal of Mechanical Science* 43, 2597-2611.

Ma, H. and Qin, Q.H. (2005). "A second-order scheme for integration of one-dimensional dynamic analysis." *Computers and Mathematics with Applications* 49, 239-252.

Ma, L.S. and Wang, T.J. (2003). "Nonlinear bending and post-buckling of a functionally graded circular plate under mechanical and thermal loadings." *International Journal of Solids and Structures* 40, 3311-3330.

Malik, M. and Bert, C.W. (1996). "Implementing multiple boundary conditions in the DQ solution of high-order PDE's: Application to free vibration of plates." *International Journal of Numerical Methods in Engineering* 39, 1237-1258.

Malik, M. and Bert, C.M. (1998). "Three-dimensional elasticity solutions for free vibrations of rectangular plates by the differential quadrature method." *International Journal of Solids and Structure* 35, 229-318.

Mitchell, A.R. and Wait, R. (1977). *The Finite Element Method in Partial Differential Equations*, Wiley, Chichester.

Moradi, S. and Taheri, F. (1998). "Differential quadrature approach for delamination buckling analysis of composites with shear deformation." *AIAA Journal* 36, 1869-1873.

Muskhelishvili, N.I. (1953). *Some Basic Problems of the Mathematical Theory of Elasticity,* Noordhoff, Groningen.

Nakasone, Y., Nishiyama, H. and Nojiri, T. (2000). "Numerical equivalent inclusion methods: a new computational method for analyzing stress fields in and around inclusions of various shapes." *Materials Science and Engineering* A 285, 229-238.

Newman, J.N. (1977). *Marine Hydrodynamics,* MIT Press, Cambindge Massachusetts.

Noda, N. (1991). "Thermal stresses in materials with temperature-dependent properties." *Applied Mechanics Review* 44, 383-397.

Noor, A.K and Burton, W.S. (1990). "Assessment of computational models for multilayered composite shells." *Applied Mechanics Review* 43, 67-97.

Pradhan, S.C., Loy, C.T., Lam, K.Y. and Reddy, J.N. (2000). "Vibration characteristics of functionally graded cylindrical shells under various boundary conditions." *Applied Acoustics* 61, 111-129.

Press, W.H, Flannery, B.P., Teukolsky, S.A. and Vetterling, W.T. (1986). *Numerical Recipes: The Art of Scientific Computing,* Cambridge University Press, Cambridge.

Quan, J.R. and Chang, C.T. (1989a). "New insights in solving distributed system equations by the quadrature method - I. Analysis." *Computational Chemical Engineering* 13, 779-788.

Quan, J.R. and Chang, C.T. (1989b). "New insights in solving distributed system equations by the quadrature method-II. Application." *Computational Chemical Engineering* 13, 1017-1024.

Reddy, J.N., Wang, C.M. and Kitipornchai, S. (1999). "Axisymmetric bending of functionally graded circular and annular plates." *European Journal of Mechanics A-Solids* 1, 185-199.

Schinzinger, R. and Laura, P.A.A. (2003). *Conformal Mapping: Methods and Applications,* Dover, New York.

Schultz, M.H. (1973). *Spline Analysis,* Prentice-Hall, Englewood Cliffs, New Jersey.

Schumaker, L.L. (1981). *Spline Functions: Basic Theory,* Wiley, New York.

Shen, H.S. (2002). "Postbuckling analysis of axially-loaded functionally graded cylindrical shells in thermal environments." *Computational Science and Mechanics* 62, 977-987.

Shu, C. and Richards, B.E. (1992). "Application of generalized differential quadrature to solve two-dimensional incompressible Navier-Strokes equations." *International Journal of Numerical Methods in Fluids* 15, 791-798.

Shu, C., Khoo, B.C., Chew, Y.T. and Yeo, K.S. (1996a). "Numerical studies of unsteady boundary layer flows past in an impulsively started circular cylinder by GDQ and GIQ approaches." *Computer Methods in Applied Mechanics and Engineering* 135, 229-241.

Shu, C., Chew, Y.T., Khoo, B.C. and Yeo, K.S. (1996b). "Solutions of three-dimensional boundary layer equations by global methods of generalized differential-integral quadrature." *International Journal of Numerical Methods for Heat and Fluid Flow* 6, 61-75.

Shu, C. (2000a). *Differential Quadrature and Its Applications in Engineering,* Springer, Berlin.

Shu, C. (2000b). "Analysis of lliptical waveguides by differential quadrature method." *IEEE Transactions on Microwave Theory and Techniques* 48, 319-322.

Shu, C., Chen, W., Xue, H. and Du H. (2001). "Numerical study of grid distribution effect on accuracy of DQ analysis of beams and plates by error estimation of derivative approximation." *International Journal of Numerical Methods in Engineering* 51, 159-179.

Shu, C., Yao, Q., Yeo, K.S. and Zhu, Y.D. (2002a). "Numerical analysis of flow and thermal fields in arbitrary eccentric annulus by differential quadrature method." *Heat and Mass Transfer* 38, 597-608.

Shu, C., Yao, Q. and Yeo, K.S. (2002b). "Block-marching in time with DQ discretization: An efficient method for time-dependent problems." *Computer Methods in Applied Mechanics and Engineering* 191, 4587-4597.

Shu, C. and Kha, A. (2002). "Numerical simulation of natural convection in a square cavity by simple-generalized differential quadrature method." *Computers and Fluids* 31, 209-226.

Shu, C., Ding, H. and Yeo, K.S. (2004). "Solution of partial differential equations by a global radial basis function-based differential quadrature method." *Engineering Analysis with Boundary Elements* 28, 1217-1226.

Sod, G.A. (1978). "A survey of several finite difference methods for systems of hyperbolic conservation laws." *Journal of Computational Physics* 27, 1-31.

Smith, B.T., Boyle, J.M., Garbow, B.S., Ikebe, Y., Klema, V.C. and Moler, C.B. (1976). Matrix Eigensystem Routines-EISPACK Guid, Springer-Verlag, Berlin.

Smith, G.D. (1985). *Numerical Solution of Partial Differential Equations: Finite Difference Methods,* Clarendon Press, Oxford.

Smith, W.E. (1966). "Computation of pitch and heave motions for arbitrary ship forms." *International Ship Progress* 14, 155.

Striz, A.G., Chen, W.L. and Bert, C.W. (1994). "Static analysis of structures by the quadrature element method." *International Journal of Solids and Structure* 31, 2807-2818.

Takahashi, H. (2005). "Complex function theory and numerical analysis." *Publ. RIMS, Kyoto University* 41, 979-988.

Tanigawa, Y. (1995). "Some basic thermoelastic problems for nonhomogeneous structural materials." *Applied Mechanics Review* 28, 377-389.

Timoshenko, S.P. and Goodier, J.N. (1970). *Theory of Elasticity,* 3rd ed., McGraw-Hill, New York

Touloukian, Y.S. (1967). *Thermophysical Properties of High Temperature Solid Materials,* Macmillian, New York.

Wang, Y. (2001). *Differential Quadrature Method & Differential Quadrature Element Method—Theory and Practice,* Ph.D. Dissertation, Nanjing University of Aeronautics and Astronautics, China (in Chinese).

Wang, X. and Bert, C.W. (1993). "A new approach in applying differential quadrature to static and free vibrational analyses of beams and plates." *Journal of Sound and Vibration* 162, 566-572.

Wang, X., Gu, H. and Liu, B. (1996). "On Buckling Analysis of Beams and Frame Structures by the Differential Quadrature Element Method." *Proceedings of Engineering Mechanics* 1, 382-385.

Wang, X.W. and Gu, H.Z. (1997). "Static analysis of frame structures by the differential quadrature element method." *International Journal of Numerical Methods in Engineering* 40, 759-772.

Wang, X.W., Liu, F., Wang, X.F. and Gan, L.F. (2005). "New approaches in application of differential quadrature method to fourth-order differential equations." *Communications in Numerical Methods in Engineering* 21, 61-71.

Washizu, K. (1982). *Variational Methods in Elasticity and Plasticity,* Pergamon Press, New York.

Weaver, Jr. W., Timoshenko, S.P. and Young, D.H. (1990). *Vibration Problems in Engineering,* 5th ed., Wiley, New York.

Wu, C.P., Tarn, J.Q. and Chi, S.M. (1996a). "Three-dimensional analysis of doubly curved laminated shells." *Journal of Engineering Mechanics ASCE* 122, 391-401.

Wu, C.P., Tarn, J.Q. and Chi, S.M. (1996b). "An asymptotic theory for dynamic response for doubly curved laminated shells." *International Journal of Solids and Structures* 26, 3813-3841.

Wu, C.P. and Chiu, S.J. (2001). "Thermoelastic buckling of laminated composite conical shells." *Journal of Thermal Stresses* 24, 881-901.

Wu, C.P and Tsai, Y.H. (2004). "Asymptotic DQ solutions of functionally graded annular spherical shells." *European Journal of Mechanics A-Solids* 23, 283-299.

Wu, T.Y. and Liu, G.R. (2000). "Application of generalized differential quadrature rule to sixth-order differential equations." *Communications in Numerical Methods in Engineering* 16, 777-784.

Wu, T.Y. and Liu, G.R. (2001). "Free vibration analysis of circular plates with variable thickness by the generalized differential quadrature rule." *International Journal of Solids and Structures* 38, 7967-7980.

Yang, J. and Shen, H.S. (2001). "Dynamic response of initially stressed functionally graded rectangular thin plates." *Computers and Structures* 54, 497-508.

Yu, Y.Y. (1957). "Axisymmetrical bending of circular plates under simultaneous action of lateral load, force in the middle plane, and elastic foundation." *Journal of Applied Mechanics* 24, 141-143.

Zhang, J. and Katsube, N. (1995). "A hybrid finite element method for heterogeneous materials with randomly dispersed elastic inclusions." *Finite Element Analysis in Design* 19, 45-55.

Zhang, Y.Y. (2003). *Development of Differential Quadrature Methods and Their Applications to Plane Elasticity,* Master Thesis, National University of Singapore.

Zhang, Y.Y., Zong, Z. and Liu, L. (2007). "Complex differential quadrature method for two-dimensional potential and plane elastic problems." *Ship and Offshore Structures* 2, 1-10.

Zhong, H. (1998). "Elastic torsional analysis of prismatic shafts by differential quadrature method." *Communications in Numerical Methods in Engineering* 14, 195-208.

Zhong, H. (2000). "Triangular differential quadrature." *Communications in Numerical Methods in Engineering* 16, 401-408.

Zhong, H. (2001). "Triangular differential quadrature and its application to elastostatic analysis of Reissner plates." *International Journal of Solids and Structures* 38(16), 2821-2832.

Zhong, H.Z. (2002). "Application of triangular differential quadrature to problems with curved boundaries." *Communications in Numerical Methods in Engineering* 18, 633-643.

Zhong, H.Z. and He, Y.H. (2003). "A note on incorporation of domain decomposition into the differential quadrature method." Communications in Numerical Methods in Engineering 19, 297-306.

Zhong, H., Hua, Y. and He, Y.H. (2003). "Localized triangular differential quadrature." *Numerical Methods for Partial Differential Equations* 19, 682-692.

Zhong, H. (2004). "Spline-based differential quadrature for fourth order differential equations and its application to Kirchhoff plates." *Applied Mathematical Modelling* 28, 353–366.

Zhong, H. and Guo, Q. (2004). "Vibration analysis of rectangular plates with free corners using spline-based differential quadrature." *Shock and Vibration* 11, 119-128.

Zhong, H.Z. and Lan, M.Y. (2006). "Solution of nonlinear initial-value problems by the spline-based differential quadrature method." *Journal of Sound and Vibration* 296, 908-918.

Zhong, W.X. (1994). "On the precise time-integration method for structural dynamics, (in Chinese)." *Journal of Dalian University of Technology* 34, 131-136.

Zhong, W.X. and Williams, F.W. (1994). "A precise time step integration method, In Proceedings of Institution of Mechanical Engineers, Part C." *Journal of Mechanical Engineering Science* 208, 427-430.

Zhou, D., Cheung, Y.K., Au, F.T.K. and Lo, S.H. (2002). "Three-dimensional vibration analysis of thick rectangular plates using Chebyshev polynomial and Ritz method." *International Journal of Solids and Structures* 39, 6339-6353.

Zhu, Z.H. and Meguid, S.A. (2000). "On the thermoelastic stresses of multiple interacting inhomogeneities." *International Journal of Solids and Structures* 37, 2313-2330.

Zong, Z. and Lam, K.Y. (2002). "A localized differential quadrature method and it application to the 2D wave equations." *Computational Mechanics* 29, 382-391.

Zong, Z. (2003a). "A complex variable boundary collocation method for plane elastic problems." *Computational Mechanics* 31, 284-292.

Zong, Z. (2003b). "A variable order approach to improve differential quadrature accuracy in dynamic analysis." *Journal of Sound and Vibration* 266, 307-323.

Zong, Z. (2004). "Solving three-dimensional heterogeneous potential problems by use of multi-stage differential quadrature." *Journal of Beijing Institute of Industry* 30, 160-168.

Zong, Z. (2005). "Comments on 'A variable order approach to improve differential quadrature accuracy in dynamic analysis' Author's reply." *Journal of Sound and Vibration* 280, 1151-1153.

Zong, Z., Lam, K.Y. and Zhang, Y.Y. (2005). "A multi-domain differential quadrature approach to plane elastic problems with material dis-continuity." *Mathematical and Computer Modelling* 41, 539-553.

Zong, Z. (2006). *Information-Theoretic Methods for Estimating Complicated Probability Distributions*, Elsevier Science Publisher, Amsterdam.

Index

Gaussian points, 190

Accuracy, 2, 44, 94, 114, 163, 195, 233
δ-approach, 31
Analytic, 41, 273
Approximation, 1
Asymptotic, 123

Basis, 3
Beam, 17, 61
Bending, 17
Biharmonic, 56
Block marching, 19
Boundary condition, 17, 51
Boundary effect, 166
Boundary node, 43
Built-in, 35

Casti, 1
Chebyshev node, 73, 169
Chebyshev nodes, 6
Circulation, 79
Complex, 22, 41, 210
Complex analysis, 41
Complex DQ, 41
Complex function, 49
Complex plane, 41
Complex polynomial, 45
Concentrated force, 60
Conformal mapping, 42
Consolidation, 146
Continuous, 1, 66, 164
Convergence, 10
Coordinate transformation, 210
Coordinate-based, 240
Core node, 169
Cortical node, 163

CST, 124
Curvilinear triangle, 91

Derivative, 13, 43, 86
Differential quadrature (DQ), 1
Direct DQ, 17
Dirichlet, 48
Disk, 60
Domain, 2, 16, 18, 41, 85, 165, 195, 234, 273
Dynamic numerical instability, 163

Elastic, 41, 101, 127, 195, 248
Ellipse, 51
Error, 5, 52, 100, 167, 233, 272

First-order, 13, 44, 130, 164, 195, 235
Flow, 23, 77, 110
Fourier expansion basis, 9
Function, 1, 41, 86, 113, 164, 198

Gauss elimination, 17, 50, 109, 197, 265
Gauss point, 1
Gauss quadrature, 1
Grid point, 1, 44, 233

Hole, 65
Holomorphic, 45

Interpolation, 4, 41, 180, 195, 236
Interpolatory approximation, 4
Irrotational flow, 81

Jacobian, 94

Lagrange interpolation, 5